AMERICAN MUSEUM OF NATURAL HISTORY 

Center for Biodiversity and Conservation
New Directions in Biodiversity Conservation

Eleanor J. Sterling, Series Editor

The books in this series are based on annual symposia presented by the American Museum of Natural History's Center for Biodiversity and Conservation and partners. Each symposium reviews a topic critical to biodiversity and conservation and provides diverse perspectives by scientists, resource managers, policymakers, and others.

Nature in Fragments: The Legacy of Sprawl, edited by Elizabeth A. Johnson and Michael W. Klemens

Conservation Genetics in the Age of Genomics

New Directions in Biodiversity Conservation

Edited by George Amato,
Oliver Ryder, Howard Rosenbaum,
and Rob DeSalle

Conservation Genetics in the Age of Genomics

COLUMBIA UNIVERSITY PRESS NEW YORK

COLUMBIA UNIVERSITY PRESS
Publishers Since 1893
New York Chichester, West Sussex

Library of Congress Cataloging-in-Publication Data
Conservation genetics in the age of genomics / edited by George Amato . . . [et al.].
p. ; cm. — (New directions in biodiversity conservation)
Based on papers from two symposia, one held in San Diego and sponsored by the San Diego Zoological Park, and the other held in New York and sponsored by the Wildlife Conservation Society and the American Museum of Natural History Center for Biodiversity and Conservation.
Includes bibliographical references and index.
ISBN 978-0-231-12832-2 (cloth : alk. paper)
1. Ecological genetics—Congresses. 2. Genetic resources conservation—Congresses. 3. Population genetics—Congresses. I. Amato, George. II. San Diego Zoo. III. Wildlife Conservation Society (New York, N.Y.) IV. Center for Biodiversity and Conservation (American Museum of Natural History) V. Series: New directions in biodiversity conservation.
[DNLM: 1. Conservation of Natural Resources—methods—Congresses. 2. Genetics—ethics—Congresses. 3. Biodiversity—Congresses. 4. Extinction, Biological—Congresses. 5. Genetic Techniques—Congresses. QU 450 C755 2009]

QH456.C667 2009
576.5′8—dc22

2009005325

Columbia University Press books are printed on permanent and durable acid-free paper.

This book is printed on paper with recycled content.
Printed in the United States of America

c 10 9 8 7 6 5 4 3 2 1

References to Internet Web sites (URLs) were accurate at the time of writing. Neither the author nor Columbia University Press is responsible for URLs that may have expired or changed since the manuscript was prepared.

CONTENTS

List of Illustrations vii

List of Tables viii

Foreword: Sydney Brenner, The Continuity of Genomes and Genetic Resources for the New Century ix

Acknowledgments xi

General Introduction xiii

Part I. Perspectives on the Union of Conservation and Genetics **1**

1 The Expansion of Conservation Genetics 3
Rob DeSalle and George Amato

2 Conservation Genetics and the Extinction Crisis: A Perspective 25
William Conway

3 Moving Toward a More Integrated Approach 34
George Amato

Part II. Conservation Genetics in Action: Assessing the Level and Quality of Genetic Resources in Endangered Species **39**

4 Neutral, Detrimental, and Adaptive Variation in Conservation Genetics 41
Philip W. Hedrick

5 Stopping Evolution: Genetic Management of Captive Populations 58
Robert C. Lacy

6 The Emerging Theme of Ocean Neighborhoods in Marine Conservation 82
Stephen R. Palumbi

7 Genetic Data and the Interpretation of Restoration Priorities of the *Cicindela dorsalis* Say Complex (Coleoptera: Carabidae): The Components of Conservation Genetics Revisited 96
Paul Z. Goldstein

8 Range Collapse, Population Loss, and the Erosion of Global Genetic Resources 106
James P. Gibbs

Part III. Saving Genetic Resources 113

9 Biodiversity, Conservation, and Genetic Resources in Modern Museum and Herbarium Collections 115
Robert Hanner, Angélique Corthals, and Rob DeSalle

10 Banking of Genetic Resources: The Frozen Zoo at the San Diego Zoo 124
Leona G. Chemnick, Marlys L. Houck, and Oliver A. Ryder

11 The Role of Cryopreserved Cell and Tissue Collections for the Study of Biodiversity and Its Conservation 131
Vitaly Volobouev

12 The Silent Biodiversity Crisis: Loss of Genetic Resource Collections 141
Deborah L. Rogers, Calvin O. Qualset, Patrick E. McGuire, and Oliver A. Ryder

13 Who Owns the Ark, and Why Does It Matter? 160
Cathi Lehn, Rebecah Bryning, Rob DeSalle, and Richard Cahoon

Part IV. Genomic Technology Meets Conservation Biology 167

14 Conservomics? The Role of Genomics in Conservation Biology 169
George Amato and Rob DeSalle

15 Genomics and Conservation Genetics 179
Judith A. Blake

16 Crop Transgenes in Natural Populations 187
Norman C. Ellstrand

17 The Role of Assisted Reproduction in Animal Conservation 199
Anne McLaren

18 Conservation and Cloning: The Challenges 204
Ian Wilmut and Lesley Paterson

Part V. Policy, Law, and Philosophy of Conservation Biology in the Age of Genomics 211

19 Can Our Laws Accommodate the New Conservation Genetics? 213
Gerald J. Flattmann Jr., Barbara A. Ruskin, and Nicholas Vogt

20 The Import of Uncertainty 224
Sandra D. Mitchell

Further Reading 235
List of Contributors 237
Index 239

ILLUSTRATIONS

1.1. Approaches to diagnosing conservation units in nature 8
1.2. Conservation priority areas, based on level of human disturbance 9
1.3. Whale conservation genetics 10
1.4. Scientifically managed parrot breeding programs 13
4.1. A neighbor-joining tree of red wolf MHC alleles 50
4.2. The heterozygosity at individual amino acid positions in the red wolf 51
5.1. Number of markhor in North American zoos 64
5.2. Number of Arabian oryx in North American zoos 65
5.3. Loss of gene diversity under 3 genetic management strategies in populations simulated for 60 years 70
5.4. Accumulation of inbreeding under 3 genetic management strategies in populations simulated for 60 years 71
5.5. Loss of founder alleles at a hypothetical locus under 3 genetic management strategies in populations simulated for 60 years 71
5.6. Reduction in number of recessive lethal alleles per animal under 3 genetic management strategies in populations simulated for 60 years 72
5.7. Effective population size (N_e) achieved under 3 genetic management strategies 73
5.8. Effect of increasing percentage of sires unknown in the pedigree on the effective population size 74
6.1. Offshore eddies near Heceta Head, Oregon trap nutrients 84
6.2. Number of invertebrates from the West Coast of the United States that have larvae with various pelagic periods 85
6.3. Tracks of open ocean drifters moving along the Florida current and into the Bahamas 87
6.4. Cline along the California coast in the frequency of DNA sequence variants 89
6.5. Settlement of limpets declines with increasing distance from the borders of a marine reserve 91
10.1. The Frozen Zoo repository 125
14.1. Diagrammatic representation of the roles of genomics in modern biology 170

TABLES

1.1. The roles of conservation genetics 5

4.1. The proportion of Florida panthers with developmental anomalies 46

4.2. The amount of observed neutral variation and predicted detrimental and adaptive variation 52

8.1. Relative contribution to erosion of global genetic diversity of loss of populations 109

10.1. Species Survival Plan (SSP) and Europäische Erhaltzungszucht Programm (EEP) mammals represented in the Frozen Zoo 126

12.1. The spectrum of genetic resources and their associated modes of conservation 143

12.2. Examples of organizations that maintain genetic resource collections 144

16.1. The world's 13 most important food crops in terms of area planted 189

16.2. Spontaneous hybridization, weed evolution, and extinction risk for the world's most important food crops and their wild relatives 190

Sydney Brenner

FOREWORD: The Continuity of Genomes and Genetic Resources for the New Century

Genomics promises to solve previously intractable problems in the biology of complex organisms, and in this context I believe genomics will be of high utility to conservation biology. Two important components of modern genomics can enhance the expansion of modern biology to address these important biological problems. The first concerns a new paradigm for studying organisms at a variety of levels gleaned from genomic studies. This paradigm uses the evolutionary process and evolutionary principles. The second concerns the potential availability of genetic resources from as broad an array of organisms as possible. Both of these components are relevant to conservation biology initiatives in the age of genomics.

In the study of evolution, access to the genomes of animals of the past is restricted, except for what can occasionally be gleaned from their hard parts, which are generally preserved in rocks. However, a complex record of the past is preserved in an organism's DNA. Putting some thought into what genomic information can tell us about life on this planet is of utmost importance. Because DNA carries not only the blueprint for organisms but also a record of the past, we should do everything possible to preserve these genomes and hence one aspect of the history of life on this planet.

Several initiatives exist to enhance and implement the collection and storage of genomes of a vast number of organisms, whether as actual nucleic acid samples or simply as DNA sequences. It is not entirely unreasonable to think that in the future (perhaps 50 to 100 years from now) we might be able to read the information in these genomes so as to be able to reconstruct them, either physically or computationally.

One of the great pieces of evidence to come from modern genomics for evolutionary studies is that the genes from many different organisms have a common evolutionary origin. For instance, whether you look at a bacterium or a human being, the genes specifying a protein such as triose phosphate isomerase in these two organisms have a common evolutionary ancestor. Furthermore, it is highly probable that one could take the yeast gene for triose phosphate isomerase and put it in a lobster, and it would work. This is highly probable because we know we can combine the yeast and the lobster proteins and get a functional entity out of them. That sort of experiment suggests that we can use new technology to implement a different approach to study how the biodiversity we observe every day on the planet arose.

If an organism develops a new muscle, it doesn't start from scratch and reinvent all the components of a muscle in order to provide for that new muscle. It is more likely that the genes involved in muscle formation alter not their normal function but rather the location or timing of their expression. So the regulation of genes, that is, how much of each gene is expressed, where it is expressed, and how long it is expressed, is essential information that determines whether an organism has one kind of morphology or behavior. Natural selection can be a tremendous driving force in making large organism-level changes involving a small number of mutations at the genomic level.

What do we know about these control mechanisms? We know that they exist. But we don't know very much about them, and even though the full human genome has been sequenced, most of the information from these DNA sequences is still meaningless. Unfortunately, a computer program does not exist that can look at the human genome and determine when a particular gene is functionally important. An approach that has arisen out of modern genomics called inverse genetics promises to expand comparative biology in a direction never before imagined.

Classical genetics took organisms of the same genotype and then made changes in them, called mutations, that were then studied gene by gene. In other words, genetic researchers looked at small changes in a sea of uniformity. Inverse genetics, on the other hand, looks at constancy in a sea of noise.

One of the major goals of modern genomics is to understand the function of genes. In order to make sense of the changes and hence the function of genes in the genome of two lineages, an examination of the constant parts of their genomes in comparison to the sea of noise is essential. As many evolutionary comparisons as possible are necessary for the full extent of the approach to be efficient. Even as far away from the human lineage as the fish lineage is, the neuroendocrine system and many parts of the early embryology are much the same in the two organisms. Organisms that are considered more closely related will also share common genomic elements as well as common embryology and neurology. For instance, by putting fish genes into mice and assaying whether the transferred genes function the same as mouse genes, we confirm common functions for any gene in the genomes of organisms being compared. So it is of utmost importance to have available as many organismal genomes or genetic resources as possible.

Many examples in molecular biology demonstrate that with the new genomics we can do a new kind of genetics that amounts to crossing fish and mice and looking at recombinants. Experiments have recently been done on controlling elements in lymphocytes, with the same specific results. The advancing technology opens up the possibility of analyzing sequences of evolutionarily distant species, and the availability of as many genomes from as many taxa as possible is of paramount importance. The more combinations that can be made, the richer the comparisons will become and the more information we will obtain about gene regulation. This approach to studying genomes makes DNA from all kinds of organisms valuable.

As someone who has been involved in analyzing genomes for some time, I have concluded that of the vast sums being spent in this field by both governments and the private sector, a significant percentage should be used to encourage scientists to look at this field from an evolutionary perspective, of which conservation biology and genetic resource conservation are also a part. In addition, it is important that we move toward preserving the genomes of organisms in the living world in such a form and of sufficient quality that we can use it to enrich our knowledge of life on this planet and so our descendants can experience the vast variety of life on this planet in the future.

ACKNOWLEDGMENTS

We wish to thank the participants of the San Diego and New York symposia. Equally important to the execution of these meetings was the institutional support for the symposia. At the American Museum of Natural History (AMNH) we wish to thank the Korein Foundation; the Lewis and Dorothy Cullman Program for Molecular Systematics Studies; the Sackler Institute for Comparative Genomics; Ellen Futter, president; Dr. Michael Novacek, provost of science; and Eleanor Sterling, director of the Center for Biodiversity and Conservation for their support of the New York meeting. We wish to thank our colleagues and students in the joint Wildlife Conservation Society–AMNH Conservation Genetics Program at the Center for Biodiversity and Conservation for their hard work, attention to detail, and dedication to the field of conservation genetics. Without the students in this program and other conservation genetics programs throughout the world, conservation genetics in the age of genomics would mean little. These students will extend the application of genomic technology to conservation into the critical century ahead.

GENERAL INTRODUCTION

This collection of essays is the result of two meetings held on opposite coasts of the United States, in San Diego and New York. Both symposia addressed the role of expanding technology in the conservation of endangered species, and both were sponsored by preeminent institutions with a focus on biodiversity and wildlife conservation. The San Diego meeting was sponsored by the San Diego Zoological Park, and the New York meeting was sponsored by the Wildlife Conservation Society and the American Museum of Natural History Center for Biodiversity and Conservation. During the two meetings, speakers discussed the role of expanding genomic technologies in the future of conservation genetics. Topics included the impact of databases, sequencing technology, gamete repositories, and genetic modification of domestic species on the field of conservation genetics. Although the future of conservation biology was thoroughly examined through the lens of genetics, a more important issue concerned connecting genetics with broader issues of ecology, biodiversity, human history, landscape changes, and species recovery.

After these meetings, the organizers asked contributors to produce essays on their talks, and the past three years have been spent compiling these essays. Here the essays are organized into five sections that emerged from the talks presented at these two meetings. Although the meetings were not planned with each other in mind, their goals and focus were complementary. Part I, "Perspectives on the Union of Conservation and Genetics," includes three papers that give the reader an introduction to problems in conservation biology and genetics and a historical perspective on their roles, opportunities, and challenges. Part II, "Conservation Genetics in Action: Assessing the Level and Quality of Genetic Resources in Endangered Species," focuses on the use of genetics in conservation biology, with papers from leaders in the field of genetics and its application to conservation biology. The five papers in this section give a broad overview of the kinds of studies modern genetics can approach in conservation biology.

Part III, "Saving Genetic Resources," has five papers that discuss the role of biological collections in conservation biology. This role has expanded in the past decade because of advances in genomic and reproductive technology, and these papers cover the range of collection types, from zoo-based to museum- and herbarium-based collections to electronic genomic database structures. Part IV, "Genomic Technology Meets Conservation Biology," has five papers that discuss the promise and pitfalls of expanding technology in conservation. These papers attempt to explain how far the application of technology to conservation can take the field in the twenty-first century. Part V, "Policy, Law, and Philosophy of Conservation Biology in the Age of Genomics," consists of two papers that discuss the role of philosophy and law in conservation thinking and the underlying policy decisions that must be considered in the new age of genomics and conservation.

We wish to direct the uninitiated reader to the following Web sites, which contain an online glossary and clearly written and illustrated descriptions of the utility of modern genetics and genomic approaches to conservation biology. The Web sites also include a list of addresses that are relevant to the subject of conservation and genetics. We particularly want to thank Mary Egan, Daniela Calcagnato, and Cathi Lehn for their diligent work in putting together the Web-based materials.

Conservation Genetics Sessions Guide:
http://symposia.cbc.amnh.org/archives/conservation-genetics/session.html

Symposium Resources:
http://symposia.cbc.amnh.org/archives/conservation-genetics/cgrsrcs.html

Glossary:
http://symposia.cbc.amnh.org/archives/conservation-genetics/glossary.html

George Amato, Oliver Ryder, Howard Rosenbaum, and Rob DeSalle

Conservation Genetics in the Age of Genomics

Part I

Perspectives on the Union of Conservation and Genetics

Conservation genetics has seen an expansion of goals and objectives over the past decade. This came about mostly as a result of the infusion of genomic technology into how we examine the genetics of natural and captive populations. Connecting the original goals of conservation genetics with the expanded potential of genomics is an important process, and the three chapters in this section attempt to make this connection. As an introductory chapter, we reprint from *Nature Review Genetics* a review by two of the organizers of the New York meeting (Rob DeSalle and George Amato). This chapter describes the incorporation of molecular information into conservation biology and discusses how such approaches have contributed to the expansion of conservation genetics. In chapter 2 William Conway provides an impassioned perspective on the real problems in conservation biology that can be addressed using genetic techniques. Most important, Conway suggests that conservation genetics must focus on area and ecosystem issues; focusing on broken ecosystems and marginalized species can be the best approach to this larger conservation biology goal. In the third chapter, George Amato examines the current state of conservation genetics and biology and concludes that a new paradigm for connecting ecological, life history, and area-based issues with genetics is essential for the field to remain vibrant and useful to conservation.

Rob DeSalle and George Amato

Adapted with permission from DeSalle, R. and G. Amato. 2004. The expansion of conservation genetics. *Nature Reviews. Genetics* 5:702–712.

The Expansion of Conservation Genetics

Conservation biology has been accurately described as a crisis discipline. Much like the human disease crisis disciplines of HIV biology and cancer biology, conservation biology requires an immediate understanding of the patterns and processes that make it such a critical subject. The urgency of the crises that are the subject of conservation biology are manifest in the large number of species facing imminent extinction. In fact, the last two centuries of human activity have been described as one of the most severe periods of mass extinction of all time, challenging even the most extreme periods of extinction in the long past as a period in which the majority of living species on Earth perish (Jablonski, 1986; Lande, 1988; Soulé, 1986). In addition, because conservation biologists have to make rapid decisions based on currently available data, there is a heightened sense of crisis in the discipline.

Crisis disciplines often see periods of expansion of the toolbox used to address the dilemmas posed by the forces causing the crises. Many of these tools are added to help researchers cope with the immediacy of the problems they are facing and allow them to rapidly and efficiently collect data on problems needing quick and even immediate attention. In conservation biology these tools have advanced accordingly. Some examples of this toolbox expansion include the use of global positioning systems (GPSs), satellite-based imaging of areas, the inclusion of mathematical advances in conservation theory, and perhaps the most visible: genetics. One of the factors involved in the inclusion of genetics in conservation biology over the past decade has been the proliferation of technology in genomics, systematics, and population biology, and conservation geneticists have scrambled to keep up with the pace of progress.

This essay attempts to articulate the current structure and scope of conservation genetics and to demonstrate the utility of this structure in modern conservation biology. The efficiency of this structure in conservation decision making is increased by the expanded genomic technologies in data acquisition, storage, and analysis. The improved precision and quantity of data on endangered species that can now be used in conservation genetics allows us to examine issues of captive species breeding, species boundary problems, and conservation forensics, three topics that will be the focus of this review. Another emerging area of conservation genetics concerns studies that combine genetic methods with ecological and landscape approaches. Such studies can provide conservation biologists with a much more accurate picture of the complex systems they work on.

Conservation biology, and conservation genetics with it, is expanding to include many subdisciplines and approaches, including genomics and high-throughput methods of data acquisition, and this expansion is better enabling the field to address the urgent problems involved in managing endangered species and critical areas. Significant challenges remain, and these include understanding that the context dependency of conservation decisions can be clarified by genetic information. Genetics can provide conservationists with unprecedented precision and elucidate the genetic parameters on which many decisions are made. Most importantly, conservation genetics must be placed in the context of the difficulties of working across political boundaries, amid economic challenges, and in the face of the complexity of using science to inform management decisions.

The Scope of Conservation Genetics

The expansion of the scope of conservation genetics over the past decade has been brought about by the infusion of high-throughput DNA sequencing and genotyping technology. Conservation geneticists have followed the trends of technique usage outlined by Schlotterer (2004) for population genetics as a whole. These techniques are well suited to enhancing our understanding of the evolutionary processes of endangered populations and species. An examination of the intensity of the use of various techniques applicable at the population level (Schlotterer, 2004) in animals indicates that although some techniques, such as random amplification of polymorphic DNA (RAPD), minisatellites, restriction fragment length polymorphism (RFLP), and allozymes, are either extinct or sparingly used in population-level studies, four methods—amplified fragment length polymorphism (AFLP), DNA sequencing, single nucleotide polymorphism (SNP) analysis, and microsatellites—make up most of the techniques used at this level for animals, and plant studies have continued to use AFLP, RAPD, and another DNA variation technique called inter–simple sequence repeat (ISSR; Godwin et al., 1997). The discovery of pattern in conservation genetics has also followed the development of techniques in systematics incorporating high-throughput sequencing and SNP analysis. On the whole, then, conservation geneticists have closely followed the development of genetic and genomic technology and have voraciously incorporated SNPs (Aitken et al., 2004; Brumfield et al., 2003; Zhang and Hewitt, 2003) and microsatellite variation as the major tools of their trade and are looking for more rapid and efficient screening techniques for genetic variability. Finally, some conservation geneticists have contemplated the utility of microarrays and quantitative trait mapping in conservation biology (Gibson, 2002; Purugganan and Gibson, 2003).

The scope of genetics in conservation biology is diverse and has been addressed in several publications (see table 1.1; Avise, 2004; Avise and Hamrick, 1996; Frankham et al.,

Table 1.1 The roles of conservation genetics are listed in the first column. The middle column indicates whether the listed role uses pattern or process from the genetic inference. The last column indicates the subdiscipline of evolutionary biology—systematics or population genetics—that is the major source of techniques to accomplish the role.

Role in conservation	Pattern or process	Subdiscipline
Minimize inbreeding and loss of genetic variation	Process	Population genetics
Identify populations of concern	Process and pattern	Population genetics
Resolve population structure	Pattern	Population genetics
Resolve taxonomic uncertainty	Pattern	Systematics
Define management units within species	Pattern	Systematics
Detect hybridization (genetic pollution)	Process	Population genetics and systematics
Detect and define invasive species	Process	Population genetics and systematics
Define sites and genotypes for reintroduction	Process	Population genetics and systematics
Allow application to forensics	Pattern	Systematics
Estimate population size and sex ratio	Process and pattern	Population genetics
Establish parentage and pedigree analysis	Pattern	Population genetics
Understand population connectivity	Pattern	Population genetics and systematics
Aid in the management of captive populations	Process	Population genetics
Understand relationships of focal groups of taxa	Pattern	Systematics
Implement genotoxicity studies	Process	Population genetics
Enhance reproductive capacity of organisms	Process	Population genetics

Source: The list of roles is partially after Allendorf and Leary (1986).

2003; Spellerberg, 1997; Young et al., 2000). Rather than attempt a comprehensive review of these usages, we instead attempt to place the major approaches to understanding pattern and process in a conservation context. To do this we first examine the early challenges to the infusion of genetics into conservation decision making and then discuss the present and future of pattern and process discovery in conservation genetics.

Early Challenges

The expansion of the toolbox of conservation biology in the late 1980s to include molecular evolutionary genetic techniques and markers was immediately met with three challenges to the relative merit and priority of genetics in the discipline. The first was a cogent direct challenge to the relevance of genetics to demographic issues in conservation biology by Lande (1988). In a landmark paper, Lande pointed out that demographic factors (the biology of population growth and life history) were much more important in explaining extinction than any of the genetic factors that could be incorporated into a theory of conservation biology. These demographic factors were viewed as swamping out most of the genetic effects ascribed to extinction. Lande's challenge to conservation genetics was a healthy one: It opened the way for a better-defined role of the concepts of inbreeding and genetic variation in the discipline. It is also clear that population genetics without demographics in conservation usually leads to less than useful recommendations (Ballou and Lacy, 1995). However, careful integration of demographic and fine-grained genetic approaches can often allow strong conservation inferences.

The second challenge came on the pattern side of the conservation genetics toolbox. The idea of conservation units was challenged by Ryder (1986), and the term *evolutionarily significant unit* (ESU) was imbedded in the conservation genetics literature as a result of healthy debate after Ryder's original challenge to subspecies definition and, more importantly, to the utility of subspecies definitions in conservation biology (Amato, 1991; O'Brien and Mayr, 1991; Waples, 1991). This problem of unit designation in conservation is often set at the interface of population genetics and systematics, because the goal is to discover species units. These debates, at both levels of resolution, resulted in a more applied focus in conservation genetics research. Clearly articulating the goals of a specific conservation problem provided better guidance to the selection of techniques, tools, and theory.

The third challenge was a posthumously published paper by Caughley (1994) that suggested that too much focus on technical approaches to conservation (including conservation genetics) had resulted in neglect of more important issues such as habitat threat and disease. This criticism was answered by the enormous expansion in the use of conservation genetics in ecology and applied wildlife management (Hedrick et al., 1996) and the realization that genetics could aid in landscape ecology approaches.

Pattern and Process: The Targets of Conservation Genetics

The most significant result of debate on these three challenges was to define the roles of conservation genetics in elucidating genetic and evolutionary processes and delineating the patterns relevant to managing endangered populations. Understanding that genetics leads us to a better picture of pattern and process in endangered species defines the most important roles of conservation genetics. The first is to aid in a more precise description and understanding of the processes that gave rise to the current state of an endangered population or species. This role is most important because the identification of factors such as inbreeding depression, breeding effective population size, minimum viable population size, and levels of genetic variation and gene flow in natural populations (Allendorf and Leary, 1986; Ralls et al., 1988; Templeton and Read, 1984; Waples, 1989) lends greater definition to the processes that affect endangered populations and also indicates immediate genetically based responses to the detrimental effects of these processes.

The tools used to analyze intrapopulation and genealogical problems have been summarized in several publications (Glaubitz et al., 2003; Goodnight and Quellar, 1999; Petit et al., 2002; Ritland, 2000; Vekemens and Hardy, 2004). Ex situ population genealogical and inbreeding analysis is a particularly important area where these methods are applied, and several studies making recommendations concerning breeding programs have arisen from these kinds of pedigree analysis studies (Geyer et al., 1993; Lacy, 2000a; Miller et al., 2003; Russello and Amato, 2001, 2004). The urgency of making the best possible genetic decisions in breeding programs of endangered species is exemplified by several studies in which pedigrees inform matchmaking. In addition, pedigrees can be extremely important in examining life history traits in endangered wild populations (e.g., of whales).

Classic population genetic measures such as Wright's inbreeding coefficients (*F* statistics) were initially used in such studies to characterize the levels of variation and genetic contact in the populations under study, but more recently these classic measures have come under scrutiny as being imprecise (Neigel, 2002; Puragganan and Gibson, 2003; Templeton, 2002; Zhang and Hewitt, 2003). As much more molecular data became routinely available, population genetics methods such as refined F_{ST} approaches (e.g., analysis of molecular variance [AMOVA]), coalescence theory–based analyses, and related approaches have become more important for the statistical analysis of population-level variation (Excoffier et al., 1992; Schneider et al., 2000). These methods therefore have become important in revealing process phenomena in conservation genetics. Another analytical approach that has become increasingly important at the same level is population viability analysis (PVA). PVA uses models of population dynamics (sometimes incorporating genetic and pedigree information) to estimate minimal viable population sizes for threatened populations subject to a variety of conditions.

The second major area is in the delineation of appropriate units for conservation attention, an area located at the intersection of population genetics and systematics. Conservation decisions often rely on the determination of species boundaries, a very contentious subject in evolutionary and systematic biology. The contentions involved in this area of evolutionary biology, and by default in conservation biology, arise from the basic plurality of species definitions and lack of agreement on how to objectively and operationally use data to delimit species boundaries based on a particular species concept or definition (Goldstein and DeSalle, 2000; Goldstein et al., 2000). Despite the contentious nature of delimiting species boundaries, several objectively based and operational approaches do exist.

At the boundary between recently diverged species, two general systematics approaches to delimitation have been taken: character based (population aggregation analysis [PAA]; Davis and Nixon, 1992) and tree based (figure 1.1; Goldstein et al., 2000; Losos and Glor, 2003; Sites and Crandall, 1997; Sites and Marshall, 2003). The tree-based approaches (summarized in Sites and Marshall, 2003) all have the common step of producing a phylogenetic tree that indicates the position of a group of individuals relative to other groups. In these approaches, regardless of the method of tree construction, the concept of reciprocal monophyly (Moritz, 1994; Sites and Marshall, 2003) is used to delimit boundaries of entities in the analysis. Character-based approaches result in the determination of diagnostics from the attributes used to perform the PAA (see figure 1.1) and therefore are highly operational tools in conservation genetic work stemming from the unit determinations they are initially used in.

Pattern and Process: The Future

The most important advances and extensions in conservation genetics in the future will be those that incorporate pattern and process together into a cohesive approach to decision making. We can think of three major areas where this cohesion of pattern and process is being developed: nested clade analysis, cladistic diversity measures, and genetically informed demography-based approaches. Nested clade analysis is an approach that has become particularly popular (Clement et al., 2000; Crandall, 1998; Posada et al., 2000;

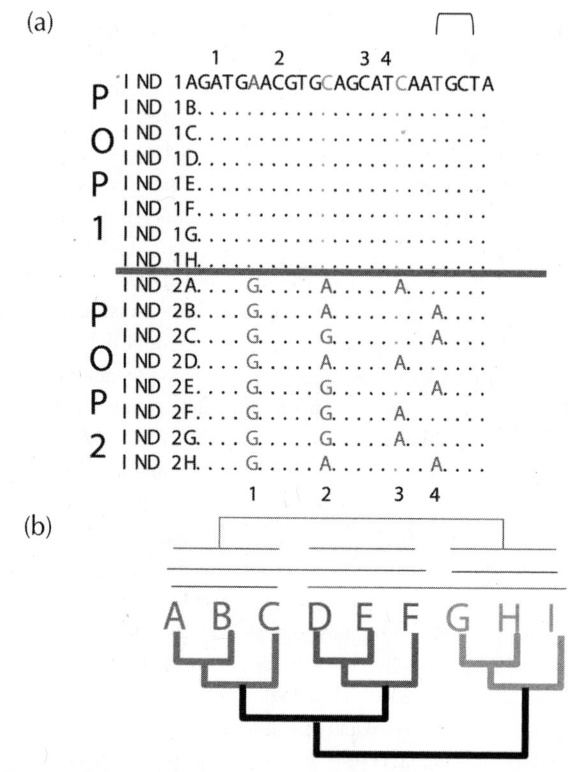

Figure 1.1 Approaches to diagnosing conservation units in nature. (a) Diagnostic character-based approach. Hypothetical DNA sequences from 16 individuals (IND) from 2 populations (POP 1 and POP 2) of organisms. The top sequence is shown for IND 1A, and dots in all sequences below indicate identity to the sequence in IND 1A. The DNA column numbered 1 indicates a DNA position in the sequence that unambiguously diagnoses POP 1 (A in position 1) as distinct from POP 2 (G in position 1). The DNA column numbered 2 indicates a DNA position in the sequence that, although polymorphic in POP 2, still diagnoses POP 1 (C in position 2) as distinct from POP 2 (either an A or G in position 1). The DNA columns numbered 3 and 4 indicate DNA positions in the sequence that are polymorphic and singly are not diagnostic. However, in combination the information in columns 3 and 4 diagnoses POP 1 (C in position 3 and T in position 4, or CT) as distinct from POP 2 (A in position 3 or 4, CA or AC). The approach described here is highly dependent on sample size of the 2 populations. **(b) Tree-based diagnostic approach.** The tree-based approach to conservation unit diagnosis depends on assaying groups of populations that are reciprocally monophyletic. The figure shows a hypothetical phylogeny of 3 groups of populations (A–C, D–F, and G–I) that illustrates the concept of reciprocal monophyly. Taken in isolation, each of the 3 groups is monophyletic (indicated by a line above the group) with respect to all other populations in the phylogeny; that is, each group includes the most recent common ancestor of that group and all its descendants. However, a larger group that includes both Group G, H, I and either Group A, B, C or Group D, E, F is paraphyletic with respect to the remaining smaller group because these groups do not contain some of the descendants of the most recent common ancestor of the populations that make up that group. However, Group G, H, I is reciprocally monophyletic to the union of Group A, B, C and Group D, E, F.

Templeton, 2002) and combines pattern and process issues. This approach uses a network constructed from genetic information that is nested according to a set of rules that results in increasingly larger nested groups, giving a detailed pattern-based way of looking at endangered organisms. The nested groups are then evaluated in the context of their geographic arrangement such that the statistical significance of geographic associations is assessed. The last step is to tie the process to the nested patterns. A statistical approach is used in a step algorithm to designate aspects of restricted gene flow, fragmentation, and range expansion, all important process factors in examining the genetics of endangered species (Crandall, 1998; Templeton, 2002).

Another approach that combines pattern and process is the cladistic diversity method, which is most useful at higher taxonomic levels. The first step in this approach is the discovery of phylogenetic patterns using genetic information and phylogenetic tree-building methods. In this approach taxa are assigned priority based on their uniqueness in a phylogenetic tree-based context and therefore have been suggested as aids in making difficult decisions about conservation priorities (Crozier, 1997; Faith, 1992a, 1992b, 2002; Nixon and Wheeler, 1992). The patterns observed in a phylogenetic tree are interpreted in a process-based context. The taxa in the tree that are basal, with few close relatives, are considered to have more cladistic diversity, and the more derived taxa, with many close relatives, are considered to have less cladistic diversity. The approach then allows an objective estimation of a cladistic diversity index that some have suggested can be useful in making conservation decisions (Crozier, 1997; Faith, 1992a, 1992b, 2002; Nixon and Wheeler, 1992).

Although genetic analysis of species and populations is a popular goal of conservation biologists, more recently there has been a growing realization that the field needs to focus on area-based recommendations (figure 1.2). Species-based approaches will

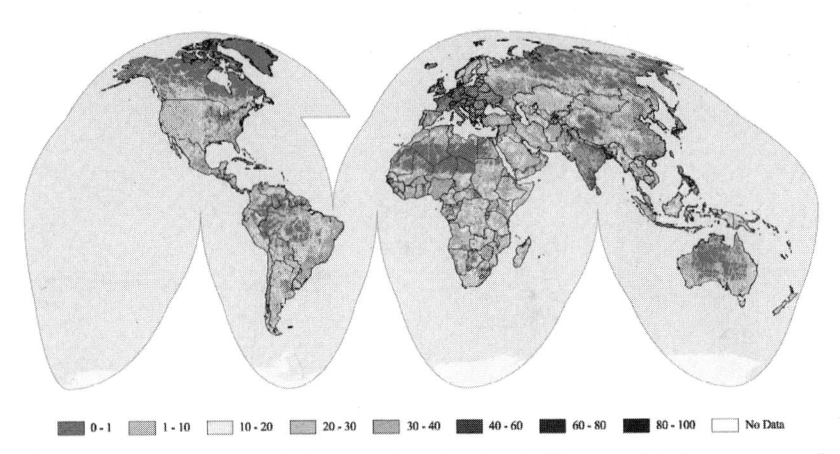

Figure 1.2 Conservation priority areas, based on level of human disturbance across all terrestrial landscapes. Conservation genetics research uses these landscape prioritization exercises to direct different efforts, focusing on large protected areas and highly fragmented habitats.

address issues relevant to the immediate success of a species, but the goal of conservation biology should be the long-term health of endangered organisms. Consequently, steps to preserve ecological or geographic areas that address these more confined species problems are the most efficient and long-term approaches to conservation. In this context, the third approach we describe as tying pattern and process together is the genetic approach or contribution to landscape ecology and areas of endemism (Brook et al., 2002; Gibbs and Amato, 2000; Losos and Glor, 2003; Moritz et al., 2000; Neel and Cummings, 2003; Palumbi, 2001; Roemer and Wayne, 2003). In particular, methods that discover genetic or genealogical patterns at these levels can assist in the assessment of connectivity in reserve formation and of the potential genetic impact of translocations and reintroductions. Marine (Palumbi, 2001) and botanical reserves (Neel and Cummings, 2003) are good examples of the incorporation of genetic information into a multidisciplinary approach to conservation management. In particular, the genetic information can be used to detail spatial and often temporal continuity of allelic composition of populations and, in addition to ecological data, can aid in selection of areas that are critical for healthy reserve systems.

Two examples of the utility of intersecting demographics and genetics concern cetaceans. Many cetacean species in the North Atlantic have undergone severe population reduction as a result of whaling. Roman and Palumbi (2003) used mitochondrial D-loop sequences and coalescence theory fit to mitochondrial DNA to estimate current and prewhaling population sizes of three North Atlantic whale species (figure 1.3a). Their study suggests that the population sizes in the prehistorical past were about 300,000 whales in each of the three species: humpback whale (*Megaptera novaeangliae*), fin whale (*Balaenoptera physalus*), and minke whale (*Balaenoptera acutorostrata*). The current population sizes are a fraction of this prewhaling population size, as shown in figure 1.3a, where the genetic estimate of the historical population size of North Atlantic humpback, fin, and minke whales is shown next to current census sizes for these species (confidence intervals are shown in light gray), and the results of this study prompted those authors to recom-

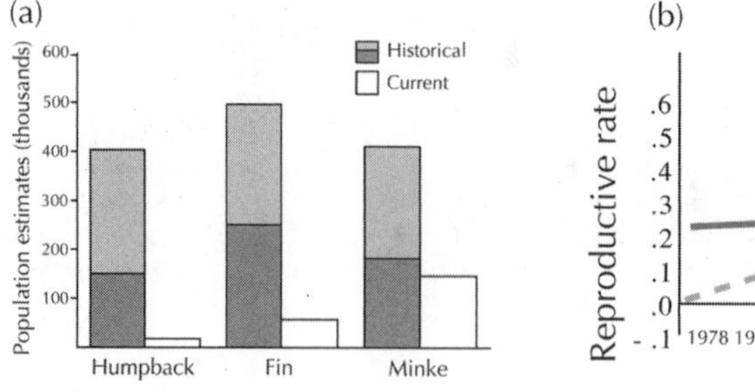

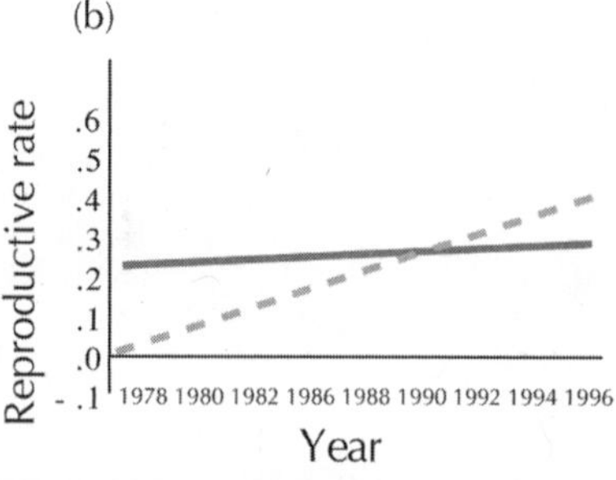

Figure 1.3 **(a)** The estimated numbers of North Atlantic humpback, fin, and minke whales in prewhaling times are shown next to current census sizes for these species (confidence intervals are shown in light gray). Redrawn from Roman and Palumbi (2003). **(b)** Regression of reproductive success of female humpback whales in 2 genealogically distinct maternal lineages. Redrawn from Rosenbaum et al. (2002).

mend that whaling of these species be further curtailed even though inferences from historical whaling records suggest that the current populations are doing well enough to sustain some harvesting.

The second study attempted to understand the life history trait evolution of humpback whales in the North Atlantic (Rosenbaum et al., 2002). This study demonstrated the potential of combining genealogy with life history traits. Figure 1.3b shows a regression of reproductive success of female humpback whales in two genealogically distinct maternal lineages. The solid line represents a clade of maternal haplotypes called the IJK lineage, and the dashed line is the BCD lineage. The study, conducted over a sixteen-year period, demonstrates the correlation of life history traits with maternal genealogy and might be used as critical information in the continuing conservation biology of this species.

Genetic Threats to Endangered Species

The number of practical applications of genetics to help manage endangered species grows larger and larger. In this section we focus on three areas of modern conservation genetics that we hope illuminate the unique problems conservationists face.

The Conservation Biologist as Matchmaker

Theory and recommendations about breeding genetics using pedigrees and genetic analysis to direct breeding efforts of endangered species have been a mainstay of modern conservation biology (Lacy, 2000a). Although managers in zoos and aquariums have long struggled with the challenges of maintaining viable ex situ populations, it wasn't until the last twenty years that genetic-based tools were incorporated into species survival programs (SSPs) (Ryder, 1986). Intrapopulation and interpopulation genetic variation assessments are particularly important approaches in single-species and captive population management efforts. The classic example of captive Speke's gazelles is a good indicator of the importance of considering these aspects of biology. In this example, genetic information about the captive population highlighted the urgency of depleted genetic variability in this captive population and informs approaches to breeding programs (Templeton and Read, 1984).

The tools of population genetics were used to assess levels of inbreeding (Allendorf and Leary, 1986; Ralls et al., 1988; Templeton and Read, 1984; Waples, 1989), especially in ex situ populations, and to devise methods to avoid the fitness depression caused by inbreeding. In addition, population geneticists pointed out that small population sizes (both captive and natural) tend to reduce genetic variation and therefore might decrease the ability of such populations to adapt to ecological challenges. Later it was pointed out that outbreeding or the direct mixing of individuals from genetically distinct populations could also be deleterious to the genetic health of an endangered species (Allendorf and Leary, 1986; Templeton, 1986). Perhaps the most important outcome of these approaches was the establishment of scientifically managed breeding programs (Ballou and Lacy, 1995) and the use of population genetic parameters to estimate minimum viable population sizes through PVA (Ballou and Lacy, 1995; Beissinger and McCullough, 2002; Reed et al., 2003; Ruggeiero et al., 1984).

Many of the complications of captive population matchmaking are demonstrated in recent work on Amazon parrots (Russello and Amato, 2001, 2004), where several genetic analysis techniques were extended to develop a comprehensive breeding plan. This species is a highly endangered island endemic restricted to the fragmentary oceanic tropical forest on St. Vincent, in the eastern Caribbean. Among other threats, the species is vulnerable to stochastic events such as hurricanes. An ex situ population of approximately one hundred birds was under the custodial care of the St. Vincent forestry department as part of an amnesty program on the island. Genetic relationships and even sex were unknown for the majority of the captive individuals. In order to optimize the breeding of these parrots, researchers determined the sex of all captive birds using a W chromosome–specific polymerase chain reaction (PCR) assay and genotyped the relatedness of all captive birds one to another using microsatellite technology. Relative relatedness values were established, allowing a scientifically managed program using empirical genetic data rather than the assumed unrelatedness of the founders (figure 1.4).

Other issues in conservation matchmaking concern wild populations and operate at the population or species level. Matchmaking in this context involves assessing the genetic health and integrity of an endangered population or species. Sometimes hybridization can occur between two populations or species that are conservation targets. Understanding when hybridization is a natural phenomenon or the result of anthropogenic factors is important in developing conservation strategies. Two examples of human-induced hybridization events that currently threaten species with extinction are the cases of the Simien wolf–domestic dog hybrids in Ethiopia and the Cuban crocodile–American crocodile hybridization in Cuba. Both cases involve rare, highly restricted endemics. An example of natural hybridization that complicates conservation is the case of the red wolf. The red wolf was believed to be an endangered unique taxon and is the focus of a U.S. Fish and Wildlife species recovery program. It has now been demonstrated that "red wolves" are the descendants of a natural wolf–coyote hybrid zone that occurred in the southeastern United States (Garcia-Moreno et al., 1996; Roy et al., 1994; Wayne and Jenks, 2002). Now that coyotes have naturally recolonized the east, they are interbreeding with the reintroduced "red wolves," and there is no consensus in the conservation community as to what should be done.

Uninformed Population Genetics and Bad Systematics Can Lead to Endangerment

The classic "bad taxonomy" problem in conservation genetics concerns the tuatara and the taxonomic lumping of species of these unique New Zealand reptiles (Daughtery et al., 1990). Looking superficially like lizards, tuataras are the sole surviving members of an ancient order of reptiles. For this reason they are a high priority for conservation. Mistakenly, they were lumped into a single species, endangering the various demonstrable species units that existed in this complex. In this case bad taxonomy placed several tuatara ESUs in harm's way and resulted in the extinction of species.

A potential example of bad systematics working in reverse concerns the Russian sturgeon (*Acipenser gueldenstaedtii*). DNA-based species diagnosis has been developed for

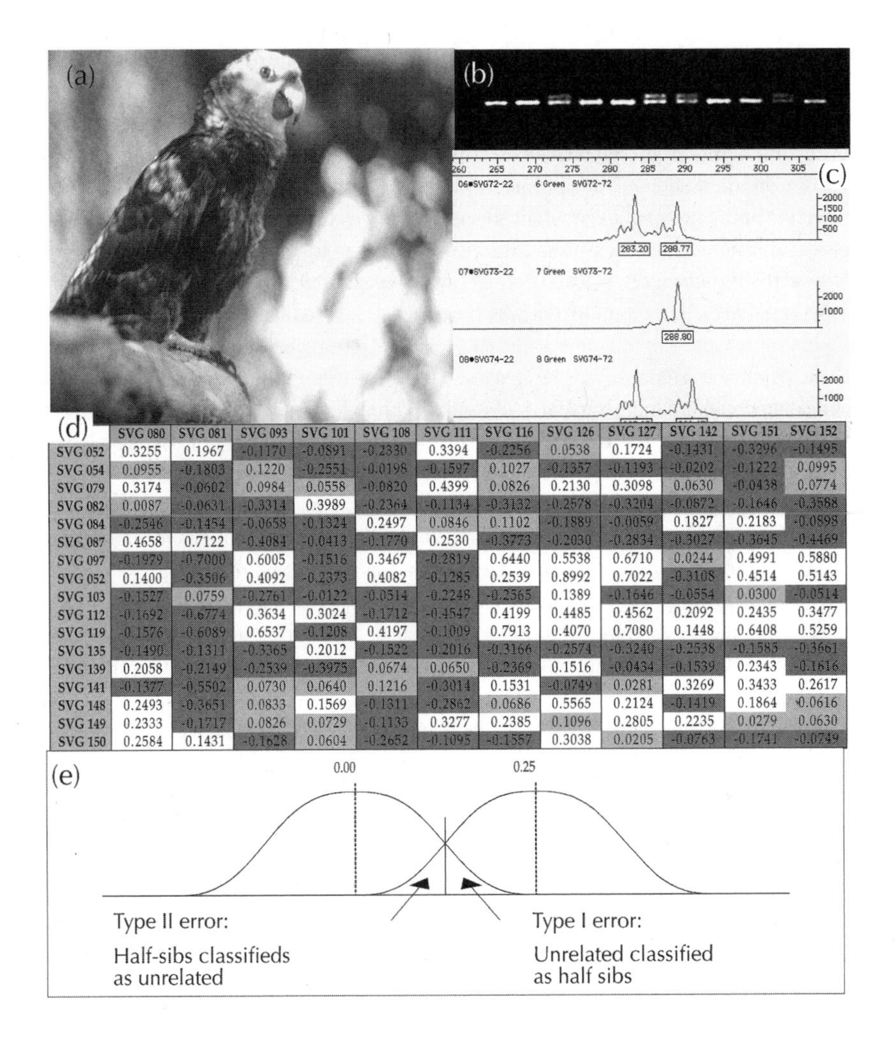

	SVG 080	SVG 081	SVG 093	SVG 101	SVG 108	SVG 111	SVG 116	SVG 126	SVG 127	SVG 142	SVG 151	SVG 152
SVG 052	0.3255	0.1967	-0.1170	-0.0891	-0.2330	0.3394	-0.2256	0.0538	0.1724	-0.1431	-0.3296	-0.1495
SVG 054	0.0955	-0.1803	0.1220	-0.2551	-0.0198	-0.1597	0.1027	-0.1357	-0.1193	-0.0202	-0.1222	0.0995
SVG 079	0.3174	-0.0602	0.0984	0.0558	-0.0820	0.4399	0.0826	0.2130	0.3098	0.0630	-0.0438	0.0774
SVG 082	0.0087	-0.0631	-0.3314	0.3989	-0.2364	-0.1134	-0.3132	-0.2578	-0.3204	-0.0872	-0.1646	-0.3588
SVG 084	-0.2546	-0.1454	-0.0658	-0.1324	0.2497	0.0846	0.1102	-0.1889	-0.0059	0.1827	0.2183	-0.0898
SVG 087	0.4658	0.7122	-0.4084	-0.0413	-0.1770	0.2530	-0.3773	-0.2030	-0.2834	-0.3027	-0.3645	-0.4469
SVG 097	-0.1979	-0.7000	0.6005	-0.1516	0.3467	-0.2819	0.6440	0.5538	0.6710	0.0244	0.4991	0.5880
SVG 052	0.1400	-0.3506	0.4092	-0.2373	0.4082	-0.1285	0.2539	0.8992	0.7022	-0.3108	0.4514	0.5143
SVG 103	-0.1527	0.0759	-0.2761	-0.0122	-0.0514	-0.2248	-0.2565	0.1389	-0.1646	-0.0554	0.0300	-0.0514
SVG 112	-0.1692	-0.6774	0.3634	0.3024	-0.1712	-0.4547	0.4199	0.4485	0.4562	0.2092	0.2435	0.3477
SVG 119	-0.1576	-0.6089	0.6537	-0.1208	0.4197	-0.1009	0.7913	0.4070	0.7080	0.1448	0.6408	0.5259
SVG 135	-0.1490	-0.1311	-0.3365	0.2012	-0.1522	-0.2016	-0.3166	-0.2574	-0.3240	-0.2538	-0.1585	-0.3661
SVG 139	0.2058	-0.2149	-0.2539	-0.3975	0.0674	0.0650	-0.2369	0.1516	-0.0434	-0.1539	0.2343	-0.1616
SVG 141	-0.1377	-0.5502	0.0730	0.0640	0.1216	-0.3014	0.1531	-0.0749	0.0281	0.3269	0.3433	0.2617
SVG 148	0.2493	-0.3651	0.0833	0.1569	-0.1311	-0.2862	0.0686	0.5565	0.2124	-0.1419	0.1864	0.0616
SVG 149	0.2333	-0.1717	0.0826	0.0729	-0.1133	0.3277	0.2385	0.1096	0.2805	0.2235	0.0279	0.0630
SVG 150	0.2584	0.1431	-0.1628	0.0604	-0.2652	-0.1095	-0.1557	0.3038	0.0205	-0.0763	-0.1741	-0.0749

Figure 1.4 A scientifically managed parrot breeding program was designed for the ex situ population of **(a)** *Amazona guildingii* in the absence of pedigree information by application of a PCR-based sexing technique and microsatellite analysis to establish relative relatedness measures. **(b)** The most obvious but previously difficult step in the process is to discover the sex of all birds in the potential breeding population. A rapid and simple PCR test exists for this task (Russello and Amato, 2001, 2004), where a single band indicates a female and a doublet indicates a male. In this figure there are 4 males and 7 females. **(c)** The second step is to characterize variability using microsatellites at a number of loci (Russello and Amato, 2004). **(d)** The third step is to construct a pairwise genetic relatedness table that indicates degree of relatedness. White boxes indicate immediate kinship or half-sibship, with light gray boxes indicating medium relatedness not discernible from half-sibship and dark gray boxes indicating that the individuals are unrelated. The final figure **(e)** shows the method behind classifying comparisons as half-sibs and unrelated.

the caviar of this fish (osetra caviar) and for all its close relatives (Birstein et al., 1998; DeSalle and Birstein, 1996). In essence, these DNA diagnostics can serve as barcodes for the twenty-four species of fish in the family Acipenseridae. Because the caviar produced by the fish in this group can be matched to species using a diagnostic approach, the Convention on International Trade in Endangered Species of Wild Fauna and Flora (CITES) listed the entire group of twenty-four species on their red book list (Ginsberg, 2002). The other two commercial caviar-producing species in this group of sturgeons (*Acipenser stellatus* and *Huso huso;* sevruga and beluga caviar, respectively) are easily diagnosed using DNA sequences, and therefore diagnosis of commercial products from these two species is not problematic. On the other hand, *A. gueldenstaedtii* is very difficult to type because of the possible existence of cryptic species in the areas where this fish is caught and used as a commercial product (Birstein et al., 2000). Through examination of diagnostic SNPs in commercial caviar, it was discovered that commercial products from Russia labeled as osetra are being typed as a species closely related to *A. gueldenstaedtii* called the Siberian sturgeon (*A. baerii*). The current DNA tests used to monitor the importation of commercial caviar might not be sensitive enough to detect the species origin of much of the osetra caviar exported from Russia. In this case the Russian fishers who rely on sturgeon catch for a livelihood are at a disadvantage as a result of errors in systematics.

The delineation of conservation units is the major underlying problem in these two examples. Several unit definitions have been coined; the most prominent are the ESU and the management unit (MU), which have been very useful in determining conservation priorities. Other units related to this boundary are semispecies, incipient species, and subspecies, of which *subspecies* often is considered a formal epithet. However, these latter designations are what we consider secondary to determination of ESUs and MUs. To further complicate matters, hybrid populations or hybrid individuals have also been a target of conservation geneticists. Hybridization complicates the delineation and identification of distinct evolutionary units in conservation genetics. In addition, the Endangered Species Act (ESA) does not protect hybrids between recognized endangered species. A good example of these complications is the gray wolf–red wolf hybrids. The hybrid origin of the endangered red wolf (Garcia-Moreno et al., 1996; Roy et al., 1994; Wayne and Jenks, 2002) has complicated the restoration and conservation of this species in nature (Hedrick, 1995). Other issues involve recognizing different units to match them with the demography and genetics of populations in other areas for reintroductions. The introduction of the Texas panther to Florida is a good example of this problem. In this example, lack of detailed consideration of the genetic match of these populations has led to decreased fitness of the population. A clearer and more extreme example of the impact of this approach is the lakeside daisy (DeMauro, 1993), where a lack of diversity at a self-incompatibility locus has prevented sexual reproduction in the last remaining Illinois population of this plant.

Forensics and Dead DNA

The final expanding area of conservation genetics concerns the ability to use unconventional sources of tissue for conservation work. Conservation biology by definition should be noninvasive. One of the major problems of doing conservation genetics at the DNA sequence level is that noninvasiveness is hard to achieve when tissues are needed to accomplish the genetic analysis (through biopsies and necropsies of animals and plant tissues). The use of PCR has opened doors to many conservation problems from a genetic perspective that would have remained closed because of the noninvasive requirement of conservation work. A broad array of unconventional sources for DNA are used in conservation work that include feces, feathers, fur, sloughed-off skin, plants from herbarium sheets, and clever direct biopsy approaches that minimize invasiveness. The number of studies using museum specimens and specimens of extinct organisms to detail the past genetics of endangered populations and species is extensive (Hoftreiter et al., 2001; Payne and Sorenson, 2002). These studies range from using hundred-year-old museum specimens, including whales (Rosenbaum et al., 2000), beetles (Goldstein and DeSalle, 2003), prairie chickens (Bellinger et al., 2003; Bouzat et al., 1998), muntjac (Amato et al., 1999), birds (Bunce et al., 2005; Fleischer et al., 2000, 2001; Glenn et al., 1999), fish (Nielsen et al., 1999), several endangered carnivores (Hoelzel et al., 1993; Leonard et al., 2000), and plants (Rogers and Bendich, 1985), to using long-extinct organisms thousands of years old, such as bears (Paabo, 2000), moas (Cooper et al., 1996), and sloths (Höss et al., 1996; Poinar et al., 2003). More recently, some studies have characterized whole mtDNA genomes of extinct organisms and extensive stretches of nuclear DNA (Poinar et al., 2006; Rogaev et al., 2006).

Other sources of tissue for DNA studies of endangered species are more purposeful and involve concerted efforts to store and archive high-quality tissues from endangered species, such as modern cryogenic techniques for the preservation of genetic resources of endangered species, gamete banks, and even potential tissues for animal cloning. Storage of plant seeds has been a tool of botanists for almost a century, and the continued attention to plant seed banks is also an important aspect of modern conservation. Because whole organism cloning from extinct animal tissues has been demonstrated as a reality (Lanza et al., 2000), high-quality freezing of tissues of endangered animals seems to be worthwhile. We emphasize here that the practice of freezing gametes and tissues for later regeneration of endangered species is not the most important or efficient current approach to endangered species management and recovery. However, there are still large problems with the viability of cloned animals, and, more importantly, the current endangered populations and the areas where they live should be the foci of our conservation efforts.

Another potential expansion of conservation genetics concerns recent efforts to form centralized DNA barcodes (Hebert et al., 2003; Stoekle, 2003) and DNA registries. Barcodes in everyday life are unique identifiers of commercial products. Such commercial barcodes identify products for retail sale or for storage and retrieval. Commercial barcodes are placed on products and, when needed, are identified using barcode readers. Simply put, DNA barcoding uses specific DNA sequences as unique identifiers of species. The dynamics of DNA sequence change have led some researchers to suggest that short DNA sequences can be used as a source of information for obtaining unique identifiers (barcodes) in organisms. DNA barcodes are developed through sequencing of specific genes such as cytochrome oxidase I for animals, or ITS2 for plants, for a reasonable number of individuals of a species. There are several ways barcodes can be read once these sequences have been obtained; multivariate statistical analysis (Hebert et al., 2003), tree building (Baker et al., 2003), and population aggregation analysis (Davis and Nixon, 1992; DeSalle et al., 2005) are the most widely used so far.

Both excitement (Hebert et al., 2003; Stoekle, 2003; Tautz et al., 2003) about and criticism (Dunn, 2003; Lipscomb et al., 2003; Seberg et al., 2003) of DNA barcoding have arisen in the scientific community. Criticism of the initiative has come mostly from classic taxonomists, with lively debate ensuing from both perspectives. We suggest that any source of character information that can be shown to be a unique identifier of a species should be considered a potential barcode. Such molecular tools as RFLP, AFLP, RAPD, microsatellites, gene order, gene presence or absence, and SNPs would be appropriate methods to discover unique identifiers for DNA barcoding. DNA barcoding can also be complemented and enriched with character information from other sources such as morphological or allozyme approaches (DeSalle et al., 2005). Several initiatives toward barcoding specific groups have already been launched; cetaceans (Baker et al., 2003) and bacteria (http://www.dsmz.de/bactnom/bactname.htm) are two prominent groups being examined using barcoding methods.

Some authors have recently pointed out that DNA fingerprints using sequences or microsatellite profiles can be used to identify individuals in populations of endangered species. Whales, in particular, have been a major focus of this approach (Cipriano and Palumbi, 1999), where samples are noninvasively taken from individuals in natural populations and a suitable number of informative microsatellite probes are used to generate a genetic profile for these individuals. The genetic profile is then stored in a database for later reference and matching to either living whales or whale tissue found in commercial situations.

DNA barcodes and DNA registries in conservation biology are useful in two main arenas. The first concerns wildlife forensics, in which they can be used to identify illegally imported animal or plant products, sometimes down to the individual (Cipriano and Palumbi, 1999). The second major use of these identifiers is in rapid assessment biodiversity studies. Identification of specimens in biodiversity studies often is tedious and time consuming. It is possible that DNA barcodes can facilitate more rapid and precise identification of species in these biodiversity studies. Expansion of the genetic toolbox to these important new initiatives will be of great assistance to conservation geneticists.

The Context Dependency of Conservation Genetics

The goals of conservation biology are highly context dependent. Much thought concerning the specific goals of genetic and ecological analysis in conservation biology has been couched in terms of naturalness and natural imperatives (Angermeier, 2000). These concepts are important in assessing the role of genetic, demographic, and ecological information in decision making. An important consideration is that genetic and demographic information represents only a snapshot in time, and in such genetic studies, that snapshot can be useful only if a frame of reference is available. This frame of reference in conservation, placed in the context of naturalness, would result in the establishment of benchmarks (Haila et al., 1997; Hunter, 1996). Such benchmarks would establish conditions for various times in the past as natural states for those periods. One problem to consider in this way of looking at naturalness is the determination of the most appropriate benchmark to adopt in making conservation decisions: prehuman, pre–human civilization, pre-Columbian (for North and South American initiatives), preindustrialization, or other. Establishing genetic benchmarks at these various stages is difficult, but not impossible, and has been accomplished in several cases (Goldstein and DeSalle, 2003; Paabo, 2000; Rosenbaum et al., 2000). An extremely important role of genetics will be to establish current population genetic and systematic profiles for endangered species with an eye on the information as a benchmark for conservation work in the future.

Future Challenges

As conservation expands, many challenges confront the discipline. These challenges include integrating genetic information with other biological and nonbiotic factors and, perhaps more important, using the results of conservation genetics studies to implement successful conservation strategies. A significant element of this effort involves using landscape dynamics and area-based conservation as part of the framework. These challenges concern what can be done to enhance the utility and efficiency of conservation genetics in making decisions about areas, landscapes, and species and also incorporating the genetic and biological issues into a coherent theory of conservation decision making. In the area of expanding technology and scientific infusion, we suggest that collections of organisms such as those in museums, herbaria, zoological parks, and botanical gardens are essential. The utility of museums and herbaria in a large number of studies to date indicates the importance of collections in establishing a baseline for genetic comparisons. We have an opportunity to store high-quality genetic resources for a large number of endangered species (Morin, 2000), if only to monitor the effects of our current conservation efforts. Studies using such methods to go back in time has been extremely useful in deciphering the effects of land use, species protection, and conservation efforts and the negative effects of human use of biotic resources. Continued collecting and, more importantly, preservation of tissues as high-quality sources of DNA and proteins will greatly enhance our ability to monitor endangered populations in the future. In this context it is important that we rethink current restrictions on the collection and storage of biodiversity, an issue that is complicated by national and international legal issues.

There is a great need to enhance the ability of conservation geneticists to noninvasively collect, archive, and use genetic resources of as broad an array of biodiversity as possible. This enhanced effort would entail a reconsideration of current restrictions conservation scientists and policymakers must conform to, such as CITES, ESA, and Convention of the Parties regulations. Because conservation biology is a crisis discipline, rapid biodiversity assessment will become more and more important in modern conservation decision making. Methods for the rapid detection and assessment of biodiversity are essential. It seems to us that DNA identification methods such as barcoding are an efficient and potentially effective way of assessing biodiversity of complex communities. Genetic and genomic approaches such as DNA barcoding and high-throughput population censusing will enhance this aspect of modern conservation biology. We do not suggest that these high-throughput methods replace classic taxonomy but rather that the DNA-based methods augment rapid and automated identification and assessment of biodiversity. Another essential area for expansion is in the development of genetically focused analytical tools such as quantitative genetic approaches (Moran, 2002) and Bayesian (Bertotorelle et al., 2004) methods in important conservation decisions.

Most important is the recognition that conservation decisions depend on a number of factors beyond the scientific. Some of these include economic, social, and cultural factors that should be considered in the light of informed biological information. Therefore, one of the major challenges to modern-day conservation genetics is the more efficient and better defined incorporation of genetics into conservation decision making in the context of the complexities of social, cultural, and political issues. Some conservationists have cogently argued that scientific issues should not be confused with social and legal issues (Freyfogle and Lutz, 2002; Sanderson et al., 2002) and that scientific terminology and inferences in conservation decisions (especially landscape management) should be explicit about the scientific content used in decision making. Infusion of scientific terminology and the scientific process in conservation decision making should be made only when the implications of such use are clearly articulated and impacts precisely assessed. The future of conservation biology will certainly need a broader theory for handling genetic information from population genetics and systematics and the interface with ecology. Such a formalized theory would go a long way in making the utility or context dependency of conservation genetics information clear and precise and enhance such information as a tool in conservation decisions at all levels.

Glossary

Areas of endemism: Geographic areas where maximal numbers of endemic species exist.

Breeding effective population size: The number of individuals that make up the breeding population in an idealized population.

Cladistic diversity: A measurement that ranks areas for biodiversity conservation priorities based on information in cladograms or phylogenetic trees. Also called phylogenetic diversity.

Convention on International Trade in Endangered Species of Wild Fauna and Flora (CITES): A convention established in 1973 among participating nations to restrict the international commerce of plant and animal species harmed by international trade.

Endangered Species Act (ESA): A U.S. congressional act established in 1973 that articulates guidelines and rules for the protection of species on the brink of extinction.

Evolutionarily significant unit (ESU): A population of organisms that is reproductively isolated from other populations of the same species and represents an important component in the evolutionary legacy of the species.

Evolutionary novelty: Heritable new attribute that is the result of phylogenetic and developmental factors and that leads to adaptive radiation of the taxa in which it arose.

Inbreeding depression: Reduction in fitness or vigor caused by one or more generations of inbreeding.

Incipient species: A newly formed species pair.

Management unit (MU): A population, stock, or group of stocks of a species that are aggregated for the purposes of achieving a desired conservation objective.

Minimum viable population size: An estimate of the smallest number of individuals in a population capable of maintaining that population, without significant manipulation.

Natural; naturalness: A thing is natural if it is not made by humans. Naturalness is the degree to which something is natural.

Natural imperative: An essential and urgent duty to restore some aspect of the environment to its "natural" state.

Paraphyletic: A term applied to a clade of organisms that includes the most recent common ancestor of all of its members but not all the descendants of the most recent common ancestor.

Population aggregation analysis (PAA): A straightforward criterion for demarcating a phylogenetic break between aggregates of individuals. An aggregate is said to be distinct when an attribute or combination of attributes in one aggregate is fixed and different from an attribute or combination of attributes in another aggregate.

Population viability analysis: The process of identifying threats faced by a species and incorporating these threats into an estimation of the likelihood of persistence of a species for a given time into the future.

Species survival program (SSP): A program established to ensure the survival of selected wildlife or captive species.

Subspecies: A physically distinct subunit of a species.

Systematics: The field of biology concerned with the diversity of life on Earth. Systematics is usually viewed as having two components: phylogenetics and taxonomy.

Wright's inbreeding coefficients: Measures of inbreeding in a subpopulation first devised by Sewall Wright to describe the amount of homozygosity in a population caused by inbreeding. Measures of inbreeding at different hierarchical levels of comparison can be obtained using the approach.

Bibliography

Aitken, N., S. Smith, C. Schwarz, and P. A. Morin. 2004. Single nucleotide polymorphism (SNP) discovery in mammals: a targeted-gene approach. *Molecular Ecology* 13(6):1423–1431.

Allendorf, F. W. and R. F. Leary. 1986. Heterozygosity and fitness in natural populations of animals. In *Conservation biology*, ed. M. E. Soulé, 57–76. Sunderland, Mass.: Sinauer.

Amato, G. 1991. Species hybridization and protection of endangered animals. *Science* 253: 250–251.

Amato, G., M. G. Egan, G. B. Schaller, R. H. Baker, H. C. Rosenbaum, W. G. Robichaud, et al. 1999. The rediscovery of Roosevelt's barking deer (*Muntiacus rooseveltorum*). *Journal of Mammalogy* 80:639–643.

Angermeier, P. L. 2000. The natural imperative for biological conservation. *Conservation Biology* 14:373–381.

Avise, J. C. 2004. *Molecular markers, natural history and evolution*. Sunderland, Mass.: Sinauer.

Avise, J. and P. Hamrick. 1996. *Conservation genetics: Case histories from nature.* New York: Chapman & Hall.

Baker, S., M. L. Dalebout, S. Lavery, and H. A. Ross. 2003. www.DNA-surveillance: Applied molecular taxonomy for species conservation and discovery. *Trends in Ecology and Evolution* 18:271–272.

Ballou, J. D. and R. C. Lacy. 1995. In *Population management for survival and recovery*, ed. J. Ballou, M. E. Gilpin, and T. J. Foose, 76–111. New York: Columbia University Press.

Beissinger, S. R. and D. R. McCullough. 2002. *Population viability analysis* Chicago: The University of Chicago Press.

Bellinger, M. R., J. A. Johnson, J. Toepfer, and P. Dunn. 2003. Loss of genetic variation in greater prairie chickens following a population bottleneck in Wisconsin USA. *Conservation Biology* 17:717–724.

Bertotorelle, G., M. Bruford, C. Chemini, C. Vernesi, and H. C. Hauffe. 2004. New flexible Bayesian methods to revolutionize conservation genetics. *Conservation Biology* 18: 584–585.

Birstein, V. J., P. Doukakis, and R. DeSalle. 2000. Caviar species identification and polyphyletic structure of Russian sturgeon. *Conservation Genetics* 1:81–88.

Birstein, V. J., P. Doukakis, B. Sorkin, and R. DeSalle. 1998. Population aggregation analysis of three caviar-producing species of sturgeons and implications for the species identification of black caviar. *Conservation Biology* 12:766–775.

Bouzat, J. L., H. A. Lewin, and K. N. Paige. 1998. The ghost of genetic diversity past: Historical DNA analysis of the greater prairie chicken. *The American Naturalist* 152:1–6.

Brook, B. W., D. W. Tonkyn, J. J. O'Grady, and R. Frankham. 2002. Contribution of inbreeding to extinction risk in threatened species. *Conservation Ecology* 6:6–28.

Brumfield, R. T., P. Beerli, and D. A. Nickerson. 2003. The utility of single nucleotide polymorphisms in inferences of population history. *Trends in Ecology and Evolution* 18:249–256.

Bunce, M., M. Szulkin, H. R. Lerner, I. Barnes, B. Shapiro, A. Cooper, et al. 2005. Ancient DNA provides new insights into the evolutionary history of New Zealand's extinct giant eagle. *PLoS Biology* 3(1):e9.

Caughley, G. 1994. Directions in conservation biology. *Journal of Animal Ecology* 63: 215–244.

Cipriano, F. and S. R. Palumbi. 1999. Genetic tracking of a protected whale. *Nature* 397: 307–308.

Clement, M., D. Posada, and K. A. Crandall. 2000. TCS: A computer program to estimate gene genealogies. *Molecular Ecology* 9:1657–1659.

Cooper, A., J. Rhymer, H. F. James, S. Olson, C. E. McIntosh, M. D. Sorenson, et al. 1996. . Ancient DNA and island endemics. *Nature* 381:484.

Crandall, K. A. 1998. Conservation phylogenetics of Ozark crayfishes: Assigning priorities for aquatic habitat protection. *Biological Conservation* 84:107–117.

Crozier, R. H. 1997. Preserving the information content of species: Genetic diversity, phylogeny, and conservation worth. *Annual Review of Ecology and Systematics* 28:243–268.

Daugherty, C. H., A. Cree, J. M. Hay, and M. B. Thompson. 1990. Neglected taxonomy and continuing extinctions of tuatara (*Sphenodon*). *Nature* 347:177–179.

Davis, J. I. and K. C. Nixon. 1992. Populations, genetic variation, and the delimitation of phylogenetic species. *Systematic Biology* 41:421–435.

DeMauro, M. M. 1993. Relationship of breeding system to rarity in the lakeside daisy (*Hymenoxys acaulis* var. *glabra*). *Conservation Biology* 7:542–550.

DeSalle, R. and V. Birstein. 1996. PCR analysis of black caviar. *Nature* 381:197–198.

DeSalle, R., M. G. Egan, and M. Siddall. 2005. The unholy trinity: Taxonomy, species delimitation and DNA barcoding. *Philosophical Transactions of the Royal Society of London B: Biological Sciences* 360(1462):1905–1916.

Dunn, C. P. 2003. Keeping taxonomy based in morphology. *Trends in Ecology and Evolution* 18:270–271.

Excoffier, L., P. E. Smouse, and J. M. Quattro. 1992. Analysis of molecular variance inferred from metric distances among DNA haplotypes: Application to human mitochondrial DNA restriction data. *Genetics* 131:479–491.

Faith, D. P. 1992a. Conservation evaluation and phylogenetic diversity. *Biological Conservation* 61:1–10.

——. 1992b. Systematics and conservation: On predicting the feature diversity of subsets of taxa. *Cladistics* 8:361–373.

——. 2002. Quantifying biodiversity: A phylogenetic perspective. *Conservation Biology* 16:248–252.

Fleischer, R. C., S. L. Olson, H. F. James, and A. C. Cooper. 2000. Identification of the extinct Hawaiian eagle (*Haliaeetus*) by mtDNA sequence analysis. *Auk* 117:1051–1056.

Fleischer, R. C., C. L. Tarr, H. F. James, B. Slikas, and C. E. Mcintosh. 2001. Phylogenetic placement of the Po'ouli, *Melamprosops phaeosoma,* based on mitochondrial DNA sequence data and osteological characters: *Studies in Avian Biology* 22:98–103.

Frankham, R., J. D. Ballou, and D. A. Briscoe. 2003. *Introduction to conservation genetics.* Cambridge, Mass.: Cambridge University Press.

Freyfogle, E. T. and N. J. Lutz. 2002. Putting science in its place. *Conservation Biology* 16:863–873.

Garcia-Moreno, J., M. S. Roy, E. Geffen, and R. K. Wayne. 1996. Relationships and genetic purity of the endangered Mexican wolf based on analysis of microsatellite loci. *Conservation Biology* 10:376–389.

Geyer, C. J., O. A. Ryder, L. G. Chemnick, and E. A. Thompson. 1993. Analysis of relatedness in the California condors, from DNA fingerprints. *Molecular Biology and Evolution* 10:571–589.

Gibbs, J. P. and G. Amato. 2000. Genetics and demography in turtle conservation. In *Turtle conservation,* ed. M. W. Klemens. Washington, D.C.: Smithsonian Institution Press.

Gibson, G. 2002. Microarrays in ecology and evolution: A preview. *Molecular Ecology* 11: 17–24.

Ginsberg, J. 2002. CITES at 30, or 40. *Conservation Biology* 12:1184–1191.

Glaubitz, J. C., E. Rhodes Jr., and J. E. Dewoody. 2003. Prospects for inferring pairwise relationships with single nucleotide polymorphisms. *Molecular Ecology* 12:1039–1047.

Glenn, T. C., W. Stephan, and M. J. Braun. 1999. Effects of a population bottleneck on whooping crane mitochondrial DNA variation. *Conservation Biology* 13:1097–1107.

Godwin, I. D., E. A. B. Aitken, and L. W. Smith. 1997. Application of inter-simple sequence repeat (ISSR) markers to plant genetics. *Electrophoresis* 18:1524–1528.

Goldstein, P. Z. and R. DeSalle. 2000. Phylogenetic species, nested hierarchies, and character fixation. *Cladistics* 16:364–384.

Goldstein, P. Z. and R. DeSalle. 2003. Calibrating phylogenetic species formation in a threatened species using DNA from historical specimens. *Molecular Ecology* 12:1993–1998.

Goldstein, P. Z., R. DeSalle, G. Amato, and A. Vogler. 2000. Conservation genetics at the species boundary. *Conservation Biology* 14:120–131.

Goodnight, K. F. and D. C. Quellar. 1999. Computer software for performing likelihood tests of pedigree relationship using genetic markers. *Molecular Ecology* 8:1231–1234.

Haila, Y., P. J. Comer, and M. Hunter Jr. 1997. A "natural" benchmark for ecosystem function. *Conservation Biology* 11:300–305.

Hebert, P. D. N., A. Cywinska, S. L. Ball, and J. R. deWaard. 2003. Biological identifications through DNA barcodes. *Proceedings of the Royal Society of London* B 270:313–322.

Hedrick, P. W. 1995. Gene flow and genetic restoration: The Florida panther; a case study. *Conservation Biology* 9:996–1007.
Hedrick, P. W., R. C. Lacy, F. W. Allendorf, and M. E. Soulé. 1996. Directions in conservation biology: Comments on Caughley. *Conservation Biology* 10:1312–1320.
Hoelzel, A. R., R. J. Halley, S. J. O'Brien, C. Campagna, B. Arnbom, B. LeBoeuf, et al. 1993. Elephant seal genetic variation and the use of simulation models to investigate historical population bottlenecks. *Journal of Heredity* 84:443–449.
Hofreiter, M., D. Serre, H. N. Poinar, M. Kuch, and S. Paabo. 2001. Ancient DNA. *Nature Reviews. Genetics* 2:353–359.
Höss, M., A. Dilling, A. Currant, and S. Pääbo. 1996. Molecular phylogeny of the extinct ground sloth *Mylodon darwinii*. *Proceedings of the National Academy of Sciences USA* 93:181–185.
Hunter, M. Jr. 1996. Benchmarks for managing ecosystems: Are human activities natural? *Conservation Biology* 10:695–697.
Jablonski, D. 1986. Background and mass extinctions: The alternation of macroevolutionary regimes. *Science* 231:129–133
Lacy, R. C. 2000a. Should we select genetic alleles in our conservation breeding programs? *Zoo Biology* 19:279–282.
——. 2000b. Structure of the VORTEX simulation model for population viability analysis. *Ecological Bulletins* 48:191–203.
Lande, R. 1988. Genetics and demography in biological conservation. *Science* 241:1455–1460.
Lanza, R. P., J. B. Cibelli, F. Diaz, C. T. Moraes, P. W. Farin, C. E. Farin, et al. 2000. Cloning of an endangered species (*Bos gaurus*) using interspecies nuclear transfer. *Cloning* 2:79–90.
Leonard, J. A., R. K. Wayne, and A. Cooper. 2000. Population genetics of Ice Age brown bears. *Proceedings of the National Academy of Sciences USA* 97:1651–1654.
Lipscomb, D., N. Platnick, and Q. Wheeler. 2003. The intellectual content of taxonomy: A comment on DNA taxonomy. *Trends in Ecology and Systematics* 18:64–65.
Losos, J. B. and R. E. Glor. 2003. Phylogenetic comparative methods and the geography of speciation. *Trends in Ecology and Evolution* 18:220–227.
Miller, C. R., J. R. Adams, and L. P. Waits. 2003. Pedigree-based assignment tests for reversing coyote (*Canis latrans*) introgression into the wild red wolf (*Canis rufus*) population. *Molecular Ecology* 12:3287–3299.
Moran, P. 2002. Current conservation genetics: Building an ecological approach to the synthesis of molecular and quantitative genetic methods. *Ecology of Freshwater Fish* 11:30–55.
Morin, P. A. 2000. Genetic resources: Opportunities and perspectives for the new century. *Conservation Genetics* 1:271–275.
Moritz, C. 1994. Defining "evolutionarily significant units" for conservation. *Trends in Ecology and Evolution* 9:373–375.
Moritz, C., J. L. Patton, C. J. Schneider, and T. B. Smith. 2000. Diversification of rainforest faunas: An integrated molecular approach. *Annual Review of Ecology and Systematics* 31:533–563.
Neel, M. and M. P. Cummings. 2003. Genetic consequences of ecological reserve design guidelines: An empirical investigation. *Conservation Genetics* 4:427–439.
Neigel, J. E. 2002. Is FST obsolete? *Conservation Genetics* 3:167–173.
Nielsen, E. E., M. M. Hansen, and V. Loeschcke. 1999. Genetic variation in time and space: Microsatellite analysis of extinct and extant populations of Atlantic salmon. *Evolution* 53:261–268.
Nixon, K. C. and Q. D. Wheeler. 1992. Measures of phylogenetic diversity, In *Extinction and phylogeny*, ed. M. J. Novacek and Q. D. Wheeler, 216–234. New York: Columbia University Press.

O'Brien, S. J. and E. Mayr. 1991. Bureaucratic mischief: Recognizing endangered species and subspecies. *Science* 251:1187–1188.

Paabo, S. 2000. Of bears, conservation genetics and the value of time travel. *Proceedings of the National Academy of Sciences USA* 97:1320–1321.

Palumbi, S. R. 2001. The ecology of marine protected areas. In *Marine community ecology*, ed. M. D. Bertness, S. D. Gaines, and M. E. Hay, 509–530. Sunderland, Mass.: Sinauer.

Payne, R. B. and M. D. Sorensen. 2002. Museum collections as sources of genetic data. *Bonner Zoologische Beiträge Band* 51.

Petit, E., F. Balloux, and L. Excoffier. 2002. Mammalian population genetics: Why not Y? *Trends in Ecology and Evolution* 17:28–35.

Poinar, H., M. Kuch, G. McDonald, P. Martin, and S. Paabo. 2003. Nuclear gene sequences from a late Pleistocene sloth coprolite. *Current Biology* 13:1150–1502.

Poinar, H. N., C. Schwarz, J. Qi, B. Shapiro, R. D. Macphee, B. Buigues, et al. 2006. Metagenomics to paleogenomics: Large-scale sequencing of mammoth DNA. *Science* 311(5759):392–394.

Posada, D., K. A. Crandall, and A. R. Templeton. 2000. GeoDis: A program for the cladistic nested analysis of the geographical distribution of genetic haplotypes. *Molecular Ecology* 9:487–488.

Purugganan, M. and G. Gibson. 2003. Merging ecology, molecular evolution, and functional genetics. *Molecular Ecology* 12:1109–1112.

Ralls, K., J. D. Ballou, and A. Templeton. 1988. Estimates of lethal equivalents and the cost of inbreeding in mammals. *Conservation Biology* 2:40–56.

Reed, D. H., J. J. O'Grady, B. W. Brook, J. D. Ballou, and R. Frankham. 2003. Estimates of minimum viable population size for vertebrates and factors affecting those estimates. *Biological Conservation* 113:23–34.

Ritland, K. 2000. Marker-inferred relatedness as a tool for detecting heritability in nature. *Molecular Ecology* 9:1195–1204.

Roemer, G. W. and R. K. Wayne. 2003. Conservation in conflict: A tale of two endangered species. *Conservation Biology* 17:1251–1260.

Rogaev, E. I., Y. K. Moliaka, B. A. Malyarchuk, F. A. Kondrashov, M. V. Derenko, I. Chumakov, et al. 2006. Complete mitochondrial genome and phylogeny of Pleistocene mammoth *Mammuthus primigenius*. *PLoS Biology* 4(3):e73.

Rogers, S. O. and A. J. Bendich. 1985. Extraction of DNA from milligram amounts of fresh, herbarium and mummified plant tissues. *Plant Molecular Biology* 5:69–76.

Roman, J. and S. R. Palumbi. 2003. Whales before whaling in the North Atlantic. *Science* 301:508–510.

Rosenbaum, H. C., M. G. Egan, P. Clapham, R. Brownell, S. Malik, M. Brown, et al. 2000. The utility of museum specimens in right whale conservation genetics. *Conservation Biology* 14:1837–1842.

Rosenbaum, H. C., M. T. Weinrich, S. A. Stoelson, J. P. Gibbs, C. S. Baker, and R. DeSalle. 2002. The effect of different reproductive success on population genetic structures: Correlations of life history with matrilines in humpback whales of the Gulf of Maine. *Journal of Heredity* 93:389–399.

Roy, M. S., D. J. Girman, A. C. Taylor, and R. K. Wayne. 1994. The use of museum specimens to reconstruct the genetic variability and relationships of extinct populations. *Experientia* 50:551–557.

Ruggiero, L. F., G. D. Hayward, and J. R. Squires. 1984. Viability analysis in biological evaluations: Concepts of population viability analysis, biological population, and ecological scale. *Conservation Biology* 8:364–372.

Russello, M. and G. Amato. 2001. Application of a non-invasive, PCR-based test for sex identification in an endangered parrot, *Amazona guildingii. Zoo Biology* 20:41–45.

Russello, M. and G. Amato. 2004. Ex situ management in the absence of pedigree information: Integration of microsatellite-based estimates of relatedness into a captive breeding strategy for the St. Vincent Amazon parrot (*Amazona guildingii*). *Molecular Ecology* 13(9):2829–2840.

Ryder, O. A. 1986. Species conservation and systematics: The dilemma of the subspecies. *Trends in Ecology and Evolution* 1:9–10.

Sanderson, E. W., M. Jaiteh, M. A. Levy, K. H. Redford, A. V. Wannebo, and G. Woolmer. 2002. The human footprint and the last of the wild. *BioScience* 52:891–904.

Schlotterer, C. 2004. The evolution of molecular markers: Just a matter of fashion? *Nature Reviews. Genetics* 5:63–69.

Schneider, S., D. Roessli, and L. Excoffier. 2000. *Arlequin: A software for population genetics data analysis. Ver. 2.000.* Geneva: Genetics and Biometry Lab, Department of Anthropology, University of Geneva.

Seberg, O., C. J. Humphries, S. Knapp, D. W. Stevenson, G. Petersen, N. Scharff, et al. 2003. Shortcuts in systematics? A commentary on DNA-based taxonomy. *Trends in Ecology and Systematics* 18:63–64.

Sites, J. W. and K. A. Crandall. 1997. Testing species boundaries in biodiversity studies. *Conservation Biology* 11:1289–1297.

Sites, J. and J. C. Marshall. 2003. Delimiting species: A Renaissance issue in systematic biology. *Trends in Ecology and Evolution* 18:461–471.

Soulé, M. E. 1986. *Conservation biology: The science of scarcity and diversity.* Sunderland, Mass.: Sinauer.

Spellerberg, I. 1997. *Conservation biology*. Harlow, U.K.: Longman.

Stoeckle, M. 2003. Taxonomy, DNA, and the bar code of life. *BioScience* 53:2–3.

Tautz, D., P. Arctander, A. Minelli, R. H. Thomas, and A. P. Vogler. 2003. A plea for DNA taxonomy. *Trends in Ecology and Evolution* 18:70–74.

Templeton, A. R. 1986. Coadaptation and outbreeding depression. In *Conservation biology: The science of scarcity and diversity,* ed. M. E. Soulé, 105–116. Sunderland, Mass.: Sinauer.

——. 2002. Nested clade analyses of phylogeographic data: Testing hypotheses about gene flow and population history. *Molecular Ecology* 7:381–397.

——. 2004. Statistical phylogeography: Methods of evaluating and minimizing inference errors. *Molecular Ecology* 13:789–809.

Templeton, A. R. and B. Read. 1984. Factors eliminating inbreeding depression in a captive herd of Speke's gazelle. *Zoo Biology* 3:177–199.

Vekemens, X. and O. J. Hardy. 2004. New insights from fine-scale spatial genetic structure analyses in plant populations. *Molecular Ecology* 13:921–935.

Waples, R. S. 1989. A generalized approach for estimating effective population size from temporal changes in allele frequency. *Genetics* 121:379–391.

——. 1991. Pacific salmon, *Oncorhynchus* spp., and the definition of "species" under the Endangered Species Act. *Marine Fisheries Review* 53:2–11.

Wayne, R. K. and S. M. Jenks. 2002. Mitochondrial DNA analysis implying extensive hybridization of the endangered red wolf *Canis rufus*. *Nature* 351:565–568.

Young, A. G., D. Boshier, and T. Boyle. 2000. *Forest conservation genetics.* Collingwood, Victoria, Australia: CSIRO Publishing.

Zhang, D. X. and G. M. Hewitt. 2003. Nuclear DNA analyses in genetic studies of populations: Practice, problems and prospects. *Molecular Ecology* 12:563–584.

William Conway

2

Conservation Genetics and the Extinction Crisis

A Perspective

Max Frisch (1957) observed, "Technology is a way of organizing the world so that man does not have to experience it." Nevertheless, these are heady times for genomics. The significance of the sequencing of the human genome has been compared with our first landing on the moon. New computational power has brought genetic insights to our perception of species survival (Lacy, 1993). Embryo transfer, cloning, artificial insemination, and a host of other techniques offer a new level of control of animal reproduction. Serendipitously, all this is taking place at a time when wildlife populations are facing a desperate extinction crisis. How, then, are the new developments in genetics and reproductive science to be tied to real-world conservation, to its scale, and to wild species?

Under the title "Noah's New Ark," *Time* magazine recently reported the cloning of a threatened Southeast Asian species of wild cattle called the gaur (*Bos gaurus*). The line about cloning that caught my eye was, "If it lives up to its billing, it could produce potentially unlimited numbers of endangered creatures" (Bird, 2001:1). We have those already; the author meant something else. If we divert attention from preserving nature to reproductive technologies, with dubious genetic justifications, we neither enhance the public's understanding of the extinction crisis nor provide much help.

It is tempting to consider the sickening decline of wild animals and plants a kind of disease. Can it not be cured with science and technology? After all, we defeated smallpox. No natural case of this disease has developed anywhere in the world since Ali Maalin developed the rash in Merka Town, Somalia, on October 26, 1977. But we cannot even hope that a comparable cure for the disappearance of tigers and turtles or cranes and crocodiles will ever be found, as one may be for cancer or AIDS. Compared with saving nature, curing human disease is child's play. The immense scale of wildlife extinction does not seem to be understood, nor does humanity's responsibility.

Even if we knew all that there is to know about the genetics of wildlife, we could not save a single species in nature unless we could influence the sovereignty of human social and political forces. And yet so much of wildlife dear to us is at stake because it is so necessary to us and so close to us.

We share 98 percent of our DNA with chimpanzees. Matt Ridley (1999:125) observed, "If you held hands with your mother, and she held hands with hers and she with hers, the line would stretch only from New York to Washington before you were holding hands with your common ancestor with chimpanzees." This emphasizes the fundamental nature of our biological bonds.

People Pressure

Although human biomass is already about 600 billion pounds, we are growing by 85 million people each year, and ecological constraints are beginning to affect our numbers. More than 3 billion people worldwide are said to be malnourished. Per capita grain production has been diminishing since 1983. Over the past decade, per capita cropland has declined by 20 percent and fish production by 7 percent (Pimental et al., 1999). Is it unreasonable to fear that producing more food for humans will produce more humans and the need for more food, even as we consume more land, more forest, and more wildlife habitat every day?

Environmental Destruction

Most terrestrial animals and plants are found in forests, especially tropical forests, and the plight of almost all tropical forests is desperate. Only about 7.5 million square kilometers of tropical evergreen forests remains. If their deforestation continues at the rate that held between 1979 and 1989, the last tropical forest tree might fall before 2100 (Terborgh, 1999). For example, the Brazilian Amazon contains about 40 percent of the world's remaining tropical rainforest and plays a vital role in maintaining biodiversity, regional hydrology, climate, and terrestrial carbon storage. It also has the world's highest rate of absolute forest destruction, currently averaging nearly 2 million hectares per year (Laurance et al., 2001).

The good news is that the world's wealthiest nations have recently decided to channel $340 million into Amazon forest conservation. The bad news is that Brazil has announced planned rainforest developments of more than $40 billion (Laurance et al., 2001) and that tropical forest animals are even more endangered than these figures suggest. The Wildlife Conservation Society's John Robinson and his colleagues estimate the bushmeat kill in the Brazilian Amazon at 67,000 to 164,000 metric tons each year: monkeys, peccaries, tapirs, and toucans. It is a figure dwarfed by that in the tropical forests of Africa, now annually more than 1 million metric tons: boiled gorillas, roasted chimpanzees, and salted elephants (Robinson et al., 1999). In Cameroon, for example, more than 800 gorillas were killed and cooked in 1998. Their bodies sold for about forty dollars each. These are intelligent, gentle creatures. Their DNA differs little from ours. We are eating members of our own family.

Superstitious, cruel, and needless international commerce, cloaked in the garments of traditional culture and inexcusable fashions, is equally depressing, with Indian tigers killed for their parts for elderly Chinese men and Tibetan antelopes slaughtered to make $2,000 Shahtoosh scarves for society's wealthiest.

Almost an eighth of the world's remaining wild species of birds and nearly a fifth of the mammals, 5 percent of the fish, and 8 percent of terrestrial plants are seriously threatened with extinction (Heywood and Watson, 1995). Almost all big animals are in trouble: the big cats, storks and cranes, parrots, pythons, antelopes, the great apes, elephants, and rhinos. Fewer than 5,000 tigers remain in nature, fewer than 400 Sumatran rhinos, perhaps 600 mountain gorillas, fewer than 100 Philippine eagles, and a handful of Hawaiian monk seals. The African violet, so common in our parlors, is nearly gone in Africa. Ninety percent of the black rhinos have been killed in the past eighteen years, one-third of the world's 266 turtle species are now threatened with extinction, and 10 of the 17 species of penguins are declining.

A Plague of Towers: Domestic and Exotic Animals

Bushmeat hunters are not common in the United States, but tower builders and cat lovers are everywhere. Tall communication towers kill as many as 40 million birds of some 230 species each year, mostly at night. There are now nearly 50,000 such towers and 10,000 more in planning (Holden, 2001; Winegrad, 2000). House cats are even more destructive.

Free-ranging domestic cats in Wisconsin alone are killing about 39 million birds each year (Coleman et al., 1997). Nationwide, free-ranging cats probably kill more than a billion small mammals and hundreds of millions of birds each year. Worldwide, cats are thought to have been responsible for the diminution of more bird species than any other cause except habitat destruction. In the United States, they are contributing to the endangerment of such birds as least terns, piping plovers, and loggerhead shrikes.

Beyond cats, humanity's vast herds of domestic animals have become a plague to wildlife, devastating habitat and spreading disease, such as anthrax, foot and mouth, rinderpest, and distemper. Bovine tuberculosis has spread from domestic cattle to wild buffalo, and thence to lions, cheetahs, kudus, and baboons in South Africa, and it threatens wood bison in Canada. The World Resources Institute reports that the average population of domestic cattle in the years 1990 through 1992 was 1.3 billion (and each cow produces raw waste equivalent to that of 22 human beings); there were 1.8 billion sheep and goats, 900 million pigs, 100 million equines, 200 million buffalo, and 17.2 billion chickens.

The only hope for many island creatures, whether they be Galápagos giant tortoises, Jamaican iguanas, or Hawaiian honeycreepers, is the extermination of introduced goats, rabbits, pigs, dogs, mongooses, rats, and cats. Some native species, their numbers disproportionately enlarged by their ability to live on human garbage, have become equally destructive. For example, herring (*Larus argentatus*) and black-backed gulls (*Larus marinus*) have become predators of terns in the United States, and superabundant kelp gulls (*Larus dominicanus*) are killing cormorants and terns in Argentina.

Amazingly, we continue to distribute an unrelenting stream of destructive exotics. They range from zebra mussels in the Great Lakes to European boars in California and

Scottish deer in Argentina—to say nothing of what has happened in the famously disastrous situations in New Zealand, Hawaii, and Australia. In California's San Francisco Bay, an average of one new species has been established every thirty-six weeks since 1850, every twenty-four weeks since 1970, and every twelve weeks in the last decade (Eisenrink, 1999; Kaiser, 1999). In the United States, it is estimated that invading exotic species cost us about $137 billion annually (Pimental et al., 1997, 2000).

A far-ranging study has concluded that more than 40 percent of Earth's total terrestrial photosynthetic productivity is being appropriated by human beings (Vitousek et al., 1986). We also consume 25 to 35 percent of the primary productivity of the ecosystems of the continental shelves (Roberts, 1997) and use 54 percent of all runoff in rivers, lakes, and other accessible sources of fresh water (Postel et al., 1996). Yet we have invested only 4 to 6 percent of the land and 0.5 percent of the marine realm in nature protection (Freese, 1998). This is the background of the extinction crisis.

Caring for Small Populations: Megazoos, Miniparks

Stuart Pimm (1999:853) observed, "When it comes to protecting ecosystems, larger is better than smaller, connected is better than fragmented and natural is better than managed." However, in the foreseeable future, the survival of more and more species will have to depend on the care of habitat fragments, the enhancement of marginal habitat, and the restoration of some of that which has been lost. Many species are destined to survive, if at all, in small, disjunct populations in areas with limited carrying capacity and no room to grow. Their survival will depend on human care.

This means that we will have to agree what species we want to care for, where, and in what numbers, and we will have to control the numbers of predators or competitors that threaten them, and their diseases. Their survival will require the creation of genetic and demographic models with which managers must work. The task is so immense that we will have to focus on the well-being of particular kinds, especially "landscape species" (Lambeck, 1997; Robinson, 1999). Ultimately, we must develop new ways of thinking about species–area relationships (see May and Stumpf, 2000). Like it or not, much wildlife conservation will become an aggregation of miniparks and megazoos (Conway, 1989).

Will genetic change affect captive animal populations or small reserves, reducing the ability of some species to survive? The answer is "yes," but the problem is not just change but the direction of change and its time scale. Animals bred in zoos to sustain maximum heterozygosity, or preserved as frozen zygotes, may almost be said to be genetically fossilized or obsolescent, adapted to environments that may no longer exist. Those in undersized reserves are no better off (Conway, 1995b). What does this mean for reintroductions?

Reintroduction

Initially, rare animal reintroductions from captive collections were discouraging, much more so than translocations (Beck et al., 1994). After all, unless the factors that resulted in a species' decline in the first place have been improved, there is no place for reintroduction. Of course, some genetic change will be less relevant if both the reintroduction habitat and its diseases and predators have changed. Besides, reintroduction efforts usually stimulate habitat restoration and protection, some species are more elastic than we thought, and we are getting better at it.

The reintroduction of the American bison (*Bison bison*) in the West, shortly after the turn of the century, by the Wildlife Conservation Society's Bronx Zoo, was among the earliest. There are now more than 150,000 bison in parks and reserves—and on menus. The recent reestablishment of the peregrine falcon (*Falco peregrinus*) has been particularly successful and logistically complex. In fact, since 1986, twenty-one of twenty-eight reintroductions of raptors resulted in the establishment of viable breeding populations. Most used captive-bred young (Cade, 2000), and some are extraordinary.

For example, in 1974 only four Mauritius kestrels (*Falco punctatus*) were left in nature. By 1984, captive breeding, clutch-doubling, predator control, and the like had made possible the release of nearly 400. There are now at least 600 birds in the wild (Jones et al., 1999). But there is also the case of the Arabian oryx. Shot to near extinction by local hunters, they were nearly lost. Then, beginning in the 1960s, captive stocks were gradually increased. At last, with support from the sultan of Oman, they were reintroduced on the Arabian Peninsula in 1982. Soon more than 400 of these magnificent antelope were running free. Last year, a quarrel broke out between local tribes, and the oryx were slaughtered (M. R. Stanley-Price, personal communication, 2004). Scarcely 100 were rescued and returned to captivity. Nevertheless, reintroduction attempts are becoming more common.

Between 1900 and 1992, attempts were made to reintroduce only two species of invertebrates, but in 1998 alone, nineteen were reintroduced. The comparable figure for fish reintroductions from 1900 to 1992 is nine, but eleven were reintroduced in 1998 alone. For reptiles and amphibians, there were twenty-two reintroductions from 1900 to 1992 and forty-two in 1998; for birds, fifty-four and sixty-nine, respectively; and for mammals, thirty-nine from 1900 to 1992 and then seventy-seven in 1998 alone (Maunder et al., 1999). Almost all of these efforts have brought new attention to habitat protection.

Most reintroduced animals will have to live in an altered environment where predation and behavioral challenges are dealt with by monitoring and management (Conway, 1995a, 1995b). Studies of moose (*Alces alces*) predation by reintroduced wolves (*Canis lupus*) in Wyoming, by the Wildlife Conservation Society's Joel Berger, have shown that a single generation was enough time for naive populations, which had had no contact with wolves for seventy-five years, to get wise to them and react appropriately (Berger et al., 2001; Gittleman and Gompper, 2001).

Climate Change

Naive or not, almost all wildlife is threatened by the menace of global warming. Because future climatic changes will involve both anthropogenic and natural forces, they may be rapid and large and will affect huge human populations. Although some societies may be able to cope, there will be unprecedented social disruptions. Most plants and wild animals will be unable to follow shifting climate zones, blocked by human developments. Climate is expected to warm most dramatically in the arctic and boreal zones and change least in the biodiverse tropics. However, human land use remains the major driver of habitat change and extinction, and it is taking place most rapidly in the tropics (Sala et al., 2000; Weiss and Bradley 2001).

Conservation Strategies

What conservation strategies might be helped by new genetic knowledge or reproductive technologies? Will cloning prove important?

Dolly the sheep was cloned as the sole result of 277 nuclear transfer–derived embryos produced from 400 oocytes recovered surgically from donor ewes. Dolly was the only successful result. Obviously, we know far too little of the reproductive biology of most endangered species and have far too few appropriate surrogates for such remarkably invasive clone or embryo transfer procedures (Loskutoff, 2001). Consider the number of species that need help and the scale of the problem.

Nonetheless, there are situations in which assisted reproduction, beyond the low-tech methods of artificial incubation, egg and chick switches, and clutch-doubling commonly used with birds, can be useful. Artificial insemination has long been important in endangered crane and raptor propagation and, recently, in breeding black-footed ferrets (*Mustela nigripes*). Genome resource banking and "frozen zoo" sperm banks might provide alternatives to natural breeding when animal pairs are incompatible. They might allow posthumous reproduction of genetically valuable animals and offer insurance against catastrophic loss. The ability to move sperm rather than animals between parks or zoos may become of increasing genetic importance (Goodrowe, 2001). However, given the scale of the global extinction crisis (there are nearly 1,000 threatened species in the United States alone), the conservation implications of high-tech assisted reproductive technology are extremely modest (Asa, 2001).

But modest or not, genetic understandings and applications are fundamental to the maintenance and reconstruction of small populations. Its insights are essential to the evaluation of founder stock for reintroduction, dispersal effectiveness, genetic drift, minimum viable populations, and translocation strategies—and hence management methods. Moreover, genetics has become a forensic science, helping to determine what a species is or was and where it came from. From as little as a restaurant scrap, geneticists can distinguish an endangered species of whale or monkey. From parrot feathers or gorilla dung, they can reveal family relationships with surprising accuracy.

However, when it comes to dung, trained scenting dogs have proven 100 percent accurate in distinguishing kit fox scats (*Vulpes macrotis mutica*) from those of coyote (*Canis latrans*), skunk (*Mephitis mephitis*), and badger (*Taxidea taxus*) (Smith et al., 2001), whereas identifications based on mtDNA analysis by geneticists cost about forty dollars per sample. Perhaps some analysts should be trained to follow their noses (Smith et al., 2001).

Conclusions

Today, almost every educated person understands that the preservation of biodiversity is fundamental to the maintenance of healthy ecosystems, the stewardship of natural resources, the realization of human potential and enjoyment, and a sustainable society.

Everyone also knows that few decision makers take any of this seriously and that not one nation on Earth has made preservation of its biological environment a budgetary priority. At the Kyoto Conference on climate change, for example, the United States agreed to cut carbon dioxide emissions by 7 percent from their 1990 levels by 2010. Between 1990 and 1998, emissions grew 11.5 percent, and the United States has discarded the agreement.

Over the long term, nature conservation and economic development are not in conflict. Over the short term and on the local level, there are huge conflicts. No one has figured out how to get rich from mountain gorillas, whooping cranes, black-footed ferrets, snow leopards, or furbish louseworts. For most people, wildlife conservation seems a luxury. Technological fixes and easy ways out dominate our attention.

Disconnects between scientific understanding and productive applications to global problems are common. But if the overall task of wildlife conservation is to save functioning ecosystems and wildlife populations, perhaps a particular focus of conservation genetics must be on helping to save broken ecosystems and marginalized species. Conservation genetics can help us care for small populations that will otherwise be lost.

The news that a population is facing genetic drift or founder effect will rarely win protection for a species or its habitat. That takes inspiration, education, and on-the-ground involvement. But we must not allow the public to delude itself about the scale of the extinction crisis or the difficulty of its resolution. There are no quick technological fixes. To achieve major wildlife conservation, we must orchestrate a continuity of political commitment that is wholly foreign to human affairs. It will take the work of thousands of people and the understanding of millions to make a difference. If we want to save wildlife, we will have to save habitat.

Acknowledgments

I thank George Amato, James Doherty, and John Robinson for helpful information and comments.

References

Asa, C. 2001, February. The future of reproductive biology in zoos and in conservation. *Communiqué,* pp. 19–21.

Beck, B. B., L. G. Rapaport, M. R. Stanley Price, and A. C. Wilson. 1994. Reintroduction of captive-born animals. In *Creative conservation: Interactive management of wild and captive animals. Proceedings of the Sixth World Conference on Breeding Endangered Species,* ed. G. Mace, P. Olney, and A. Feistener, 265–286. London: Chapman and Hall.

Berger, J., E. Swenson, and I.-L. Persson. 2001. Recolonizing carnivores and naïve prey: Conservation lessons from Pleistocene extinctions. *Science* 291:1036–1039.

Bird, M. 2001, January 8. Noah's new ark. *Time,* pp. 30–31.

Cade, T. J. 2000. Progress in translocation of diurnal raptors. In *Raptors at Risk,* ed. R. D. Chancellor and B.-U. Meyburg, 343–372. New York: WWGBP/Hancock House.

Coleman, J., S. A. Temple, and S. R. Craven. 1997. *Cats and wildlife*. Odanah: Great Lakes Indian Fish and Wildlife Commission and the University of Wisconsin.

Conway, W. 1989. *Miniparks and megazoos: From protecting ecosystems to saving species*. The Thomas Hall Lecture. St. Louis, Mo.: Washington University.

——. 1995a. Foreword. Altered states: Population viability analyses, planning and caring for wildlife in parks. In *Population management for survival & recovery,* ed. J. Ballou, M. Gilpin, and T. Foose, xiii–xix. New York: Columbia University Press.

——. 1995b. Wild and zoo animal interactive management and habitat conservation. *Biodiversity and Conservation* 4:573–594.

Eisenrink, M. 1999. Biological invaders sweep in. *Science* 285:1834–1836.

Freese, C. H. 1998. *Wild species as commodities: Managing markets and ecosystems for sustainability.* Washington, D.C.: Island Press.

Frisch, M. 1957. *Homo Faber.* New York: Harvest.

Gittleman, J. and M. Gompper. 2001. The risk of extinction: What you don't know will hurt you. *Science* 291:997–999.

Goodrowe, K. 2001, February. The role of genome resource banking in wildlife conservation programs. *Communiqué,* pp. 13–14.

Heywood, V. and R. Watson, eds. 1995. *Global biodiversity assessment 4.* Cambridge: Cambridge University Press.

Holden, C. 2001. Curbing tower kill. *Science* 291:2081.

Jones, C., K. Swinnerton, J. Hartley, and Y. Mungroo. 1999. The restoration of the free-living populations of the Mauritius kestrel (*Falco punctatus*), pink pigeon (*Columba mayeri*) and echo parakeet (*Psittacula eques*). In *Seventh World Conference on Breeding Endangered Species: Linking Zoo and Field Research to Advance Conservation,* ed. T. Roth, W. Swanson, and L. Blattman, 77–86. Cincinnati: Cincinnati Zoo and Botanical Garden.

Kaiser, J. 1999. Stemming the tide of invading species. *Science* 285:1836–1841.

Lacy, R. C. 1993. Vortex: A computer simulation model for population viability analysis. *Wildlife Research* 20:45–65.

Lambeck, R. 1997. Focal species: A multi-species umbrella for nature conservation. *Conservation Biology* 14:849–856.

Laurance, W. F., M. A. Cochrane, S. Bergen, P. M. Fearnside, P. Delamônica, S. Barber, et al. 2001. The future of the Brazilian Amazon. *Science* 291:438–439.

Loskutoff, N. 2001, February. Giving nature a helping hand. *Communiqué,* pp. 5–6:43.

Maunder, M., M. Stanley-Price, and P. Soorae. 1999. The role of in-country ex-situ facilities in supporting species and habitat recovery: Some perspectives from East Africa. In *Seventh World Conference on Breeding Endangered Species: Linking Zoo and Field Research to Advance Conservation,* ed. T. Roth, W. Swanson, and L. Blattman, 31–47. Cincinnati: Cincinnati Zoo and Botanical Garden.

May, R. and M. Stumpf. 2000. Species–area relations in tropical forests. *Science* 290:2084–2086.

Pimental, D., O. Bailey, P. Kim, E. Mullaney, J. Calabrese, L. Walman, et al. 1999. Will limits of the earth's resources control human numbers? *Environment, Development and Sustainability* 1:19–39.

Pimental, D., L. Lach, R. Zuniga, and D. Morrison. 2000. Environmental and economic costs of nonindigenous species in the United States. *BioScience* 50(1):53–65.

Pimental, D., C. Wilson, C. McCullum, R. Huang, P. Dwen, J. Flack, et al. 1997. Economics and environmental benefits of biodiversity. *BioScience* 47(11):747–757.

Pimm, S. L. 1999. Seeing both the woods and the trees. *Nature* 402:853–854.

Postel, S. L., G. C. Daily, and P. R. Ehrlich. 1996. Human appropriation of renewable fresh water. *Science* 271:785–787.

Ridley, M. 1999. *Genome*. New York: HarperCollins.
Roberts, C. M. 1997. Ecological advice for the global fisheries crisis. *T.R.E.E.* 12:35–38.
Robinson, J. 1999. *Biodiversity conservation at the landscape scale*. Unpublished manuscript. Bronx, N.Y.: Wildlife Conservation Society.
Robinson, J., K. Redford, and E. Bennett. 1999. Wildlife harvests in logged tropical forests. *Science* 284:595–596.
Sala, O., F. S. Chapin III, J. J. Armesto, R. Berlow, J. Bloomfield, R. Dirzo, et al. 2000. Global biodiversity scenarios for the year 2100. *Science* 287:1770–1774.
Smith, D. A., K. Ralls, B. Davenport, B. Adams, and J. E. Maldonado. 2001. Canine assistants for conservationists. *Science* 291:435.
Terborgh, J. 1999. *Requiem for nature*. Washington, D.C.: Island Press.
Vitousek, P., P. R. Ehrlich, A. H. Ehrlich, and P. A. Matson. 1986. Human appropriation of the products of photosynthesis. *BioScience* 36:368.
Weiss, H. and R. Bradley. 2001. What drives societal collapse? *Science* 291:609–610.
Winegrad, G. 2000. New study documents towers killing birds. *Bird Calls* 4(1):1.

George Amato

3

Moving Toward a More Integrated Approach

The debate over reductionism versus holism has been ubiquitous in the different disciplines within biology (Hagen, 1989). This debate dominated my intellectual environment while I was being trained as a geneticist and conservation biologist in the early stages of genetic studies of populations. In the 1970s we explored population biology and genetics using the "cutting-edge" technologies of isoelectric focusing. We quickly leapt to Southern blotting as we sought to understand evolutionary processes by examining and characterizing genetic variation, admittedly a bit indirectly at times.

As a contrast, consider the university where I spent most of my formative years: Yale University. Osborn Memorial Laboratory is an old, stone building, looking a bit like a medieval castle sitting at the bottom of "Science Hill," the high point of the Yale campus. High above us on the hill, casting an enormous shadow, was the modern, fourteen-story brick-and-glass monument to biological reductionism: Kline Biology Tower. It was here that the secrets of living systems were being revealed by examination of each individual molecule. Those of us who continued to hold faith in our core belief that, as Dobzhansky said, "Nothing in biology makes sense except in the light of evolution," criticized the small-mindedness of the Kline Biology Tower's reductionist approach while adopting their techniques as ours and watching most of the funding go to them. Although it remains unclear how much we are a product of our environment, it certainly plays a significant role in how we view the world. And this experience certainly shapes my view of how the field of conservation genetics needs to evolve.

The ever-expanding effect humans have on the environment and the concomitant extinction crisis are the reason why many of us turned our interests in the natural world to a calling in conservation. Although we often talk about this as some sort of objective, reasoned decision, it can honestly be viewed only as a subjective appreciation for living systems that persist in less human-dominated landscapes. However, the current trend is obvious to anyone who shares this admittedly subjective appreciation and concern. In chapter 2, Bill Conway provides us with the facts and figures—as he has so often in the past—that reinforce the bleak future that confronts us. And yet so many of us feel compelled to marshal a challenge to these trends, although I often wonder whether the notion that it is within our power to alter this future is just another example of the remarkable hubris that seems to define our species.

As everyone who has thought about and discussed conservation observes, it will take changes and commitments from nearly every element of human society to implement meaningful and sustained conservation. As is natural, those who respond to the calling look to their own interests and talents as they seek to contribute to this effort. Conservation biology arose out of the notion that traditional disciplines in the biological sciences could provide important insights and tools for conservation. Although I believe that most people think of science as a set of facts that change with sufficient frequency to cast doubt on their veracity, scientists tend to be committed to the discipline and process of science as one that has enormous explanatory power and is grounded in sufficient objective and repeatable rules to distinguish it from cultural superstition and other nonscience systems. Because the goal is to conserve living systems, it seemed especially logical to conclude that science would be a necessary, if not sufficient, component of a successful conservation effort. Those of us who had an interest in and understanding of population genetics and evolutionary biology combined with a conservation ethic saw the emergence of a new discipline, conservation genetics, as a specific and obvious example of the value of this new conservation biology paradigm.

Populations that were declining in numbers while becoming increasingly fragmented would be subject to the effects that result when genetic variation declines, individuals become inbred, and natural patterns of dispersal and recruitment are disrupted. Using theoretical and experimental methods from these established disciplines, scientists (including many who contributed to this volume) helped conservation managers by documenting and exposing intrinsic genetic threats in specific endangered species, by more accurately identifying the natural distribution of genetic variation both below and above the species level, and by providing insights into the complexity of these living systems. Interestingly to many conservation geneticists, this discipline has been viewed as a high-tech approach by other conservationists and even by other conservation biologists. In fact, conservation genetics has seemed always to be associated with a controversy pitting overstatements of its importance by its proponents against statements such as Michael Soulé's comment at the San Diego Symposium in 2000 "that there is no example of a successful, high-tech approach to conservation." This debate within the conservation community was elevated almost to Civil War proportions by Grahame Caughley's posthumously published paper relegating the argument to his division of "declining" and "small" population paradigms. The implication was that a focus on molecular genetic analysis and examination of intrinsic biological factors was of little or no importance in comparison to examining the broader anthropogenic reasons for the species' decline (Caughley, 1994). This argument, taken a bit out of context, became a classic straw man debate on both sides, when it is obvious that to understand and ameliorate the threats, it was necessary to examine and research each conservation challenge as a single problem rather than to artificially divide it into two competing paradigms.

This brings us to the direction in which conservation genetics needs to move if it is to contribute more substantively to biodiversity conservation. What we need to do is move toward a greater integration with the other disciplines of conservation biology and conservation management. And we need to continue and expand it in a holistic framework. In the debate over reductionism versus holism in ecology, the notion that ecosystems have emergent properties that could not be understood only through examination of their components drove the holistic approach (DeSalle and Amato, 2004). Similarly, conservation issues involving such complex systems will necessitate even greater attention to the relationships between biological and nonbiological phenomena in a way that is challenging to traditional science. However, this is not a call for less detailed or less basic research. On the contrary, we will need to renew our efforts to apply the highest-quality scientific research to these complex problems while attempting to elucidate and modify human effects on these systems. We start down a slippery slope when we try to artificially divide applied and basic conservation research. We must resist the tempting but simplistic call for cookie cutter approaches based on the conservation paradigm of the week. It reduces our opportunity to make the novel discoveries we need to better understand the impact of our actions on these complex systems.

In the age of genomics, we find both opportunities and challenges to moving in this direction. The opportunities lie primarily in an increased capacity to rapidly accumulate data to test the numerous hypotheses and answer the questions confronting conservationists. Other opportunities include the development of some novel ways of manipulating living systems. However, there are a great many challenges. In terms of data collection, one of the challenges will be to keep the endangered forest in sight as we genetically characterize every single tree. Another challenge is to avoid letting the press and public's fascination with genetic manipulation lead to a false sense of optimism and hope for a technological solution to this extinction crisis. It is also important that we avoid allowing technological manipulation of individuals to further separate them from the environments in which they are uniquely connected and defined. The overriding message is that context and connection are the necessary framework. We have now sequenced the human genome—a technologically remarkable achievement—but we remain challenged by what the sequences do not yet reveal. Back in the Kline Biology Tower at Yale University, one of my former colleagues, Frank Ruddell, is beginning to think that some of the answers lie in what we used to call junk DNA (Shashikant et al., 1998), representing a clearly less reductionist approach. Whatever genomics has to offer to conservation genetics, it is not a call to further reductionism.

I remain optimistic that we are beginning to head in the right direction. In recent years we have seen a merging of conservation genetics and ecology in the new field of molecular ecology. Using molecular markers to better understand the life history traits of organisms that are difficult to observe directly provides a very powerful way of understanding behavior, ecology, and genetics simultaneously. Similarly, molecular forensics applied to the trade in wildlife integrates an understanding of human behavior and the law while giving insights into species distributions and other biological information.

In our own conservation genetics program in New York, we use very different approaches than when I began fifteen years ago. It is more difficult to administratively wrap hands around our program in this less reductionist approach because a few of the lines are a little more blurred, as I strongly feel that they should be. I've never considered that the program was precisely a genetics program or that it should be. I understood why the Wildlife Conservation Society was more comfortable describing my work as genetics (which was murky enough), even though I always felt that I had been trained as and was an evolutionary biologist. Our program has become an authentic combination of population biology, genetics, systematics, molecular ecology, and forensics. Our staff, postdoctoral fellows, and graduate students come from varied backgrounds, but nearly all have authentic field research experience and excellent laboratory skills and are broadly trained in interdisciplinary approaches in biology. They are encouraged to develop conservation programs that bridge the basic research and science with a conservation management and training focus. Our experiences and successes have taught us to resist the institutional push to reduce this work to more service-oriented, uncreative, routine analyses that ultimately will not result in high-quality work or synthetic understandings of biological processes necessary for conservation.

A few brief examples illustrate this approach. The Wildlife Conservation Society has developed a major cetacean conservation program (jointly with the American Museum of Natural History) that includes molecular genetics, population biology and demography, geographic information systems, remote tracking, local training, graduate student training, capacity building, and ecotourism (Rosenbaum and Collins, 2004). Another example of this synthesis is provided by the use of novel microsatellite markers for the endangered St. Vincent Amazon parrot. Working with Bronx Zoo staff and resource managers on St. Vincent and the Grenadines, researchers used genetic analysis of captive animals to manage the ex situ population. This research has also revealed illegal smuggling of birds, included ecological studies in the eastern Caribbean, and is in part a collaborative, long-term, on-the-ground conservation effort with local nongovernment organizations and other nongovernment partners (Russello and Amato, 2001, 2004a). These two examples reflect how our program has tried to encourage and train a new generation of conservation geneticists.

In all our current projects, we seek to find greater understanding or greater application to conservation. In the last couple of years we have received a lot of press attention for our work describing new species of deer from Southeast Asia, in collaboration with George Schaller and Alan Rabinowitz, in our field research program (Amato et al., 2000; Rabinowitz et al., 1998). However, this was never a research project about deer taxonomy and systematics, nor was it about the excitement of new discovery. It was a well-articulated and novel attempt to add rigor to prioritizing areas for conservation based on measures of species richness and levels of endemism. We sought to demonstrate that linking molecules and systematics with biological surveys could result in an enhanced protected area strategy (Amato et al., 1999a).

Finally, we have found that in a holistic approach it is important to develop meaningful strategic partnerships. Our primary one has been our long-term collaborative relationship with the American Museum of Natural History and the Molecular Systematics Program. Also, the museum's Center for Biodiversity and Conservation has provided support for additional collaboration with museum scientists and staff. Our program's links with Yale, Columbia, New York, and Fordham universities add immeasurably to these efforts while providing conservation context and experiences for students in these programs. Although

I have used our own program to illustrate these goals, I do want to acknowledge that we have been inspired by and learned from many creative colleagues, especially those who have contributed to this volume.

As conservation genetics continues to evolve as a discipline, we need to be open to new approaches representing greater integration with ecology and other subdisciplines of biology, greater integration with the abiotic issues in conservation, and a holistic, system-based approach using all the innovative techniques synthesized from our past successes and failures. We must avoid intramural wrestling matches while developing collegial, strategic partnerships. In this way, we position ourselves to better understand these living systems, the effects of human behavior on these systems, and the options that are open to us to conserve significant elements of these systems.

References

Amato, G., M. G. Egan, and G. Schaller. 1999a. A new species of muntjac, *Muntiacus putaoensis* (Artiodactyla: Cervidae) from northern Myanmar. *Animal Conservation* 2:1–7.

Amato, G., M. G. Egan, and G. B. Schaller. 2000. Mitochondrial DNA variation in muntjac: Evidence for discovery, rediscovery, and phylogenetic relationships. In *Antelopes, deer, and relatives: Fossil record, behavioral ecology, systematics, and conservation,* ed. E. S. Vrba and G. B. Schaller, 245–249. New Haven, Conn.: Yale University Press.

Amato, G., M. G. Egan, G. B. Schaller, R. H. Baker, H. C. Rosenbaum, W. G. Robichaud, et al. 1999b. The rediscovery of Roosevelt's barking deer (*Muntiacus rooseveltorum*). *Journal of Mammalogy* 80(2):639–643.

Caughley, G. 1994. Directions in conservation biology. *Journal of Animal Ecology* 63:215–244.

DeSalle, R. and G. Amato. 2004. The expansion of conservation genetics. *Nature Reviews. Genetics* 5:702–712.

Hagen, J. B. 1989. Research perspectives and the anomalous status of modern ecology. *Biology and Philosophy* 4:433–455.

Rabinowitz, A. A., G. Amato, and U. Saw Tun Khaing. 1998. Discovery of the black muntjac, *Muntiacus crinifrons (Artiodactyla, Cervidae)*, in north Myanmar. *Mammalia* 62(1):105–108.

Rosenbaum, H. C. and T. Collins. 2004. The ecology, population characteristics and conservation efforts for humpback whales (*Megaptera novaeangliae*) on their wintering grounds in the coastal waters of Gabon. In *Natural history of the Gamba complex,* ed. A. Alonso and P. Campbell, 175–187. Washington, D.C.: Smithsonian Institution Press.

Russello, M. A. and G. Amato. 2001. Application of a noninvasive, PCR-based test for sex identification in an endangered parrot, *Amazona guildingii*. *Zoo Biology* 20:41–45.

Russello, M. A. and G. Amato. 2004a. Ex situ population management in the absence of pedigree information. *Molecular Ecology* 13:2829–2840.

Russello, M. A. and G. Amato. 2004b. A molecular phylogeny of *Amazona*: Implications for Neotropical parrot biogeography, taxonomy and conservation. *Molecular Phylogenetics and Evolution* 30:421–437.

Shashikant, C. S., C. B. Kim, M. A. Borbely, W. C. H. Wang, and F. H. Ruddle. 1998. Comparison studies on mammalian Hoxc8 early enhancer sequence reveal a baleen whale–specific deletion of a *cis*-acting element. *Proceedings of the National Academy of Sciences USA* 95:15446–15451.

Part II

Conservation Genetics in Action

Assessing the Level and Quality of Genetic Resources in Endangered Species

The broad range of specific approaches to conservation genetics is a result of continual incorporation of new techniques and theory into the field. In this section several genetic approaches are articulated. Philip Hedrick points out that expanding genomic technology has given us a unique and novel view of genome dynamics in particular neutral and adaptive variants. This perspective is of great importance in conservation genetics because the information from the model organism genome sequencing will provide a richer background knowledge of genome dynamics. Next, Robert Lacy describes the utility of genetics in managing captive populations. In particular, Lacy explains that so-called low-tech genetic approaches such as inferences from animal breeding via pedigree analysis are valid and important to conservation biology. Stephen Palumbi examines the current status of many marine ecosystems as conservation subjects. Palumbi suggests that by combining a wide array of modern techniques, including the integration of geographic information and DNA sequence tagging, we can meet the goal of marine conservation more easily. Paul Goldstein describes the plight of the endangered tiger beetle (*Cicindela dorsalis*). In so doing he describes a framework for species recognition using gene sequences and explains the importance of species recognition in modern conservation biology. In the final chapter, James Gibbs examines the erosion of global genetic resources. According to Gibbs, the bottom line is that more attention must be paid to population loss in order to stop the trend of eroding genetic resources in conservation. The chapters in this section should provide the reader with a sense of the broad range of conservation genetics questions currently being examined.

Philip W. Hedrick

4

Neutral, Detrimental, and Adaptive Variation in Conservation Genetics

Because of the new molecular genetic information from various genome projects, new applications and insights for genetic studies in endangered species are being developed. Neutral variants are generally used for conservation applications and for estimating evolutionary parameters, and with highly variable loci, greater availability of markers, or extensive sequence data, these approaches should become much more informative. Detrimental and adaptive variations are also important in conservation genetics, but identification and characterization of such variations are more difficult. Neutral variants might be used to identify adaptive variants, but the overlay of different mutational processes and selective regimes suggests that great caution should be used in making such predictions.

Although other forces are of primary concern for avoiding extinction of most endangered species, for long-term persistence genetics has been a focus of conservation effort. Part of this emphasis on genetics resulted from the extensive coverage in the early edited volumes on conservation biology (Soulé, 1986; Soulé and Wilcox, 1980) and the detailed introduction of many aspects of evolutionary genetics to conservation (Frankel and Soulé, 1981; Schonewald-Cox et al., 1983). Various researchers thought that the measurement of genetic variation in a population genetics context would permit direct predictions about the history and status of endangered species. The framework of population genetics theory furnished an enticing and elegant approach to interpreting the measured amounts of genetic variation and predicting the future effects of evolutionary factors and management strategies.

However, the recommendations from population genetics were general and essentially amounted to avoiding inbreeding and maintaining genetic variation, with some exceptions (Hedrick and Miller, 1992). In some cases, these recommendations were consistent with management ideas from other risk avoidance approaches (e.g., avoid a prolonged low population number, split the population into subpopulations). However, the recommendations were generally vague and often focused on an expectation of what might occur for a typical, neutral polymorphic gene. Recently, the application of new molecular techniques has made examination of genetics feasible for many more endangered species, and the level of genetic analysis has become much more sophisticated. For example, instead of a few polymorphic loci, many highly polymorphic loci may be available so that questions that at one time were not completely resolved by molecular genetic surveys can be examined definitively.

These advances have generally applied to genetic variations that are thought to be neutral and consequently are used as markers of various evolutionary phenomena. Of great significance in conservation are variants that are not neutral, that is, those that have either detrimental or adaptive effects. Here, I provide an overview of how these three major types of genetic variation—neutral, detrimental and adaptive—may contribute to and be used in the conservation of endangered species. Of course, whether a particular variant is neutral, detrimental, or adaptive depends on the environment, population size, genetic background, and so on. For example, a particular allele that is adaptive, such as one providing resistance to an infectious disease in one environment, could be detrimental when the pathogen is absent because of a pleiotropic cost associated with the allele. Or genetic variants that are neutral in one situation may be adaptive in another. I also consider one of the great promises of neutral molecular variation: the use of the extent and pattern of neutral variation to predict the amount and significance of detrimental and adaptive variation. Although it has not proven compelling in some cases, with loci that are more informative and a larger number of genes, in the future we might be able to better use such observed associations.

Neutral Variation

The extent and pattern of molecular variation within a population are generally consistent with neutral theory, that is, a balance predicted by a reduction in variation from genetic drift and an increase in variation from mutation (Kimura, 1985; Nei, 1987). When the population is small, even if selection is acting on the variation at a given gene, genetic drift could have a greater effect on allele frequencies than does selection. In general, neutrality of genetic variants can be assumed when the selection coefficient s (either the selective disadvantage of a detrimental allele or the advantage of an adaptive allele) is less than $1/(2N_e)$, where N_e is the effective population size (Kimura, 1985). Therefore, because endangered species generally have low effective population sizes, genetic variants are more likely to be effectively neutral in endangered than in common species. For example, if s 0.01 and the effective population size is 50, then genetic drift should play a more important role than selection in determining the fate of the variant, whereas in a population of size 500, selection should be more important than genetic drift.

Most recent conservation genetics research has focused on the use of neutral molecular markers. Molecular genetic markers hold great promise for estimating fundamental parameters or characteristics important in conservation, such as past effective population size (Garrigan et al., 2002), past bottlenecks (Luikart and Cornuet, 1998), population origin of individuals (Cornuet et al., 1999), individual inbreeding level (Ellegren, 1999;

Lynch and Ritland, 1999), and sex-specific gene flow (Latta and Mitton, 1997) or founder contributions (Carvajal-Carmona et al., 2000). Overall, neutral genetic markers have been used in conservation primarily to identify species, evolutionarily significant units, and management units (e.g., Moritz, 1999). Highly variable genetic markers, such as microsatellite loci, have allowed the quantification of patterns that are not apparent when genetic markers with less variation are used. The use of extensive sequence data or large numbers of single-nucleotide polymorphisms might also provide high genetic resolution. Molecular markers can also be used to infer the historical and geographic relationships between groups (Avise, 2000). Although the power to infer such relationships is substantial, data from ancient specimens can now provide additional insight into the relationships between contemporary groups (Barnes et al., 2002).

However, application of highly variable loci (or very large numbers of markers) must be used with some caution (Hedrick, 1999) because statistically significant differences might not reflect biologically important differences or might give a different signal than do other markers (Balloux et al., 2000). For example, it is presumed that a statistically significant difference between groups for neutral molecular markers indicates the presence of biologically important differences or is at least thought to indicate that the groups have been separated long enough for biologically important differences to accumulate. Traditional molecular markers often provided inadequate statistical power to estimate the differences between groups in endangered species because there was generally little variation for these markers. However, with the use of highly variable loci, large numbers of independent markers, or extensive sequence data, the statistical power to differentiate between groups now is often very high.

What are the possible relationships between statistical significance based on molecular markers and biological meaningfulness of comparisons between groups? First, there may be both no significant statistical and no meaningful biological difference between groups, or there may be a significant statistical difference between groups that reflects a meaningful biological difference. In both situations, statistics based on molecular markers result in an appropriate evaluation of the biological situation.

However, problems result when statistical significance does not reflect biological meaningfulness, a conflict that can occur in two basic forms. There may be no statistical significance when there are actual biologically meaningful differences between groups, or there may be statistical significance between groups when there is no meaningful biological difference. In the former instance, there may be no significant difference based on molecular genetic markers, but other adaptively important loci may be highly differentiated between populations. For example, in Scots pine (*Pinus sylvestris*) from Finland, a number of different molecular markers show very little differentiation between northern and southern populations (Karhu et al., 1996). However, many important adaptive quantitative traits show high levels of genetic differentiation between these populations in common experimental environments. In this case, the molecular data appear to be adequately reflecting the high level of gene flow in Scots pine; however, the selective forces between populations are so strong that they overwhelm the effects of gene flow and result in large adaptive genetic differences between populations. In this case, the error is not a typical false negative because the result is correct for the neutral nuclear markers. The error results from not assaying, or being able to assay, directly the genes involved in adaptation.

The final classification portends a major concern in both evolutionary and conservation biology as large numbers of highly variable markers become available in many species. In general, statistical power for determining differentiation between groups is closely related to the number of independent alleles (S. Kalinowski, personal communication; 2001), so that even for a few highly variable microsatellite loci, there may be extremely high statistical power. When there is such high statistical power, very small molecular genetic differences between groups become statistically significant. This is not a typical false positive because the differences detected are real but so small that they do not reflect biologically meaningful differences.

To determine a biologically meaningful difference, we need to define some measure or effect related to the likelihood of the accumulation of significant biological differences. One potential way to examine the relationship between biological and statistical significance is to evaluate the statistical power to detect a known biological effect. For example, the statistical power to detect a one-generation genetic bottleneck of different sizes can be compared with the ancestral population for different numbers of loci (Hedrick, 1999). In general, we need to quantify the extent of the evolutionary effect that we are able to detect with highly variable molecular markers and evaluate whether this effect is likely to have important biological consequences.

Detrimental Variation

Perhaps the most important early contribution of genetics to conservation was the recognition of the importance of inbreeding depression (for a recent review, see Hedrick and Kalinowski, 2000; see also Keller and Waller, 2002). Inbreeding depression is generally defined as a reduction in fitness (or some component of fitness) with an increase in inbreeding, an effect thought to result from increasing the homozygosity of detrimental alleles (Charlesworth and Charlesworth, 1999). The mean population fitness can also decline over time because detrimental mutations with a small selective disadvantage in a small population can become fixed, much as if they were neutral (Wang et al., 1999). It is useful to distinguish between these effects (Kirkpatrick and Jarne, 2000) and define the genetic load as the reduction in mean population fitness compared with a population without lowered fitness resulting from detrimental variation. Inbreeding depression and genetic load have been of major concern for endangered species, and inbreeding avoidance has become a priority in captive breeding.

In a large population at equilibrium, substantial standing detrimental genetic variation is expected, and consequently a large reduction in fitness is expected if inbreeding occurs. There is also little genetic load because, due to the efficacy of selection in large populations, most of the detrimental variants are at low frequency and are recessive. However, if the population declines in number, purging of detrimental variation should take place, especially for alleles of large detrimental effect, thereby reducing inbreeding depression, but some detrimental variants might become fixed, particularly those of smaller effect, causing an increase in genetic load (Wang et al., 1999). If the population remains small for an extended period, more detrimental variation could be purged, further reducing inbreeding depression, but more detrimental variants could be fixed, causing higher genetic load. Such a population might show no lowered fitness upon inbreeding, but because of fixation of detrimentals, all individuals in the population might have a low fitness, and the population might have a high genetic load.

Several caveats should be mentioned. First, during this process some populations (or even species) might become extinct, and the ones going extinct could be the ones with higher genetic load. As a result, the remaining populations might not have as high a genetic load as would be expected from the standing amount of detrimental genetic variation. Second, genetic load might be documented as a low estimate of fitness compared with other populations or might be assessed by crossing with individuals from another population and observing the fitness of their progeny compared with progeny of within-population crosses. However, making such crosses might not be possible, or the groups might differ in other characteristics (Wang, 2000).

In *Drosophila melanogaster*, approximately half the effect of inbreeding depression is thought to be from nearly recessive lethals and half from detrimentals of small effect but higher dominance (Lynch et al., 1999; Wang et al., 1999). However, *D. melanogaster* generally has a very large effective population size, and the genetic architecture of their detrimental genetic variation probably reflects that of a large population near equilibrium. Alternatively, for many endangered species, genetic drift has been important, because of either a small current population size or a history of bottlenecks. As a result, these endangered species might have a different genetic architecture with fewer segregating variants of large detrimental effect (Hedrick, 2001b), lower inbreeding depression, and perhaps higher genetic load than do species with histories of larger population size.

The magnitude and specific detrimental effects of alleles on fitness are highly variable because they depend greatly on how these genotypes interact with the environment. Recent natural experiments are generally consistent with the pattern of greater inbreeding depression in more stressful environments (Hedrick and Kalinowski, 2000). Estimates of inbreeding depression from captivity or laboratory environments are generally thought to underestimate the effects in a natural environment. For example, the effect of inbreeding on male reproductive success in wild mice under laboratory conditions was minor, whereas in seminatural conditions, inbred males had only about 20 percent of the success of outbred males (Meagher et al., 2000). In a comprehensive examination in *Drosophila*, Bijlsma et al. (1999) found that the extent of inbreeding depression was greatly increased in stressful laboratory conditions and also found a low correlation between the fitness of genetic variants in different stressful environments. This suggests that generalizations about the detrimental effects of a variant over different stressful environments may not be possible.

It has been postulated that, counterintuitively, inbreeding depression may be reduced by breeding of related individuals (Byers and Waller, 1999). However, the effects observed in Speke's gazelle (Templeton and Read, 1983), often cited as a demonstration of reduction of inbreeding depression, are consistent with a temporal change in fitness in inbred individuals and are not necessarily the result of a reduction in inbreeding depression (Kalinowski et al., 2000). Significantly, in an examination of the potential effects of purging in seventeen mammalian species (Ballou, 1997), a nonsignificant reduction in inbreeding depression in the updated Speke's gazelle captive population was found, and the inbreeding depression in the Speke's gazelle was the highest of any of the species analyzed. After a thorough theoretical examination of the factors that may influence purging of inbreeding depression, Wang (2000:84) concluded that "it is not justified to apply a breeding program aimed at purging inbreeding depression by inbreeding and selection to a population of conservation concern."

Populations of some endangered species have become so small that they have lost genetic variation and appear to have deleterious genetic variants occurring at high frequency or fixed in the population (Land et al., 2001; Madsen et al., 1999; Westemeier et al., 1998). To avoid extinction from this genetic deterioration, some populations may benefit from the introduction of individuals from related populations or even subspecies for genetic restoration, that is, elimination of deleterious variants and recovery to normal levels of genetic variation. Recent research has shown that immigration can result in genetic rescue in experimental populations (Ebert et al., 2002; Richards, 2000; Saccheri and Brakefield, 2002).

The last remaining population of the Florida panther (*Puma concolor coryi*) provides an extreme example of this phenomenon in an endangered species. This population has a suite of traits that suggests that genetic drift has fixed (or nearly fixed) the population for previously rare and potentially deleterious traits. These traits, which are found in high frequency only in the Florida panther and are unusual in other puma subspecies, include high frequencies of cryptorchidism (unilateral undescended testicles), kinked tail for the last five vertebra, cowlick on the back, and the poorest semen quality recorded in any felid (Roelke et al., 1993, table 4.1). In addition, a large survey of microsatellite loci has shown that Florida panthers have much lower molecular variation than other North American populations of pumas (Culver et al., 2000).

The potential positive and negative genetic effects of introducing individuals from genetically diverse but geographically isolated populations into apparently inbred populations were evaluated theoretically before the introduction of Texas cougars into Florida (Hedrick, 1995). Assuming 20 percent gene flow from outside in the first generation (and 2.5 percent every generation thereafter), the fitness generally improves quickly. One concern about this approach is that any locally adapted alleles may be swamped by outside

Table 4.1 The proportion of Florida panthers (sample size in parentheses) with a kinked tail, cowlick, or cryptorchidism with no Texas cougar ancestry and those with Texas cougar ancestry (F_1, F_2, backcross to Texas [BC-TX], and backcross to Florida [BC-FL]).

		Texas ancestry				
	No Texas ancestry	F_1	F_2	BC-TX	BC-FL	Total
Kinked tail	0.88 (48)	0.00 (17)	0.00 (7)	0.00 (3)	0.20 (15)	0.07 (42)
Cowlick	0.93 (46)	0.20 (10)	0.00 (5)	0.00 (1)	0.60 (5)	0.24 (21)
Cryptorchidism	0.68 (22)	0.00 (2)	0.00 (2)	--	0.00 (1)	0.00 (5)

Sources: Land et al. (2001); Roelke et al. (1993).

gene flow. However, with this level of gene flow, fitness from advantageous alleles is only slightly reduced.

A program to release females from the closest natural population from Texas was initiated in 1995 to genetically restore fitness in this population. Five of the eight introduced Texas females produced offspring with resident Florida panther males, and a number of F_1 and F_2 offspring have been produced (Land and Lacy, 2000). At this point approximately 20 percent of the overall ancestry is from the introduced Texas cougars. Of the animals with Texas ancestry, only 7 percent have a kinked tail (compared with 88 percent before), and the ones with a kinked tail are progeny from backcrosses to Florida cats (table 4.1). Similarly but not as dramatic, only 24 percent have a cowlick (compared to 93 percent before), two F_1s and three backcrosses to Florida cats. Only five males with Texas ancestry have been evaluated for cryptorchidism, and all have two descended testicles, a reduction from 68 percent to 0 percent cryptorchidism. Semen characteristics have been evaluated in only one F_1 male, and it appears to be as good as or better than the average of Texas cougars, much better than that of Florida panthers. In other words, the introduction of Texas cougars has already resulted in a substantial reduction of the frequency of the detrimental traits that have accumulated in the Florida panther.

Little progress has been made in the genetic characterization of genes with detrimental effects segregating in endangered species. However, efforts in the human and other genome projects to identify genes causing inherited disorders should provide information about homologous genes in endangered species. In addition, the ability to map genes affecting fitness-related traits portends imminent knowledge of the detailed architecture of genes affecting inbreeding depression, that is, the number and location of the genes, the distribution of their effects and their dominance, and the interaction (epistasis) of different genes.

There are documented examples of inherited disorders in captive populations of endangered species (Laikre, 1999; Ralls et al., 2000) that would have important negative consequences for animals reintroduced to the wild. Although these recessive alleles occur at a low frequency, many individuals in the population may be heterozygous for them. Management with the aim of reducing the frequency of these alleles could be difficult and must be undertaken carefully to avoid jeopardizing the remaining genetic variation in these species (Lacy, 2000). For example, to eliminate the recessive allele for chondrodystrophy, a form of dwarfism, in the California condor (*Gymnogyps californianus*) would require that more than half the population be prevented from breeding (Ralls et al., 2000). On the other hand, it appears that hereditary blindness in a captive population of gray wolves (*Canis lupus*) could be reduced without greatly influencing the genetic variation in the rest of the genome (Laikre et al., 1993). Selection will also reduce the frequency of detrimental alleles. However, if they are recessive the effect will be quite slow. If the population size is small, then some detrimentals will be effectively neutral, and purifying selection will not be effective. On the other hand, in small populations the expected frequency of recessive detrimentals of large effect is much lower than in an infinite population (Hedrick, 2001b; Wright, 1937).

Adaptive Variation

The extent and pattern of adaptive (advantageous) variation are crucial to the long-term survival of endangered species. In particular, if there is no adaptive variation in a population and it faces a new environmental challenge, such as a new disease or introduced species, it has no potential for adaptive response except from new mutations or gene flow from other taxa. However, determining the extent and pattern of adaptive variation in the present or presumed future environments is quite difficult. Potentially, experimental tests of fitness and adaptiveness in a variety of environments could be carried out, but this is difficult, even in a model organism such as *D. melanogaster* and virtually impossible in an endangered species.

The extensive molecular data available today provide potentially new ways to determine whether adaptive selection has operated in the past on a given gene and therefore whether it might operate in the future. For example, rather than being able to measure the impact of selection in a single or a few generations by determining differential viability or reproduction, one could observe the cumulative effect over many generations, or the results of a selection episode some time in the past, in analysis of DNA variation. Recent surveys estimate the number of genes that can be classified as having experienced bouts of adaptive evolution in vertebrates to be 280 of 5,305 genes examined (5.3 percent), whereas in plants 123 of 3,385 (3.6 percent) genes show a signal of adaptive evolution (Liberles et al., 2001).

Two approaches used to detect adaptive selection are comparisons of the rate of nonsynonymous substitutions (those that cause a change to an amino acid) to synonymous substitutions and the pattern of sequence variation within and between species (Hughes, 1999; Richman, 2000; Yang and Bielawski, 2000). Garrigan and Hedrick (2003) have found that the signal of adaptive selection observed by these tests usually is generated much more quickly than it is lost. For example, an observation that the rate of nonsynonymous substitutions is greater than the synonymous rate indicates only that at some time in the distant past, maybe even before the origin of the current species, adaptive selection occurred. To determine whether adaptive selection is occurring at present, direct measures of selective difference should be estimated (Garrigan and Hedrick, 2003; Hedrick, 2000).

In general, there is no obvious molecular way to differentiate alleles that are adaptive, neutral, or detrimental, except for detrimental variants that have stop codons or those for which there is detailed structural information on the molecule (Gao et al., 2001) or alleles that have quantifiable adaptive characteristics (Yokoyama et al., 1999). However, favorable alleles that have become fixed by a selective sweep are expected to show low variation at, and high linkage disequilibrium with, tightly lined markers (Hudson et al., 1997). This pattern has now proven useful in estimating the age of adaptive, polymorphic alleles at the G6PD locus in humans (Tishkoff et al., 2001) and could be used to identify alleles at other loci that have recently increased in frequency because of their adaptive importance.

The foremost example of adaptive evolution at the DNA level is the genes of the major histocompatibility complex (MHC). These genes encode proteins with a pivotal role in immunorecognition, and MHC variation appears to be an important component of pathogen resistance (Edwards and Hedrick, 1998; Hedrick and Kim, 2000; Hill, 2001). MHC genes have been characterized at the molecular level for more than two decades, and nearly every approach has been taken to examine the effects of selection on these genes. Because extensive and ongoing research on MHC evolution is carried out in a wide range of model vertebrate taxa, molecular studies of many of organisms of evolutionary and conservation significance are now possible. Before I discuss an example of MHC in red wolves, I will give a brief introduction to MHC biology, focusing on human MHC because it is the best studied and shares many characteristics with the MHCs of other vertebrates.

The human MHC region (known as HLA), which is composed of approximately 3,600 kilobases over 4 map units on the short arm of chromosome 6, was among the first large, contiguous genomic regions to be sequenced completely (Beck and Trowsdale, 2000). With 224 known genes, the HLA region is one of the most gene-dense regions of the genome and harbors the most variable functional genes yet known in humans. For example, in worldwide samples, genes HLA-A, HLA-B, and HLA-DRB1 have 243, 478, and 307 alleles, respectively (Garrigan and Hedrick, 2003). Additionally, 40 percent of the genes in the HLA region have a known immunological function, and many of these have medical significance because of their role in organ transplantation and autoimmune disease. Intensive research on the function of MHC proteins has provided the three-dimensional structure of MHC proteins, which enables the identification of amino acid residues that are involved in antigen binding (Brown et al., 1993). From a number of different tests and perspectives, it is obvious that the extent and pattern of MHC variation is, or has been at some point in the past, strongly influenced by one or more forms of balancing selection (Hedrick and Thomson, 1983; Hughes, 1999). Here I present some data from the endangered red wolf as an illustration.

The red wolf (*Canis rufus*) once ranged throughout much of the eastern United States (Nowak et al., 1995). However, in the early twentieth century, the numbers of red wolves declined dramatically as a result of eradication programs, habitat destruction, hybridization with coyotes, and parasite infestation (McCarley, 1962; Nowak, 1979). In 1967, red wolves were listed as endangered, and by 1970 they remained only in a small area of Texas and Louisiana. A captive breeding program was initiated in 1974, which now contains contributions from fourteen individuals from this remnant population, and the natural population became extinct in 1975. The captive population was used to start a reintroduced wild population in eastern North Carolina in 1987 (Phillips et al., 1995), which has now grown to approximately 100 individuals.

Figure 4.1 shows a neighbor-joining tree with the twenty-eight alleles that we have found at an MHC gene called *DRB* in red wolves, gray wolves, and coyotes. As has been found for MHC genes in other taxa, the alleles for a given taxon are dispersed throughout the phylogenetic tree. In particular, the four sequences from red wolves are widely dispersed in the tree, with high bootstrap numbers separating them. The average number of amino acid differences between the four red wolf alleles is 25 nucleotides (15.5 amino acids), and the range in difference between pairs for nucleotides (19 to 31) and amino acids (13 to 19) is not large.

To determine the probability that four such divergent alleles in red wolves could persist by chance, Hedrick et al. (2002) used a Monte Carlo simulation. The simulation

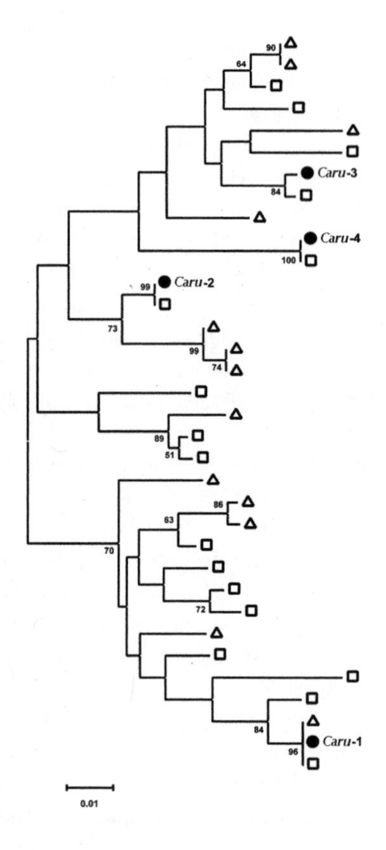

Figure 4.1 A neighbor-joining tree (with bootstrap values) showing the 4 red wolf MHC alleles, *Caru*-1 to *Caru*-4 (indicated by black circles). Also given are 17 coyote alleles (white squares) and 7 gray wolf and 6 Mexican wolf alleles (open triangles). The vertical lines indicate identical sequences found in different taxa, and the scale bar indicates the number of substitutions per site.

starting point was the twenty-eight different alleles given in figure 4.1, all with equal frequencies, and then genetic drift was allowed to reduce the allele number to four. The process was replicated 1,000 times, and the distribution of the number of pairwise amino acid differences between the remaining four alleles was determined. The mean expected from the simulation results was only 51 percent of the observed mean difference for the four red wolf alleles, and none of the simulations resulted in a value as high as that observed. In other words, it appears very unlikely that the four alleles remaining in the red wolves would be as divergent as those observed by chance, suggesting that balancing selection has favored retention of divergent MHC alleles.

Using the frequencies observed for the four alleles, we can calculate the average heterozygosity for each of the amino acid positions in the sequenced gene (figure 4.2). The highest heterozygosity was for position 28 (0.714), and eighteen amino acid positions have heterozygosities greater than 0.4. Most of the variation was concentrated in the antigen-binding site (ABS) positions (indicated by asterisks), with an average heterozygosity of 0.349, whereas the non–antigen-binding sites had a significantly lower average heterozygosity of 0.043 (12 percent of that found for ABS positions). In addition, we found a greater number of heterozygotes than expected and a higher rate of nonsynonymous than

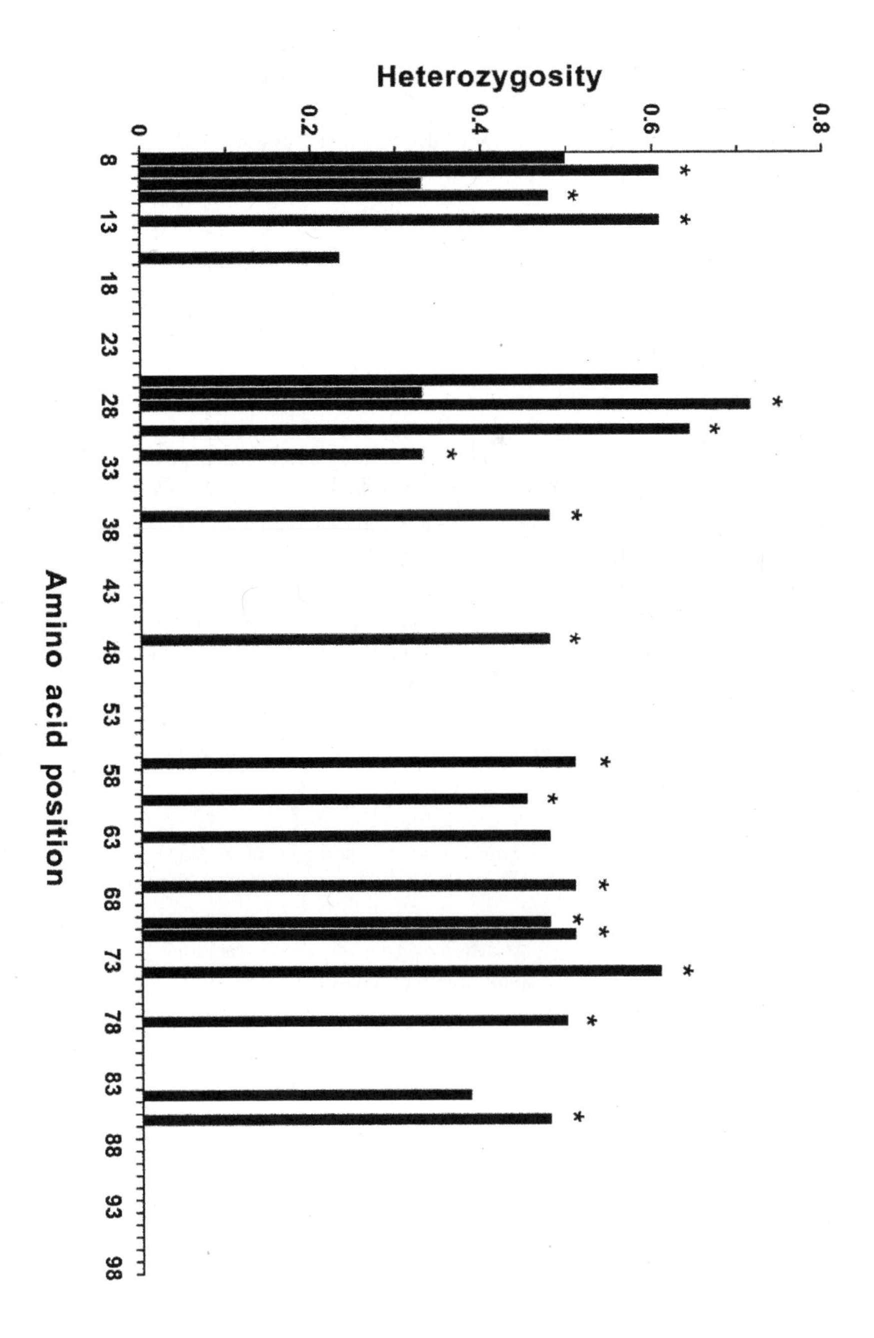

Figure 4.2 The heterozygosity at individual amino acid positions in the red wolf; asterisks indicate sites that are putative antigen binding site positions.

synonymous substitution for the functionally important ABS positions, and the distributions of alleles and amino acids at many positions were more evenly distributed than expected from neutrality. In other words, five different types of evidence support the importance of adaptive selection for this MHC gene in red wolves.

Neutral Variation as an Indicator of Detrimental and Adaptive Variation

The extent and pattern of neutral genetic variation have been used as a guide to estimating the frequency and nature of detrimental and adaptive variation. If their association was always positive and strong, it would be a good predictor of how much detrimental variation might be present that could lower fitness and how much adaptive variation might be present to deal with future challenges. With more loci and more variable loci, estimates of the extent and pattern of neutral genetic variation presumably will become more accurate. However, different evolutionary scenarios might be responsible for different amounts of neutral variation (table 4.2), and these scenarios might in turn result in different amounts or patterns of detrimental (Kirkpatrick and Jarne, 2000) or adaptive variation. Positive associations might occur when the population size has been large or small for a long time or where stochastic effects dominate the extent of genetic variation. For example, there is a high correlation of (neutral) microsatellite and (adaptive) MHC loci variation over Gila topminnow (*Poeciliposis occidentalis*) (Hedrick et al., 2001) and

Table 4.2 The amount of observed neutral variation and general predictions of the amounts of detrimental and adaptive variation under some different evolutionary scenarios.

		Observed	Predicted	
Scenario		Neutral variation	Detrimental variation	Adaptive variation
1. Equilibrium	a. Large population	High	High	High
	b. Small population	Low	Low	Low to medium (retention more than neutrality)
2. Bottleneck	a. Shortly after	Low (loss of alleles more than heterozygosity)	Low (loss of lethals more than detrimentals)	Low to medium (retention more than neutrality)
	b. Some time after	Low to high (depending on mutation rate of variants)	High (assuming high mutation rate)	Low to medium (retention more than neutrality, function of mutation rate)
	c. Mixture of separately bottlenecked populations	High (due to fixation of different alleles in different populations)	Low (due to purging of variants within each population)	Low to medium (retention more than neutrality)
3. Metapopulation	a. With extinction, recolonization, and low gene flow	Low (due to low effective population size)	Low (due to purging)	Low to medium (depending on selection model)
	b. No extinction and more gene flow	High (as if one large population)	High (as if one large population)	High (as if one large population)

Source: Hedrick (2001a).

desert bighorn sheep (*Ovis canadensis*) (Gutierrez-Espeleta et al., 2001a, 2001b) populations, two species for which stochastic factors appear to be important in the present spatial pattern of variation.

There are likely to be situations in which an inference based on a positive correlation is unfounded. For example, some time after a bottleneck there might be a negative association between neutral and detrimental variation because the amount of detrimental variation could recover faster as a result of a higher mutation rate. However, this would not be the case for all neutral variation because some genes, such as microsatellite loci, have mutation rates as high as those for detrimental variation. A negative association may also occur when separately bottlenecked populations become mixed, resulting in high neutral variation but low detrimental variation. Finally, in a metapopulation there might be a positive association between neutral and detrimental variation, but depending on the level of extinction, recolonization, and gene flow, the levels of variation could vary widely.

In some specific cases, it appears that there might be little genetic variation for some markers and substantial variation for others. For example, the cheetah (*Acinonyx jubatus*) was documented to have low genetic variation for allozymes and other genetic markers (O'Brien et al., 1985) and great concern was raised about its long-term survival. However, further studies showed that cheetahs have substantial genetic variation for microsatellites (Menotti-Raymond and O'Brien, 1995), and captive cheetahs appear to exhibit inbreeding depression for juvenile survival (Wielebnowski, 1996), indicating variation for detrimental variants. These differences were explained with scenarios that include a bottleneck or a metapopulation with a small effective population size (Hedrick, 1996) and different mutation rates for different markers.

Overall, the positive correlation between neutral and adaptive variation might not be very high. However, high neutral variation may indicate the potential for significant adaptive variation. Low neutral variation could indicate low adaptive variation, but the present population might either be well adapted or poorly adapted to its environment. Furthermore, gene flow between populations could result in low differentiation for neutral markers between populations, but there might still be strong adaptive differences in the populations. The extent and pattern of neutral variation are the result of nonselective forces and can potentially be used to identify the past importance of finite size, bottlenecks, and population structure. However, adaptive variants might differ in mutation rate from neutral markers, and the selection might result in quite different patterns of variation within and between populations. In other words, neutral variants might be used as a guide to understanding nonselective effects, but the overlay of different mutational processes and selective regimes suggests that great caution should be used in making such predictions.

Summary

Genetics probably will play an even more important role in conservation in the future. The use of information from the human and other genome projects will provide substantially more background understanding, and the appropriate use of these data should be of great benefit to conservation. In particular, techniques to screen and analyze large amounts of data, whether variations at marker loci or DNA sequence data, will be used to determine specific groups and individuals. However, underlying the high statistical power of these data, an evolutionary perspective is needed to evaluate the biological importance of these differences. In addition, these data will allow an understanding of past evolutionary events in endangered species, whether they are nonselective, such as bottlenecks or gene flow, or selective, such as detrimental mutants or adaptive variants.

References

Avise, J. 2000. *Phylogeography: The history and formation of species*. Cambridge, Mass.: Harvard University Press.

Ballou, J. D. 1997. Ancestral inbreeding only minimally affects inbreeding depression in mammalian populations. *Journal of Heredity* 88:169–178.

Balloux, F., H. Brünner, N. Lugon-Moulin, J. Hausser, and J. Goudet. 2000. Microsatellites can be misleading: An empirical and simulation study. *Evolution* 54:1414–1422.

Barnes, I., P. Matheus, B. Shapiro, D. Jensen, and A. Cooper. 2002. Dynamics of Pleistocene population extinctions in Beringian brown bears. *Science* 295:2267–2270.

Beck, S. and J. Trowsdale. 2000. The human major histocompatibility complex: Lessons from the DNA sequence. *Annual Reviews of Genomics and Human Genetics* 1:117–137.

Bijlsma, R., J. Bundgaard, and W. F. Van Putten. 1999. Environmental dependence of inbreeding depression and purging in *Drosophila melanogaster*. *Journal of Evolutionary Biology* 12:1125–1137.

Brown, J. H., T. S. Jardetzky, J. C. Gorga, L. J. Stern, R. G. Urban, J. L. Strominger, et al. 1993. Three-dimensional structure of the human class II histocompatibility antigen HLA-DR1. *Nature* 364:33–39.

Byers, D. L. and D. M. Waller. 1999. Do plant populations purge their genetic load? Effects of population size and mating history on inbreeding depression. *Annual Reviews of Ecology and Systematics* 30:479–513.

Carvajal-Carmona, L. G., R. Ophoff, S. Service, J. Hartiala, and J. Molina. 2000. Strong Amerind/white sex bias and a possible Sephardic contribution among the founders of a population in northwest Colombia. *American Journal of Human Genetics* 67:1287–1295.

Charlesworth, D. and B. Charlesworth. 1999. The genetic basis of inbreeding depression. *Genetical Research* 74:329–340.

Cornuet, J. M., S. Piry, G. Luikart, A. Estoup, and M. Solignac. 1999. New methods employing multilocus genotypes to select or exclude populations as origins of individuals. *Genetics* 153:1989–2000.

Culver, M., W. W. Johnson, J. Pecon-Slattery, and S. J. O'Brien. 2000. Genomic ancestry of the American puma (*Puma concolor*). *Journal of Heredity* 91:186–197.

Ebert, D., C. Haag, M. Kirkpatrick, M. Riek, J. W. Hottinger, and V. I. Pajunen. 2002. A selective advantage to immigrant genes in a *Daphnia* metapopulation. *Science* 295:485–488.

Edwards, S. V. and P. W. Hedrick. 1998. Evolution and ecology of MHC molecules: From genomics to sexual selection. *Trends in Ecology and Evolution* 13:305–311.

Ellegren, H. 1999. Inbreeding and relatedness in Scandinavian grey wolves *Canis lupus*. *Hereditas* 130:239–244.

Frankel, O. H. and M. E. Soulé. 1981. *Conservation and evolution*. Cambridge: Cambridge University Press.

Gao, X., G. W. Nelson, P. Karacki, M. P. Martin, J. Phair, R. Kaslow, et al. 2001. Effect of a single amino acid substitution in the MHC class I molecule on the rate of progression to AIDS. *New England Journal of Medicine* 344:1668–1675.

Garrigan, D. and P. W. Hedrick. 2003. Perspective: Detecting adaptive molecular polymorphism—Lessons from the MHC. *Evolution* 57(8):1707–1722.
Garrigan, D., P. C. Marsh, and T. E. Dowling. 2002. Long-term effective population size of three endangered Colorado River fishes. *Animal Conservation* 5:95–102.
Gutiérrez-Espeleta, G. A., P. W. Hedrick, S. T. Kalinowski, and W. M. Boyce. 2001a. Is the decline of desert bighorn sheep from infectious disease the result of low MHC variation? *Heredity* 86:439–450.
Gutiérrez-Espeleta, G. A., S. T. Kalinowski, W. M. Boyce, and P. W. Hedrick. 2001b. Genetic variation and population structure in desert bighorn sheep: Implications for conservation. *Conservation Genetics* 1:3–15.
Hedrick, P. W. 1995. Gene flow and genetic restoration: The Florida panther as a case study. *Conservation Biology* 9:996–1007.
——. 1996. Bottleneck(s) or metapopulation in cheetahs. *Conservation Biology* 10:897–899.
——. 1999. Perspective: highly variable loci and their interpretation in evolution and conservation. *Evolution* 53:313–318.
——. 2000. *Genetics of populations.* 2nd ed. Boston: Jones and Bartlett.
——. 2001a. Conservation genetics: Where are we now? *Trends in Ecology and Evolution* 16:629–636.
——. 2001b. Lethals in finite populations. *Evolution* 56:654–657.
Hedrick, P. W. and S. T. Kalinowski. 2000. Inbreeding depression in conservation biology. *Annual Review of Ecology and Systematics* 31:139–166.
Hedrick, P. W. and T. J. Kim. 2000. Genetics of complex polymorphisms: Parasites and maintenance of the major histocompatibility complex variation. In *Evolutionary genetics: From molecules to morphology,* ed. R. S. Singh and C. B. Krimbas, 204–234. Cambridge: Cambridge University Press.
Hedrick, P. W., R. N. Lee, and D. Garrigan. 2002. Major histocompatibility complex variation in red wolves: Evidence for common ancestry with coyotes and balancing selection. *Molecular Ecology* 11:1905–1913.
Hedrick, P. W. and P. S. Miller. 1992. Conservation genetics: Techniques and fundamentals. *Ecological Applications* 2:30–46.
Hedrick, P. W., K. M. Parker, and R. N. Lee. 2001. Genetic variation in the endangered Gila and Yaqui topminnows: Microsatellite and MHC variation. *Molecular Ecology* 10:1399–1412.
Hedrick, P. W. and G. Thomson. 1983. Evidence for balancing selection at HLA. *Genetics* 104:449–456.
Hill, A. V. S. 2001. The genomics and genetics of human infectious disease susceptibility. *Annual Reviews of Genomics and Human Genetics* 2:373–400.
Hudson, R. R., A. G. Saez, and F. J. Ayala. 1997. DNA variation at the *Sod* locus of *Drosophila melanogaster:* An unfolding story of natural selection. *Proceedings of the National Academy of Sciences USA* 94:7725–7729.
Hughes, A. L. 1999. *Adaptive evolution of genes and genomes.* Oxford: Oxford University Press.
Kalinowski, S. T., P. W. Hedrick, and P. S. Miller. 2000. A close look at inbreeding depression in the Speke's gazelle captive breeding program. *Conservation Biology* 14:1375–1384.
Karhu, A., P. Hurme, M. Karjalainen, P. Karvonen, K. Karkkainen, and D. Neale. 1996. Do molecular markers reflect patterns of differentiation in adaptive traits of conifers? *Theoretical and Applied Genetics* 93:215–221.
Keller, L. F. and D. M. Waller. 2002. Inbreeding effects in wild populations. *Trends in Ecology and Evolution* 17:230–241.

Kimura, M. 1985. *The neutral theory of molecular evolution*. Cambridge: Cambridge University Press.

Kirkpatrick, M. and P. Jarne. 2000. The effects of a bottleneck on inbreeding depression and the genetic load. *American Naturalist* 155:154–167.

Lacy, R. C. 2000. Should we select genetic alleles in our conservation breeding programs? *Zoo Biology* 19:279–282.

Laikre, L. 1999. Hereditary defects and conservation genetic management of captive populations. *Zoo Biology* 18:91–99.

Laikre, L., N. Ryman, and E. A. Thompson. 1993. Hereditary blindness in a captive wolf (*Canis lupus*) population: Frequency reduction of a deleterious alleles in relation to gene conservation. *Conservation Biology* 7:592–601.

Land, D., D. Shindle, M. Cunningham, M. Lotz, and B. Ferree. 2001. *Florida panther genetic restoration and management: Annual report 2000–2001*. Gainesville: Florida Fish and Wildlife Conservation Commission.

Land, E. D. and R. C. Lacy. 2000. Introgression level achieved through Florida panther genetic restoration. *Endangered Species Update* 17:100–105.

Latta, R. G. and J. B. Mitton. 1997. A comparison of population differentiation across four classes of gene markers in limber pine (*Pinus flexilis* James). *Genetics* 146:1153–1163.

Liberles, D. A., D. R. Schreiber, S. Govindarajan, S. G. Chamberlin, and S. A. Benner. 2001. The adaptive evolution database (TAED). *Genome Biology* 2:20–28.

Luikart, G. and J.-M. Cornuet. 1998. Empirical evaluation of a test for identifying recently bottlenecked populations from allele frequency data. *Conservation Biology* 12:228–237.

Lynch, M., J. Blanchard, D. Houle, T. Kibota, S. Schultz, L. Vassilieva, et al. 1999. Perspective: Spontaneous deleterious mutation. *Evolution* 53:645–663.

Lynch, M. and K. Ritland. 1999. Estimation of pairwise relatedness with molecular markers. *Genetics* 152:1753–1766.

Madsen, T., R. Shine, M. Olsson, and H. Wittzell. 1999. Restoration of an inbred adder population. *Nature* 402:34–35.

McCarley, H. 1962. The taxonomic status of wild *Canis* (Canidae) in south central United States. *Southwestern Naturalist* 7:227–235.

Meagher, S., D. Penn, J. Potts, and K. Wayne. 2000. Male–male competition magnifies inbreeding depression in wild house mice. *Proceedings of the National Academy of Sciences USA* 97:3324–3329.

Menotti-Raymond, M. and S. J. O'Brien. 1995. Evolutionary conservation of ten microsatellite loci in four species of Felidae. *Journal of Heredity* 86:319–322.

Moritz, C. 1999. Conservation units and translocation: Strategies for conserving evolutionary processes. *Hereditas* 130:217–228.

Nei, M. 1987. *Molecular evolutionary genetics*. New York: Columbia University Press.

Nowak, R. M. 1979. *North American Quaternary Canis*. Lawrence: University of Kansas Museum of Natural History Monograph No. 6.

Nowak, R. M., M. K. Phillips, V. G. Henry, W. C. Hunter, and R. Smith. 1995. The origin and fate of the red wolf. In *Ecology and conservation of wolves in a changing world*, ed. L. N. Carybn, S. H. Fritts, and D. R. Seip, 409–415. Canadian Circumpolar Institute, Occasional Publication No. 35.

O'Brien, S. J., M. E. Roelke, L. Marker, A. Newman, C. A. Winkler, D. Meltzer, et al. 1985. Genetic basis for species vulnerability in the cheetah. *Science* 227:1428–1434.

Phillips, M. K., R. Smith, V. G. Henry, and C. Lucash. 1995. Red wolf reintroduction program. In *Ecology and conservation of wolves in a changing world*, ed. L. N. Carybn, S. H. Fritts, and D. R. Seip, 157–168. Edmonton, Alberta: Canadian Circumpolar Institute, Occasional Publication No. 35.

Ralls, K., J. D. Ballou, B. A. Rideout, and R. Frankham. 2000. Genetic management of chondrodystrophy in California condors. *Animal Conservation* 3:145–153.

Richards, C. M. 2000. Inbreeding depression and genetic rescue in a plant metapopulation. *American Naturalist* 155:383–394.

Richman, A. 2000. Evolution of balanced genetic polymorphism. *Molecular Ecology* 9:1953–1963.

Roelke, M. E., J. S. Martenson, and S. J. O'Brien. 1993. The consequences of demographic reduction and genetic depletion in the endangered Florida panther. *Current Biology* 3:340–350.

Saccheri, I. J. and P. M. Brakefield. 2002. Rapid spread of immigrant genomes into inbred populations. *Proceedings of the Royal Society London B* 269:1073–1078.

Schonewald-Cox, C. M., S. M. Chambers, B. MacBryde, and L. Thomas, eds. 1983. *Genetics and conservation*. Menlo Park, Calif.: Benjamin/Cummings.

Soulé, M. E., ed. 1986. *Conservation biology: The science of scarcity and diversity*. Sunderland, Mass.: Sinauer.

Soulé, M. E. and B. A. Wilcox, eds. 1980. *Conservation biology: An ecological–evolutionary perspective*. Sunderland, Mass.: Sinauer.

Templeton, A. R. and B. Read. 1983. The elimination of inbreeding depression in a captive herd of Speke's gazelle. In *Genetics and conservation,* ed. C. M. Schonewald-Cox, S. M. Chambers, B. MacBryde, and L. Thomas, 241–261. Menlo Park, Calif.: Benjamin/Cummings.

Tishkoff, S. A., R. Varkonyi, N. Cahinhinan, S. Abbes, and G. Argyropoulos. 2001. Haplotype diversity and linkage disequilibrium at human G6PD: Recent origin of alleles that confer malarial resistance. *Science* 293:455–462.

Wang, J. 2000. Effects of population structure and selection strategies on the purging of inbreeding depression due to deleterious mutations. *Genetical Research* 76:75–86.

Wang, J., W. G. Hill, D. Charlesworth, and B. Charlesworth. 1999. Dynamics of inbreeding depression due to deleterious mutations in small populations: Mutation parameters and inbreeding rate. *Genetical Research* 74:165–178.

Westemeier, R. L., J. D. Brawn, S. A. Simpson, T. L. Esker, R. W. Jansen, J. W. Walk, et al. 1998. Tracking the long-term decline and recovery of an isolated population. *Science* 282:1695–1698.

Wielebnowski, N. 1996. Reassessing the relationship between juvenile mortality and genetic monomorphism in captive cheetahs. *Zoo Biology* 15:353–369.

Wright, S. 1937. The distributions of gene frequencies in populations. *Proceedings of the National Academy of Sciences USA* 23:307–320.

Yang, Z. and J. P. Bielawski. 2000. Statistical methods for detecting molecular adaptation. *Trends in Ecology and Evolution* 15:496–503.

Yokoyama, S., H. Zhang, F. B. Radlwimmer, and N. S. Blow. 1999. Adaptive evolution of color vision of the Comoran coelacanth (*Latimeria chalumnae*). *Proceedings of the National Academy of Sciences USA* 96:6279–6284.

Robert C. Lacy

5

Stopping Evolution

Genetic Management of Captive Populations

We need to conserve the diversity of the natural world at multiple levels. Biodiversity conservation is most often discussed in terms of species loss, because the loss of unique evolutionary lineages is irreversible. Below the species level, we are also concerned with the stability and sustainability of populations within species and with the diversity of genes within populations. Population and gene diversity often have value to us directly, and they are also critical to the persistence of the species they make up (Bijlsma et al., 2000; Frankham, 1995; Frankham et al., 1999; Lacy, 1997; Saccheri et al., 1998; Taylor et al., 2002). Moving up from the level of species, we care about loss of species because they are fundamental components of ecological communities and ecosystems. Thus, we also care about conserving ecosystem integrity and function. Conservation cannot be equated with absolute preservation of types, however. Since Darwin, we have known that species, populations, genetic diversity, and ecological communities are dynamic results of evolutionary processes, not static entities. Therefore, long-term conservation of biodiversity requires conservation of evolutionary processes and protection of the entities those processes have produced and work on (Myers and Knoll, 2001; Templeton et al., 2001).

Unfortunately, conflicts sometimes occur when we are trying to preserve biodiversity while also maintaining processes. Some evolutionary and ecological processes involve or even require change over time of species compositions. Therefore, when we work to prevent the loss of biodiversity around us, we may actually need to put on temporary hold the ecological and evolutionary processes that we seek to preserve. Thus, we are often in the position of limiting the extent to which ecological and evolutionary changes are driven by current human activities in order to preserve options for a future of more natural and healthy environments. Consider that evolutionary adaptation depletes genetic variation, as less well-adapted types are winnowed out of the population. Yet the continued existence of adequate genetic variation is a prerequisite for future evolutionary change (Fisher, 1930; Robertson, 1960; Wright, 1977). In many ways, the task of conservation is to manage amounts of diversity and rates of change in a human-dominated world so that natural processes can continue to function rather than ending with catastrophic collapse.

Captive populations can serve a number of different roles in conservation efforts (IUDZG/CBSG [IUCN/SSC], 1993). Most importantly, they can be used to educate, enlighten, and enchant people about the diversity of species and their adaptations. As living museums, they also serve as invaluable resources for scientific discovery (Gansloßer et al., 1995; Hodskins, 2000). Captive populations have also served as sources for supplementation or restoration of wild populations (Wilson and Stanley-Price, 1994), as in the case of the Arabian oryx, golden lion tamarin, black-footed ferret, eastern barred bandicoot, peregrine falcon, American burying beetle, California condor, and many other species. While waiting for opportunities to reestablish healthy populations in the wild, captive populations often serve as insurance against the final loss of a species, as in the case of Przewalski's horse, Père David's deer, Hawaiian crow, addax, several species of Partula snails, and other species that are extinct or nearly so in the wild. For captive populations the relative emphasis on these various roles has been evolving (Rabb, 1994), and there has been debate about how well and in what circumstances captive populations can serve these roles (Caughley and Gunn, 1966; Snyder et al., 1996). Importantly, however, good genetic management is a prerequisite for any of these conservation goals of captive populations to be achieved. This chapter will discuss genetic management of captive populations of animals. However, many of the conservation concerns, conceptual principles, and management methods discussed here are the same or have direct parallels in plant conservation.

Most fundamentally, captive populations need to persist if they are to have a chance to serve conservation objectives. Yet some animal populations in zoos have become so inbred or otherwise compromised genetically that they are unlikely to persist longer than a few more generations (Earnhardt et al., 2001; Lacy, 2000b; Ralls and Ballou, 1983). It may be that persistence in captivity will be more likely if we artificially select the animals in order to create semidomesticated forms that are better adapted to lives in zoo enclosures. Yet in creating such populations we would destroy much of what we want to preserve about the wild species (Lacy, 2000c). Many captive populations in zoos persist, albeit with handicaps, despite substantial losses of genetic variation and accumulated inbreeding (Ballou and Ralls, 1982; Lacy et al., 1993; Ralls et al., 1988). However, in order for captive populations to be used successfully for restoration of wild populations, they must retain high fitness and the adaptability to changed or changing environments that is conferred by genetic diversity (Jiménez et al., 1994; Lacy, 1993a, 1994). Preventing adaptation to captive conditions, and preventing loss of diversity, amounts to trying to stop evolution of the populations while they are in captivity.

If we want to preserve the option to someday release animals to restock or supplement wild populations, the captive populations need not only to retain the wild characteristics but also to retain high levels of genetic variation (Arnold, 1995; Woodworth et al., 2002). It is very hard to know what environments species may need to inhabit in the future, and therefore it is hard to know what genetic characteristics they will need to survive the rigors of release and thrive. However, we do know that environments around the globe are changing at unprecedented rates. Therefore, it is almost certain that species will need to adapt as fast as or faster than ever before if they are to survive. Maximal genetic variation must be retained to provide the capacity for future evolution.

Evolution in Captive Populations: The Challenge

To keep populations that match genetically the ancestors that came from the wild, we need to stop three causes of genetic change: random genetic drift; artificial selection for specific, favored phenotypes; and unintended natural selection for traits that confer high fitness in captivity. Random genetic drift causes changes in allele frequencies and loss of polymorphism as alleles drift to fixation (Allendorf, 1986; Wright, 1931, 1948). Genetic drift can be especially rapid in captive populations because they are usually started from a few founders and maintained at small total size. Drift can be minimized by breeding strategies that maximize the effective population size (Lande and Barrowclough, 1987), and effective strategies will be described in this chapter. Modern captive breeding programs for conservation usually avoid artificial selection (Frankham et al., 1986; Lacy, 2000c), but some selection for favored phenotypes is almost inevitable if breeding pairs are selected by on-site managers rather than through analysis of the full captive pedigree. It is hard to know how strong inadvertent selection for captive-adapted traits may be (Arnold, 1995; Frankham and Loebel, 1992). Yet for reasons that are not well understood, often as many as 50 percent of captive animals do not breed, so there may be very strong selection exerted for traits that favor breeding under those conditions. The common causes of mortality (of captive or of captive bred animals) include stillbirths and neonatal deaths attributed to a failure to thrive (Ballou and Ralls, 1982), injuries from conspecifics, and old age. These sources of mortality contrast markedly with the causes of mortality in natural populations, including disease, predation, nutritional stress, and extreme environmental conditions. The captive environment is so different from the wild environment that it is hard to imagine how natural selection could not be favoring very different traits.

It has been proposed that the rapid genetic change in small, bottlenecked populations (Meffert, 1999) may be desirable to allow evolution of adaptations to captivity or other novel environments, or that we should sometimes actively select against deleterious alleles in captive populations (Bryant and Reed, 1999; Laikre, 1999). However, promoting adaptation to the captive environment would be counterproductive to many conservation objectives (Lacy, 2000c). If wild-adapted individuals are not thriving in captivity, it may be more appropriate to change the captive environment to meet their needs rather than allowing or promoting rapid evolution of the population away from the characteristics that evolved over millennia in the wild.

Stopping Genetic Change in Captive Populations: Is Biotechnology the Solution?

Powerful tools for manipulating genes may provide us with methods to achieve the cessation of evolution that we desire in our wildlife populations being conserved in captivity.

Biotechnology is being used in various ways to assist with species conservation and may increasingly be drawn on in the future as the technology improves and its costs come down closer to the realm of resources that are being applied to species conservation.

As described by others in this book (see chapters 1, 6, and 7), molecular characterization of genetic diversity can be important for identifying the appropriate population units for conservation and for identifying the population structure in natural systems. In addition, measurement of current levels of genetic diversity can provide the baseline assessment against which we should measure our success at slowing unwanted evolutionary change in intensively managed captive populations (Arnold, 1995).

Cryopreservation of gametes and embryos can provide a powerful tool for slowing evolution by allowing use of long-dead donors as the genetic parents of future generations (Ballou, 1984, 1992; Ballou and Cooper, 1992; Holt and Watson, 2002). However, availability of cryopreserved gametes does not eliminate the need for care in determining which individuals should be used as the parents of each generation. The techniques described in this chapter for selecting optimal parents for captive breeding programs are applicable to selecting gamete donors (Johnston and Lacy, 1995) and to determining the optimal use of stored gametes (Harnal et al., 2002).

It has been proposed, perhaps more often in the popular press (e.g., the recent proposals to bring the thylacine back from extinction [Meek, 2002; Smith, 2002]) than in scientific publications, that cloning may be useful in the efforts to preserve endangered species (see chapter 18). There will probably be a few cases in which nearly extinct or recently extinct species can be propagated by cloning (Cohen, 1997). Considering that the retention of adequate genetic diversity is essential to allow maintenance or restoration of healthy, resilient, and adaptable populations (Lacy, 1997), however, it is hard to see how production of many cloned genotypes will ever contribute much more to species conservation than the emergency replication of a few individual genotypes in danger of being lost from an extremely small gene pool. The potential benefits and problems of applying cloning and other reproductive technologies to conservation have been addressed by others (Critser and Prather, 2002; Loskutoff, 2002; Santiago and Caballero, 2000) and are beyond the scope of this chapter.

At the theoretical extreme of using DNA manipulations to further species conservation is the prospect of using genetic engineering to change specific genes within an organism in order to provide it with traits thought to enhance its prospects for persistence. Although there may be a philosophical contradiction in changing the genotypes of organisms in an attempt to preserve their genetic heritage, such manipulations are not fundamentally different from artificial selection programs that have been used to "improve" stocks for centuries. The problems inherent in using such techniques for conservation (Lacy, 2000c) are no different in the two cases. Unless we acquire full knowledge of the function of all the genes, the implications of the interactions between genes, the roles that variant alleles play in diverse environments, and the future environmental challenges that species will face, it may be folly to think that we could engineer genotypes that are better able to survive than those that have resulted from billions of years of natural evolution.

Many papers in this volume and from other recent meetings (Holt and Wildt, 2002) focus on the use of modern, high-tech approaches to conservation. Rather than addressing further whether biotechnology can save the world's species, I will instead explore how well certain low-tech approaches may work to achieve some of our genetic goals in species conservation. If methods derived from more traditional approaches of animal breeding can adequately protect genomes from the onslaught of human activities and the threats to persisting in human-altered environments, then the important question may not be whether biotechnology can save species but rather whether biotechnology is a needed or even a useful adjunct to other methods available for genetic conservation.

Low-Tech Conservation Genetics

Low-tech conservation genetics relies on classic approaches of pedigree analysis and animal breeding to achieve the genetic goals of captive propagation. The scientific study of the methods of animal breeding has a long history, going back as least as far as Darwin (1868). However, some of the specific quantitative methods were only recently proposed and are still being tested. For example, most of the focus in classic animal breeding is on the development and application of breeding schemes that maximize the rate of genetic change in a population, in order to increase the production traits that are of commercial value. Yet conservation geneticists have used much of the same understanding about Mendelian and quantitative genetic processes to minimize the rate of genetic change (Arnold, 1995; Frankel and Soulé, 1980; Frankham, 1999). Interestingly, some of the pedigree techniques developed for endangered species management are being applied again to domestic livestock, in programs to preserve the unique genetic diversity present in rare breeds and landraces (Caballero and Toro, 2000).

It is perhaps a misnomer to describe modern animal breeding techniques as "low-tech." The analyses usually rely on high-speed computers, and the precision of the techniques is often enhanced by the use of molecular genetic techniques to determine parentage or other pedigree relationships. Still, it probably fits with common perception to describe pedigree analysis methods as low-tech genetic management and molecular analyses and manipulations as high-tech genomics. In addition to the obvious difference in technologies applied, there are a few fundamental differences in the assumptions and philosophies behind the low-tech and high-tech approaches. The high-tech world of genomics starts with the use of molecular genetics to characterize the genetic composition of individuals and populations. This genetic characterization usually focuses on a sample of a few genes or DNA segments and use sequencing, fragment length polymorphisms, single nucleotide polymorphisms, or other empirical assessments of DNA sequence or structure. In contrast, the low-tech world of pedigree analyses applies our understanding of Mendelian genetics to derive the theoretical or most probable genetic status of individuals or populations from the known pedigree relationships. This approach uses either probability calculations or results of computer simulations to derive from the pedigree structure estimates of genome-wide parameters such as mean relative heterozygosity, probabilities or proportions of alleles shared between pairs of individuals, probabilities of allele loss, and gene diversity or other estimates of diversity (Caballero and Toro, 2000; Lacy, 1995). Thus, the high-tech approach tends to be more precise, in the sense that genotypes are directly observed, but the low-tech approach may often be more accurate, in that the genetic metrics represent genome-wide probabilities or means.

Following from this difference between empirical and theoretical characterization of a population's genetic structure, the pedigree approach does not require or directly use any measurement of the starting condition of the population. Instead, it presumes that at the time we start interventive management, the population had a gene pool that was the successful product of past evolution. Pedigree analyses are then performed to determine how fast and how far the population probably diverged from the baseline of presumed genetic health (Lacy et al., 1995). After analyzing the pedigree, the low-tech approach uses extensions of classic animal breeding techniques to plan a breeding program that will minimize further genetic decay or even reverse prior changes. Instead of manipulating genes directly, we manipulate breeding pairs to achieve our genetic goal of stopping undesired evolution in the conserved population.

Although there are distinct differences between the high-tech approach of the age of genomics and the low-tech approach of pedigree analysis and animal breeding, the two approaches should be viewed as complementary rather than in conflict. The most effective genetic conservation will be achieved if we wisely combine the strengths of each, just as optimal biodiversity conservation requires attention to genetics, ecology, animal behavior, and other biological processes, as well as to politics, economics, sociology, ethics, and many other dimensions of human society. Unfortunately, there has been little collaboration and exchange between those applying molecular genetics to conservation and those who manage breeding programs for endangered species. Productive collaborations have benefited the conservation of several bird species, such as Puerto Rican parrots (Brock and White, 1992), California condors (Geyer et al., 1993), Guam rails (Haig et al., 1994), Micronesian kingfishers (Haig et al., 1995), and whooping cranes (Jones et al., 2002), and primate species, such as lion-tailed macaques (Morin and Ryder, 1991) and bonobos (Reinartz, 1977), but more such work is needed.

In order to assess what the age of genomics will contribute to species conservation, it is important to determine how effectively we can use more traditional pedigree analysis techniques to prevent genetic damage in our managed populations. Only then can we evaluate the necessity or incremental benefit of applying the power of biotechnology to the conservation problems. Most of the rest of this chapter describes ongoing work to refine and test strategies that are used to manage breeding programs for species in zoos and other captive facilities.

Does Pedigree Management Work?

Although the methods of pedigree analysis and management have continued to change and improve, we can get an indication of our ability to save the world's species (at least in the sense of preserving their genetic composition) through careful and intensive management of breeding programs by comparing the past performance of managed and unmanaged captive populations. One useful comparison is between the North American captive populations of two bovids: the markhor (*Capra falconeri*) and the Arabian oryx (*Oryx leucoryx*) (Lacy, 2000b).

Markhor are a wild goat species native to central Asia but greatly reduced in numbers because of poaching and competition with livestock. Markhor are closely related to

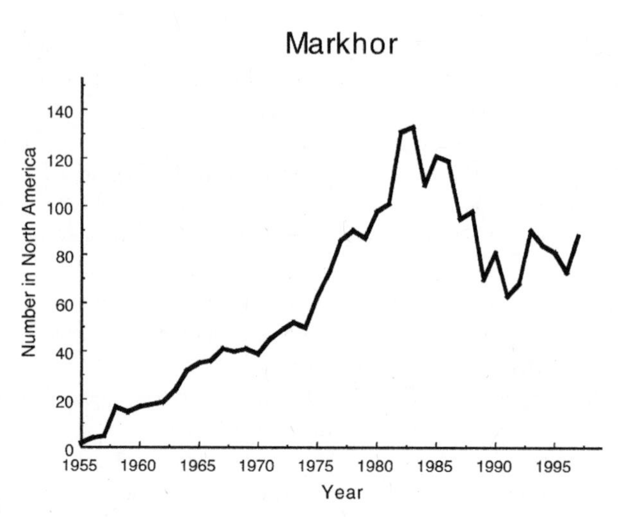

Figure 5.1 Number of markhor in North American zoos, 1955 to 1997. From Lacy (2000b).

domestic goats, can interbreed with domestic goats, and may be similar to the wild goat from which domestic goats were derived. Markhor probably harbor genes that would be valuable to agriculture. The breeding of markhor in North America zoos has not previously been managed cooperatively, and a studbook for the regional population was just recently initiated (LaBarge, 1999). As shown in figure 5.1, the population has declined about 30 percent since the mid-1980s. Much of the breeding has been due to a few prolific animals, and until recently no coordinated attempt had been made to analyze the pedigree and optimally manage the breedings. Although the captive population descends from seventeen wild-caught founders, much of the genetic variability of those wild-caught animals has been lost through the generations, with the current gene diversity (86 percent of initial levels) being that which would be represented in just 3.76 wild-caught animals; that is, the captive population has 3.76 founder genome equivalents (Lacy, 1989).

Figure 5.2 shows the population of Arabian oryx in North American zoos since 1965. Upon the realization in 1962 that the species was nearly extinct in the wild and not abundant in captivity, a group of zoos led by the Phoenix Zoo initiated a population management program of the "World Herd." The number of Arabian oryx increased rapidly and steadily under management. By 1990 the population exceeded 400 animals, after which breeding was curtailed to hold the population size at a desired level. The captive population has saved the species, as oryx have been provided for a reintroduction program that reestablished herds in the original wild habitat (Stanley-Price, 1989). Concomitant with the notable demographic success of the program, the breeding program managed to retain genetic representation from all ten of the founders of the North American population, with a high percentage (about 92 percent) of the original gene diversity still in the captive population.

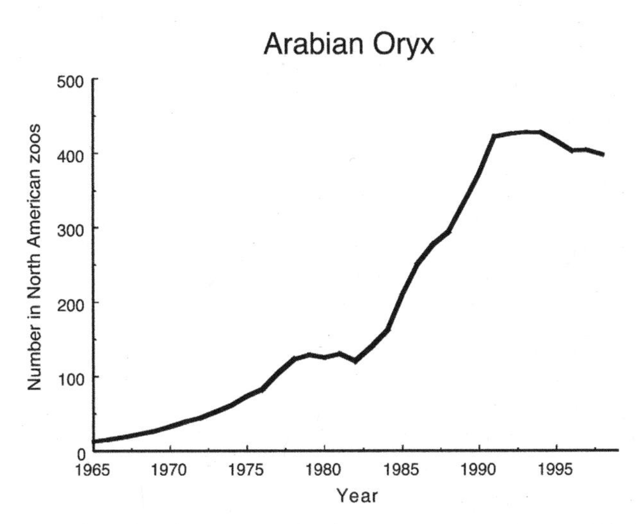

Figure 5.2 Number of Arabian oryx in North American zoos, 1965 to 1998. From Lacy (2000b).

The contrast between the genetic decay in the unmanaged population of markhor and the managed population of Arabian oryx is further reflected in the much lower ratio of effective population size to total population size (N_e/N = 0.07 vs. 0.30) and a much higher level of inbreeding (mean f = 0.19 vs. 0.07) (Lacy, 2000b). The consequence is that Arabian oryx are thriving in captivity, their breeding must be curtailed to prevent overpopulation in zoos, and they are available for restocking wild habitats, whereas markhor have an uncertain future in captivity even while the remnant wild populations are reported to be in decline.

Pedigree Management

The impact of good genetic management on not only retention of genetic variation but also on demographic growth and stability of the population and on individual animal health is apparent in the previous example and many others (Lacy, 2000b). In addition, laboratory experiments by Frankham and coworkers on *Drosophila* have confirmed that the pedigree management techniques described in this chapter do help maintain genetic variation and slow the loss of wild-adapted traits (Frankham et al., 2000; Montgomery et al., 1997).

Current population management techniques focus primarily on four measures of the genetic health of animals, their contribution to the genetic diversity of the population, and the genetic value of potential pairs to the breeding program (Ballou and Foose, 1996; Ballou and Lacy, 1995; Lacy, 1994): inbreeding, mean kinship of individuals to the population, difference between mean kinships of potential breeding partners (but see Caballero and Toro, 2000), and percentage of ancestry that is not traceable to the population founders. The avoidance of close inbreeding has been an objective of most captive breeding programs since the demonstrations by Ralls and Ballou of the National Zoo and others that inbred animals of many mammal species in zoos suffered higher mortality (Lacy et al., 1993; Ralls et al., 1979, 1988). On average, infant survival of mammals in zoos is depressed by about 15 percent for every 0.10 increment in the inbreeding coefficient, so it is not surprising that markhor, with mean $f = 0.30$, and many other species with depleted genetic variation are not thriving in captivity. Inbreeding depression has also been shown to affect many other components of fitness (Brewer et al., 1990; Lacy, 1993a; Margulis, 1998; Ryan et al., 2002) and may be stronger in wild populations than is apparent in captive populations protected from many of the stresses of life in natural environments (Jiménez et al., 1994; Keller et al., 1994; Lacy, 1997; Meagher et al., 2000; Miller, 1994). Many captive breeding programs (and most human populations) try to avoid matings between individuals related at the level of first cousins (which would produce inbred progeny with $f = 0.0625$) or closer.

Although the effects of inbreeding are widely known and most conservation biologists and geneticists would presume correctly that conservation breeding programs generally avoid inbreeding, modern pedigree management actually focuses on other aspects of genetic management more directly than on inbreeding. This is because even in the absence of the deleterious effects of close inbreeding, rapid adaptation to current captive conditions can degrade adaptations to more natural environments (Frankham and Loebel, 1992), and loss of genetic diversity in captive populations can limit future adaptive potential (Lacy, 1997). Moreover, techniques that focus on retaining genetic diversity within managed populations reduce the accumulation of inbreeding in future generations even better than do some strategies that were designed to minimize inbreeding (Ballou and Lacy, 1995).

Rather than focusing on inbreeding per se, genetic management of captive populations uses mean kinship (MK) to determine which animals to breed each year (Ballou and Lacy, 1995). The kinship coefficient between any two individuals is the probability that an allele sampled at random from one will be identical due to descent from a common ancestor with an allele sampled from the same locus in the second individual (Crow and Kimura, 1970; Malécot, 1948). The kinship between two individuals is the same as the inbreeding coefficient of any offspring they produce together. An animal's MK is the mean of its pairwise kinships to all individuals in a population (Lacy, 1995). The individuals with lowest MK are, by definition, those with the fewest close relatives. They therefore carry the least represented alleles, and they contribute the most to the population's overall gene diversity. They also have the greatest number of unrelated potential mates and therefore can most easily be used to produce noninbred, highly heterozygous progeny.

The overall population MK—the mean of all pairwise kinships—has a number of useful properties as a measure of the genetic health of a population. The population MK is the expected average inbreeding of the next generation if mating is random. Therefore, minimizing MK almost guarantees that inbreeding will be minimized in future generations. One minus MK is equal to the proportional gene diversity (the mean heterozygosity expected under Hardy–Weinberg equilibrium), expressed as a fraction of the gene diversity of the source population from which the founders were sampled. Therefore, minimizing MK is equivalent to maximizing gene diversity. Gene diversity is expected to be approximately proportional to the additive genetic variation in quantitative traits (Falconer and Mackay, 1996). Therefore, minimizing MK also maximizes the variation on which selection can act and maximizes the rate of adaptive response to future selective pressures. While maximizing future evolutionary potential, selecting breeders with the lowest MK will minimize or even reverse past selection in the captive population. This is because the lineages that contributed the fewest progeny in prior generations (perhaps because they have traits not well adapted to the captive environment) will be preferentially chosen for increased breeding in future generations. Overall, production of offspring based on minimizing MK will maximize gene diversity, minimize the rate of adaptation to the captive environment, minimize the rate of random genetic drift (and therefore, by definition, maximize the effective population size, N_e), and minimize long-term accumulation of inbreeding. Using simulations, Ballou and Lacy (1995) demonstrated the effectiveness of using MK to guide captive breeding, relative to other breeding strategies that have been proposed.

The optimality of an MK-guided breeding strategy is predicated on the assumption that animals selected for breeding will produce the desired progeny. Yet animals often differ substantially in their ability to reproduce, either because of their age or because of differences in fertility. If many animals in a lineage or kinship group have low prospects for breeding success, then optimal genetic conservation will require that extra progeny be produced by the relatives that can breed, in order to avoid a reduction or loss of alleles represented uniquely in the lineage. To account for variable chances of reproductive success in genetic management, a weighted MK, called kinship value (KV), can be used (Ballou and Lacy, 1995; Lacy, 1995). The KV of an animal is its weighted MK, where the weight is the mean reproductive value (V_x, the expected future reproduction for an animal of age x [Fisher, 1930]) of each of the kin. KV therefore discounts kin that have low potential for breeding because they are very young (if the probability of not reaching reproductive age is substantial) or old and maximally weights kin that are beginning their peak breeding years. Breeding the animals that have lowest KV should maximize gene diversity in the future if breeding pairs reproduce in accord with their V_x expectations.

Although individuals can be prioritized for breeding based on their MK or KV rank in the population, once the animals to be bred are selected, the optimal male–female combinations for pairings must be determined. It is at this stage that managers of captive populations avoid pairing closely related animals that would produce unacceptably inbred progeny. In addition, pairing males and female of similar MKs (i.e., pairing the best male with the best females, second best with second best, etc.) allows more effective genetic management in future generations (Lacy, 1994). This is because the most valuable (rarest) alleles in the pool of sires are combined with the most valuable alleles in the dams. That allows increased production simultaneously from both the paternal and maternal lineages. On the other hand, if valuable paternal lineages are linked with less valuable maternal lineages (or the reverse), then it becomes impossible to preferentially increase representation of one lineage in the population without also increasing the other to the same extent. Caballero and Toro (2000) questioned how pairings of genetically valuable sires and dams could be optimally used. Apparently they were thinking only of breeding programs in which a single offspring would be produced at each round of breeding. Instead, the optimal number of progeny to be produced per pairing can be determined through an iterative process of creating hypothetical progeny from the current highest-priority parents, updating the kinship matrix with each planned offspring (Lacy, 1995).

The fourth measure of genetic value used to manage captive breeding programs is the proportion of each individual's ancestry that is unknown (not traceable to the founders of the captive population). Unfortunately, it is not uncommon for some sires or dams in a captive breeding program to be unknown. This happens because of a lack of recordkeeping, because of acquisition of animals from sources (e.g., animal dealers and confiscations of illegally held animals) that are not part of the managed breeding program, and because of animals that are kept in multimale, multifemale social groups that preclude definitive observation of paternity or sometimes even maternity. Formulas are available for estimating kinships, inbreeding, and related measures for animals with partly known ancestries (Ballou and Lacy, 1995), but those methods assume that the genetic value of the unknown part of an animal's genome is equal to the value of the portion of the genome descended through traceable lineages. For that reason, captive breeding programs for conservation usually assign lower priority for breeding to animals that have partly unknown ancestries (Willis, 2001). Willis (1993) examined the relative costs and benefits to gene diversity and inbreeding avoidance of including or excluding animals of unknown ancestry. He found that the trade-offs can be complex but generally favor the inclusion of animals with unknown ancestry if the gene diversity of the population is very low and their exclusion if it is high.

Testing the Effectiveness of Conservation Breeding Strategies

It is important to know how effective these strategies are in achieving the genetic objective of minimizing evolutionary change in wildlife while they are in captivity. Some of the strategies described here, such as the use of MK to prioritize animals for breeding, have become widely used in conservation breeding programs. Others, such as using KV, have been proposed but have been applied only rarely. There is a diversity of opinions among practitioners about how effectively we are conserving genetic diversity and about the value of applying increasingly refined genetic management. There is also uncertainty about the robustness of the methods when some pedigree data are missing or are in error.

To test how well the MK and KV strategies perform in stopping evolution, I merged two computer programs traditionally used for different aspects of population analysis and management: the VORTEX simulation software for population viability analysis (Lacy, 1993b, 2000d; Miller and Lacy, 1999) and the GENES program for pedigree analysis (Lacy, 1999). VORTEX simulates population dynamics, under the assumptions of life history characteristics (e.g., age of breeding, maximum age, breeding structure, and variance in breeding success), mean birth and death rates, fluctuations in those rates, uncertainty in which animals live and which are successful as breeders, negative impacts of inbreeding on survival, initial age and sex structure, and limitations on the numbers of animals that can be supported. The simulation is used most often to assess threats to the viability of small, isolated populations of animals in the wild (Lacy, 2000a), and to compare likely effects of proposed conservation actions (Lacy, 1993/1994). The GENES program is used widely to calculate genetic parameters (e.g., inbreeding, kinships, gene diversity) from pedigrees and to guide selection of breeders in captive breeding programs for wildlife species.

VORTEX is used to project the future fates of simulated populations, whereas GENES is used to analyze the pedigree through previous generations in order to plan the next generation. Combining the two programs allows experimentation with different strategies for selecting animals to breed while monitoring of the fate of the simulated population under the uncertainties that face real populations. To link the two programs, VORTEX was modified to produce at each year of the simulation a list of the extant population and its pedigree. This pedigree was read by GENES, which then generated a prioritized list of animals to be paired for breeding. VORTEX used this list (rather than the default random breeding) to continue with the simulation. Each year, the VORTEX program outputs statistics describing the demographic and genetic health of the population: population size, gene diversity, mean inbreeding, number of founder alleles still present, and frequency of deleterious recessive alleles. In contrast to prior simulation tests of genetic management strategies (Ballou and Lacy, 1995), this approach incorporates the uncertainty about which of the paired animals will be successful breeders and which will survive. In contrast to prior uses of VORTEX to examine genetic change in conserved populations (e.g., Lacy and Lindenmayer, 1995), this approach allows specification of the strategy for selecting breeding pairs each generation.

The life history characteristics used for initial testing of genetic management strategies were chosen to model a typical medium-sized mammal or bird. The species was specified to begin breeding at age three years, to senesce at age ten, and to have a polygynous breeding system, with 75 percent of pairings producing offspring (±5 percent *SD* variation between years), 50 percent litters of one and 50 percent litters of two, 25 percent (±5 percent *SD* annual variation) infant mortality, 7 percent (±2 percent *SD* annual variation) annual mortality after the first year, and a mean of 3.14 recessive lethal alleles per founder. These demographic rates result in a population with a projected mean annual population growth rate (in the absence of random fluctuations) of 15 percent, moderate fluctuations in population growth, and a mean generation time of about six years. The populations were started with twenty-five animals, with a limit of fifty imposed by truncation density dependence. Populations were simulated for sixty years, or about ten generations, and results were averaged over 100 iterations of the simulations. Three breeding strategies were compared: pairing animals with the lowest MK, pairing animals with the lowest KV, and random breeding. In all three cases, pairs with inbreeding coefficients greater than 0.20 were eliminated from consideration. (Results were trivially different when simulations were run with stricter avoidance of inbreeding.) For the MK and KV strategies, after each pair was selected from the top of the list, it was presumed that they would produce a litter of one or two offspring, the kinship matrix and MKs were then updated to reflect the addition of the new offspring, and then the next pair was chosen. This iterative recalculation of the kinships after each pairing is selected is commonly used to guide breeding programs in zoos (Ballou and Lacy, 1995; Johnston and Lacy, 1995). It adjusts kinships of all unused

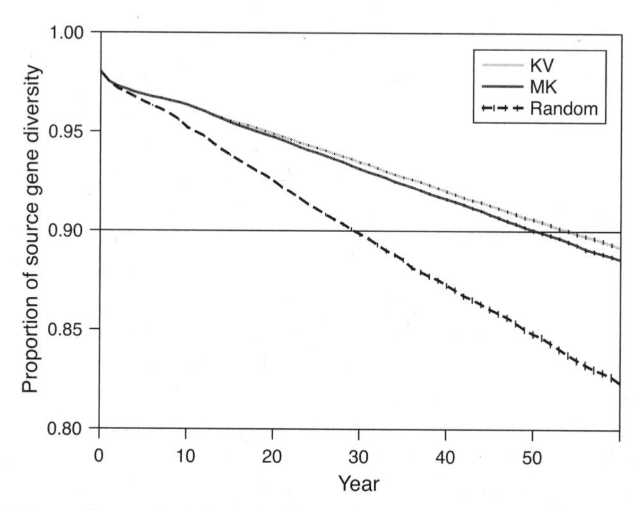

Figure 5.3 Loss of gene diversity under 3 genetic management strategies in populations simulated for 60 years. The horizontal line indicates the minimum level of gene diversity (90% of that in the wild source population) set as a goal for many long-term captive breeding programs for threatened species. The small vertical error bars are standard errors of the means across simulation iterations.

potential breeders for the prior selection of any of their kin, and it generates automatically the number of repeated breedings desired for each male.

Figure 5.3 shows the average losses of gene diversity under the three management strategies. The horizontal line shows the 90 percent criterion that is often used to define acceptable losses of gene diversity (Soulé et al., 1986). The managed populations maintained diversity much better than did the randomly paired or unmanaged populations. This demonstrates that the genetic management strategies perform well even when there is a lot of intrinsic uncertainty regarding births and deaths. The graph also shows that the KV strategy does slightly better than does the MK strategy, although the benefit is not noticeable until several generations have elapsed. Species with different life histories and demographic rates would show different amounts of benefit of the KV and MK management strategies, but my exploration of a few different cases consistently showed patterns like that shown in figure 5.3.

Figure 5.4 shows how effectively the three strategies minimize the accumulation of inbreeding. It shows that the mean level of inbreeding in the managed populations reaches the level of first-cousin matings, shown by the horizontal line, in about eight generations, as opposed to less than five generations with random breeding.

Figure 5.5 shows the loss of unique alleles in the three cases. The losses are precipitous for a few generations, but that is an artifact of the model, in which initially all twenty-five founder animals are assumed to contain two unique alleles at each locus. The rate of loss of alleles is highly dependent on starting allele frequencies, and the divergence between the lines would be less for alleles that were initially common in the population (Fuerst and Maruyama, 1986), but the simulation does show that genetic management retains more of

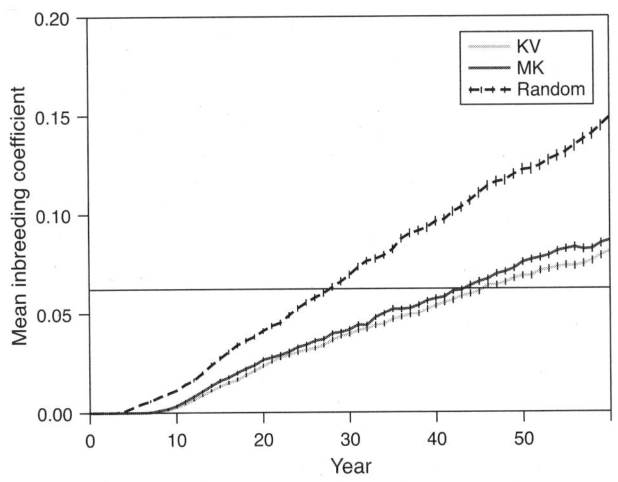

Figure 5.4 Accumulation of inbreeding under 3 genetic management strategies in populations simulated for 60 years. The horizontal line shows the maximum level of inbreeding ($f = 0.0625$) considered acceptable in many captive breeding programs.

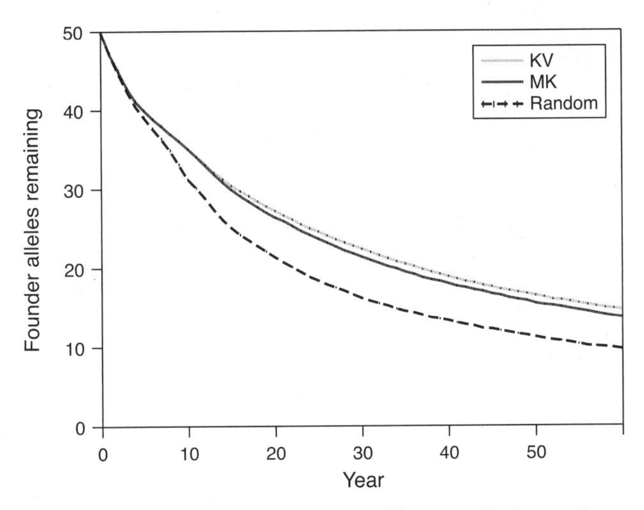

Figure 5.5 Loss of founder alleles at a hypothetical locus under 3 genetic management strategies in populations simulated for 60 years. Each of the 25 founders of the simulated population starts with 2 unique alleles.

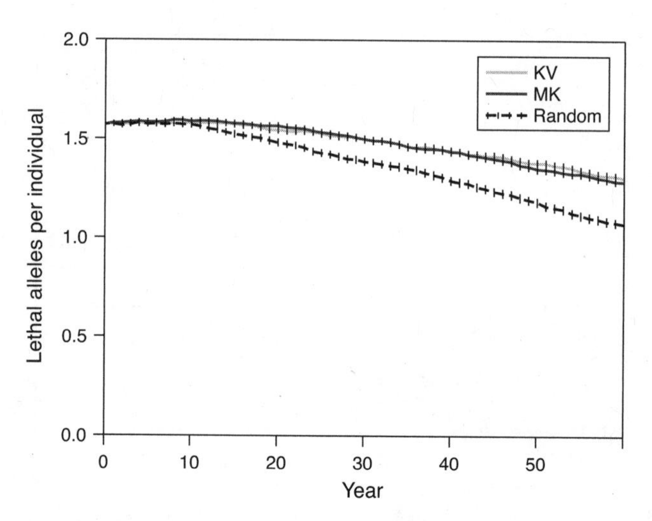

Figure 5.6 Reduction in number of recessive lethal alleles per animal under 3 genetic management strategies in populations simulated for 60 years. The population was initiated with an average of 1.57 lethal alleles per founder.

the initial alleles. As with gene diversity and inbreeding, the KV strategy is slightly better than the MK strategy.

Figure 5.6 shows the reduction in recessive lethal alleles that occurs as natural selection eliminates inbred homozygotes from the populations. At first, it might seem problematic that the managed populations retain their recessive deleterious alleles longer than does a randomly breeding population. Yet this is a desired result. An allele that is deleterious in a captive setting may very well be advantageous in a more natural environment. Thus, the loss of deleterious alleles in captivity mirrors the rate at which the populations would become adapted to captivity or domesticated. The simulation results show that the management strategies slow this rate of adaptation to the captive environment, although they cannot stop it entirely.

The rates of loss of genetic diversity from the populations can be converted to effective population sizes because the effective size of a population is defined to be the size of an ideal population that would lose diversity at the rate observed in the focal population. The rate of loss of gene diversity from a population with effective size N_e over t generations is given by $G_t = G_0\,[1 - 1/(2N_e)]^t$ (Crow and Kimura, 1970). Thus, the effective size can be calculated as $N_e = 0.5/[1 - (G_t/G_0)^{(1/t)}]$. Using this conversion, figure 5.7 shows the effective population sizes that led to the losses of diversity shown in figure 5.3. For the life history that was modeled, about thirty of the fifty animals would be adults. The effective size of the randomly breeding population is depressed a little below thirty because of the effects of population fluctuations. However, the managed populations do much better. In fact, under the KV strategy, N_e is about 1.5 times the number of adults, which is an impressive performance, considering that a theoretical maximum for N_e is twice the number of adults, and that can be achieved only if there is

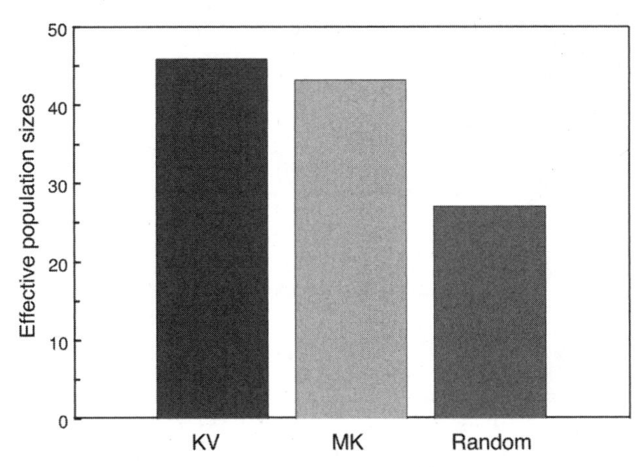

Figure 5.7 Effective population size (N_e) achieved under 3 genetic management strategies in populations with maximum census numbers of $N = 50$.

no random uncertainty with respect to which animals live and breed. Again, we see that the KV strategy slight outperforms MK.

Robustness Under Incomplete Pedigree Information

One reality of breeding programs in zoos and elsewhere is that we do not always have perfect pedigree information. As mentioned earlier, sometimes parentage data are simply not recorded. Other times, animals are kept in multimale, multifemale groups, and it is not possible to know with certainty who sired each offspring—unless you bring in the molecular geneticists. Therefore, it is important to know how robust the breeding schemes are to gaps or errors in the pedigrees. To test this, I specified in the simulation that the pedigree data given to the GENES pedigree analysis component of the program were missing information on some sires. The population simulation kept track of the real sires, so that the rate of genetic loss could be monitored, but some of those data were withheld from the algorithm that determined which animals should be paired each year.

Figure 5.8 shows the performance of KV and MK strategies under increasing percentages of unknown sires. The line shows the effective population size achieved under random breeding, which is unaffected by lack of knowledge about parentage. The KV strategy moderately outperforms the MK strategy across the levels of pedigree incompleteness that were tested. Both strategies perform well until the percentage of unknown sires exceeds about 10 percent. Even with pedigrees that are missing 20 percent of the sires, however, MK and KV breeding strategies significantly outperform random breeding. This result is perhaps surprising, given that the loss of information about the pedigree accumulates each generation. Thus, when 20 percent of sires (10 percent of the parents) are unrecorded each generation, after ten generations only 35 percent of the genes in the population ($0.35 = 0.90^{10}$) can be traced back to the founder source.

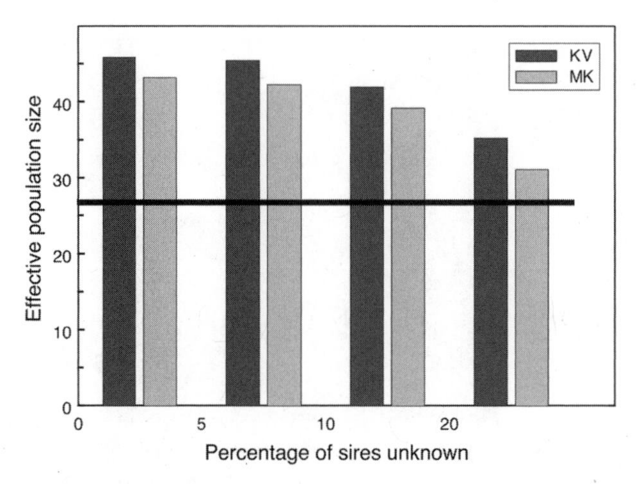

Figure 5.8 Effect of increasing percentage of sires unknown in the pedigree on the effective population size (N_e) achieved under 3 genetic management strategies in populations with maximum census numbers of $N = 50$. The horizontal line shows the N_e achieved under random breeding (which is unaffected by lack of pedigree information).

Clearly, however, full information is better, and zoos often use molecular genetic techniques to determine paternities. Figure 5.8 provides a way to assess what can be gained by filling in gaps in the pedigree. If 5 percent of the pedigree is unknown, it probably is not worth going to much trouble to resolve that uncertainty. At 10 percent, it may be worth it. At 20 percent, it probably is, because using that extra information would accomplish as much for achieving genetic goals as would increasing the population size by about one-third. The cost–benefit analysis for any given population depends on the expense of increasing the managed population size and the expense of making empirical determinations of the pedigree.

Future Trends: Merging Low-Tech Pedigree Analysis with High-Tech Molecular Genetics

Although I have contrasted the approaches of theoretical analysis and management of pedigrees with the empirical analyses afforded by molecular genetics, it is useful to consider some of the ways in which the two approaches can complement and benefit one another. First, as stated in the last section, molecular genetic analyses can be used to fill in gaps in our knowledge of the pedigrees, and this can improve the effectiveness of our breeding programs. We should also use molecular markers to empirically monitor and test how well we are doing in our quest to stop evolution in captive populations.

Standard pedigree analysis techniques start with the assumption that all founder animals are equally unrelated and not inbred. Yet in some cases we strongly suspect that some animals collected from the wild to be used to initiate a breeding program are close relatives to others. For example, the captive population of whooping cranes descends from eighty-eight eggs collected from the wild, but those eggs came from a remnant wild population that itself descended from probably just five breeding pairs in the prior generation or two (Jones et al., 2002). Jones et al. examined the relationships between the founders of the captive population by microsatellite DNA analysis and then assessed the impact on the whooping crane breeding program of considering these relationships in pedigree analysis and management.

Although most pedigree analyses begin with a kinship matrix of all zeros for between-founder kinships off the diagonal and all 0.5 for kinships of founders to themselves (indicating that they are not inbred), it is a straightforward extension of pedigree analyses to specify that the founders have some other observed kinship structure. However, caution should be exercised in using empirical data to initialize pedigree relationships. Unless the kinship structure among founders is measured with a fairly high degree of accuracy, using that information may actually degrade the effectiveness of pedigree management. It may be wise to use the relationships between founders only in cases (e.g., the whooping crane) in which there is a strong reason to believe that many close genetic relationships exist between founders.

Another suggestion that has been made regarding how molecular data could be used to augment breeding programs is to measure genotypes at some loci known to be important to fitness and manage the program to maximize retention of the valuable alleles. Two variations on this approach are worth considering. First, we could select for animals carrying alleles believed to be especially important, such as variants at the major histocompatibility complex loci (Hedrick, 2002; Hughes, 1991). Second, we could measure variation at random loci and then preferentially breed the animals that appear to carry the rarest alleles. In theory, this approach could produce a population with even more gene diversity than was present in the wild population by creating more equal allele frequencies than existed in the source population.

Although these ideas deserve more evaluation, I caution, as have others (Vrijenhoek and Leberg, 1991) that there are some potential drawbacks. First, we know very few of the many loci that might be critical to individual fitness and population viability. If we select only on the basis of the few loci about which we do know something, we are very likely to cause rapid depletion of genetic variability at other loci that may be just as important (Hedrick, 2001; Lacy, 2000c). This is especially so because the alleles that are advantageous depend on what environment the animals are in. Thus, many alleles that encode adaptations important in natural environments may be neutral or even deleterious in a specific captive environment. A strategy of preferentially breeding animals that have the rarest alleles, without trying to prejudge which alleles will be most advantageous, may have more merit than attempts to select the animal with superior alleles. However, even this strategy has risks. Initially rare alleles may have been rare for a good reason. Selecting for them may increase frequencies of mutations that were deleterious in the natural populations. I think we are on safer ground if we use strategies that attempt to minimize the rate at which the populations under our care diverge genetically from what they were before we took control of their breeding. Stopping evolution from running amok in captivity may be a better approach than trying to improve on the results of the prior evolution in wild populations.

Conclusions

The potentially rapid genetic changes that occur in managed populations of threatened species can seriously compromise the goals of conservation programs and even lead to the demise of the captive populations. Populations under careful management show better population growth and stability, maintain higher levels of individual animal health, and retain more genetic variation than do populations not under intense genetic management. Intensive genetic management will probably be essential to the long-term persistence of closed populations of the size that are typically maintained in conservation breeding programs.

Breeding strategies based on pedigree monitoring and management can be highly effective at retaining genetic variation in populations being conserved in captivity. A strategy of selecting as breeders the animals with lowest mean kinship to the managed population is nearly optimal for retaining gene diversity, avoiding loss of rare alleles, countering selection, and minimizing accumulation of inbreeding. A small further enhancement can be achieved by using weighted MKs (KVs) that consider also the likelihood that each kin will reproduce in the future. Such managed breeding programs can retain much more variation than would result from random breeding, and random breeding itself probably retains more variation than would be preserved if managers actively choose breeders based on nongenetic criteria. However, even with good genetic management, populations of the sizes that are typically maintained for conservation breeding programs by zoos and other wildlife centers are still subject to genetic drift and consequent losses of genetic diversity. Effective population sizes would need to be on the order of hundreds to keep substantial genetic change from occurring over a few generations (Lacy, 1987; Woodworth et al., 2002).

The use of MKs or KVs to prioritize animals for breeding is robust to moderate gaps in our knowledge of the pedigree. The lack of information for up to approximately 5 percent of parents does not seriously degrade our ability to provide good genetic management. Even when 10 percent of parents are not known, selection of parings based on the known portion of the pedigree provides better stabilization of the gene pool than would be obtained under a panmictic, unmanaged scheme.

Molecular genetic methods for empirical assessment of the genetic composition of individuals and the population can be used to fill in gaps in our knowledge of pedigrees and to monitor the effectiveness of conservation breeding programs. Molecular genetic analyses can also be used to assess the genetic structure of the founder population from which pedigree analysis derives.

Evolution to the highly modified and, we hope, temporary environments of conservation breeding programs meant to rescue threatened species can be substantially slowed but not stopped. Therefore, the goal of any conservation genetic program should be to ensure that the bulk of the species' population is still evolving under largely natural conditions in the wild habitats. When this is not the case, efforts must be made to restore wild populations quickly so that the duration of intensive genetic care is kept to as few generations as possible.

References

Allendorf, F. W. 1986. Genetic drift and loss of alleles versus heterozygosity. *Zoo Biology* 5:181–190.

Arnold, S. J. 1995. Monitoring quantitative genetic variation and evolution in captive populations. In *Population management for survival and recovery*, ed. J. D. Ballou, M. Gilpin, and T. J. Foose, 293–317. New York: Columbia University Press.

Ballou, J. D. 1984. Strategies for maintaining genetic diversity in captive populations through reproductive technology. *Zoo Biology* 3:311–323.

——. 1992. Potential contribution of cryopreserved germ plasm to the preservation of genetic diversity and conservation of endangered species in captivity. *Cryobiology* 28:19–25.

Ballou, J. D. and K. A. Cooper. 1992. Application of biotechnology to captive breeding of endangered species. *Symposia of the Zoological Society of London* 64:183–206.

Ballou, J. D. and T. J. Foose. 1996. Demographic and genetic management of captive populations. In *Wild mammals in captivity*, ed. D. G. Kleiman, M. E. Allen, K. V. Thompson, and S. Lumpkin, 263–283. Chicago: University of Chicago Press.

Ballou, J. D. and R. C. Lacy. 1995. Identifying genetically important individuals for management of genetic diversity in pedigreed populations. In *Population management for survival & recovery. Analytical methods and strategies in small population conservation*, ed. J. D. Ballou, M. Gilpin, and T. J. Foose, 76–111. New York: Columbia University Press.

Ballou, J. and K. Ralls. 1982. Inbreeding and juvenile mortality in small populations of ungulates: A detailed analysis. *Biological Conservation* 24:239–272.

Bijlsma, R., J. Bundgaard, and A. C. Boerema. 2000. Does inbreeding affect the extinction risk of small populations?: Predictions from *Drosophila*. *Journal of Evolutionary Biology* 13:502–514.

Brewer, B. A., R. C. Lacy, M. L. Foster, and G. Alaks. 1990. Inbreeding depression in insular and central populations of *Peromyscus* mice. *Journal of Heredity* 81:257–266.

Brock, K. M. and B. N. White. 1992. Application of DNA fingerprinting to the recovery program of the endangered Puerto Rican parrot. *Proceedings of the National Academy of Science, USA* 89:11121–11125.

Bryant, E. H. and D. H. Reed. 1999. Fitness decline under relaxed selection in captive populations. *Conservation Biology* 13:665–669.

Caballero, A. and M. A. Toro. 2000. Interrelations between effective population size and other pedigree tools for the management of conserved populations. *Genetical Research, Cambridge* 75:331–343.

Caughley, G. and A. Gunn. 1966. *Conservation biology in theory and practice*. Cambridge, Mass.: Blackwell Science.

Cohen, J. 1997. Can cloning help save beleaguered species? *Science* 276:1329–1330.

Critser, J. and R. Prather. 2002. The application and relevance of nuclear transfer technology to conservation. In *Reproduction and integrated conservation science*, ed. D. E. Wildt and B. Holt, 35–43. Cambridge: Cambridge University Press.

Crow, J. F. and M. Kimura. 1970. *An introduction to population genetics theory*. New York: Harper and Row.

Darwin, C. R. 1868. *Variation of animals and plants under domestication*. London: John Murray.

Earnhardt, J. M., S. D. Thompson, and E. A. Marhevsky. 2001. Interactions of target population size, population parameters, and program management on viability of captive populations. *Zoo Biology* 20:169–183.

Falconer, D. S. and T. F. C. Mackay. 1996. *Introduction to quantitative genetics*. Harlow, England: Longman.

Fisher, R. A. 1930. *The genetical theory of natural selection.* Oxford: Oxford University Press.

Frankel, O. H. and M. E. Soulé. 1980. *Conservation and evolution.* Cambridge: Cambridge University Press.

Frankham, R. 1995. Inbreeding and extinction: A threshold effect. *Conservation Biology* 9:792–799.

——. 1999. Quantitative genetics in conservation biology. *Genetical Research* 74:237–244.

Frankham, R., R. Hemme, O. A. Ryder, E. G. Cothran, M. E. Soulé, N. D. Murray, et al. 1986. Selection in captive environments. *Zoo Biology* 5:127–138.

Frankham, R., K. Lees, M. E. Montgomery, P. R. England, E. H. Lowe, and D. A. Briscoe. 1999. Do population size bottlenecks reduce evolutionary potential? *Animal Conservation* 2:255–260.

Frankham, R. and D. A. Loebel. 1992. Modeling problems in conservation genetics using captive *Drosophila* populations: Rapid genetic adaptation to captivity. *Zoo Biology* 11:333–342.

Frankham, R., H. Manning, S. H. Margan, and D. A. Briscoe. 2000. Does equalisation of family sizes reduce genetic adaptation to captivity? *Animal Conservation* 3:357–363.

Fuerst, P. A. and T. Maruyama. 1986. Considerations on the conservation of alleles and of genic heterozygosity in small managed populations. *Zoo Biology* 5:171–179.

Gansloßer, U., J. K. Hodges, and W. Kaumanns. 1995. *Research and captive propagation.* Fürth, Germany: Filander Verlag.

Geyer, C. J., O. A. Ryder, L. G. Chemnick, and E. A. Thompson. 1993. Analysis of relatedness in the California condors from DNA fingerprints. *Molecular Biology and Evolution* 10:571–589.

Haig, S. M., E. A. Beever, S. M. Chambers, H. M. Draheim, B. D. Dugger, S. Dunham, et al. 1994. Identification of kin structure among Guam rail founders: A comparison of pedigrees and DNA profiles. *Molecular Ecology* 3:109–119.

Haig, S. M. et al. 1995. Genetic identification of kin in Micronesian kingfishers. *Journal of Heredity* 86:423–431.

Harnal, V. K., D. E. Wildt, D. M. Bird, S. L. Monfort, and J. D. Ballou. 2002. Use of computer simulations to determine the efficacy of different genome resource banking strategies for maintaining genetic diversity. *Cryobiology* 44:122–131.

Hedrick, P. W. 2001. Conservation genetics: Where are we now? *Trends in Ecology and Evolution* 16:629–636.

——. 2002. The importance of the major histocompatibility complex in declining populations. In *Reproduction and integrated conservation science,* ed. D. E. Wildt and B. Holt, 75–85. Cambridge: Cambridge University Press.

Hodskins, L. G. 2000. *AZA annual report on conservation and science 1997–98.* Volume II. *Member institution conservation and research projects.* Silver Spring, Md.: American Zoo and Aquarium Association.

Holt, B. and P. Watson. 2002. Genetic management for inbreeding avoidance: The integrated use of genome resource banks. In *Reproduction and integrated conservation science,* ed. D. E. Wildt and B. Holt, 86–94. Cambridge: Cambridge University Press.

Holt, B. and D. E. Wildt. 2002. *Reproduction and integrated conservation science*. London: Zoological Society of London.

Hughes, A. L. 1991. MHC polymorphism and the design of captive breeding programs. *Conservation Biology* 5:249–251.

IUDZG/CBSG (IUCN/SSC). 1993. *The world zoo conservation strategy: The role of the zoos and aquaria of the world in global conservation.* Brookfield, Ill.: Chicago Zoological Society.

Jiménez, J. A., K. A. Hughes, G. Alaks, L. Graham, and R. C. Lacy. 1994. An experimental study of inbreeding depression in a natural habitat. *Science (Washington DC)* 266:271–273.

Johnston, L. A. and R. C. Lacy. 1995. Genome resource banking for species conservation: Selection of sperm donors. *Cryobiology* 32:68–77.

Jones, K. L., T. C. Glenn, R. C. Lacy, J. R. Pierce, N. Unruh, C. M. Mirande, et al. 2002. Refining the whooping crane studbook by incorporating microsatellite DNA and leg banding analyses. *Conservation Biology* 16:789–799.

Keller, L. F., P. Arcese, J. N. M. Smith, W. M., Hochachka, and S. C. Stearns. 1994. Selection against inbred song sparrows during a natural population bottleneck. *Nature* 372:356–357.

LaBarge, T. 1999. *AZA population management plan for Markhor* (Capra falconeri heptneri). Syracuse, N.Y.: Rosamond Gifford Zoo at Burnet Park.

Lacy, R. C. 1987. Loss of genetic diversity from managed populations: Interacting effects of drift, mutation, immigration, selection, and population subdivision. *Conservation Biology* 1:143–158.

——. 1989. Analysis of founder representation in pedigrees: Founder equivalents and founder genome equivalents. *Zoo Biology* 8:111–124.

——. 1993a. Impacts of inbreeding in natural and captive populations of vertebrates: Implications for conservation. *Perspectives in Biology and Medicine* 36:480–496.

——. 1993b. VORTEX: A computer simulation model for population viability analysis. *Wildlife Research* 20:45–65.

——. 1993/1994. What is population (and habitat) viability analysis? *Primate Conservation* 14/15:27–33.

——. 1994. Managing genetic diversity in captive populations of animals. In *Restoration and recovery of endangered plants and animals,* ed. M. L. Bowles and C. J. Whelan, 63–89. Cambridge: Cambridge University Press.

——. 1995. Clarification of genetic terms and their use in the management of captive populations. *Zoo Biology* 14:565–577.

——. 1997. Importance of genetic variation to the viability of mammalian populations. *Journal of Mammalogy* 78:320–335.

——. 1999. *GENES: A software package for the genetic analysis of pedigree data*. Brookfield, Ill.: Chicago Zoological Society.

——. 2000a. Considering threats to the viability of small populations. *Ecological Bulletins* 48:39–51.

——. 2000b. Management of limited animal populations. In *Bottlenose dolphin reproduction workshop report,* ed. D. Duffield and T. Robeck, 75–93. Silver Spring, Md.: AZA Marine Mammal Taxon Advisory Group.

——. 2000c. Should we select genetic alleles in our conservation breeding programs? *Zoo Biology* 19:279–282.

——. 2000d. Structure of the VORTEX simulation model for population viability analysis. *Ecological Bulletins* 48:191–203.

Lacy, R. C., J. D. Ballou, F. Princee, A. Starfield, and E. Thompson. 1995. Pedigree analysis. In *Population management for survival & recovery. Analytical methods and strategies in small population conservation,* ed. J. D. Ballou, M. Gilpin, and T. J. Foose, 57–75. New York: Columbia University Press.

Lacy, R. C. and D. B. Lindenmayer. 1995. A simulation study of the impacts of population subdivision on the mountain brushtail possum *Trichosurus caninus* Ogilby (Phalangeridae: Marsupialia), in South-eastern Australia: II. Loss of genetic variation within and between subpopulations. *Biological Conservation* 73:131–142.

Lacy, R. C., A. Petric, and M. Warneke. 1993. Inbreeding and outbreeding depression in captive populations of wild species. In *The natural history of inbreeding and outbreeding,* ed. N. W. Thornhill, 352–374. Chicago: University of Chicago Press.

Laikre, L. 1999. Hereditary defects and conservation genetic management of captive populations. *Zoo Biology* 18:81–99.

Lande, R. and G. F. Barrowclough. 1987. Effective population size, genetic variation, and their use in population management. In *Viable populations for conservation,* ed. M. E. Soulé, 87–123. Cambridge: Cambridge University Press.

Loskutoff, N. 2002. Do advanced reproductive technologies have a role in future conservation and management programmes? In *Reproduction and integrated conservation science,* ed. D. E. Wildt and B. Holt, 125–134. Cambridge: Cambridge University Press.

Malécot, G. 1948. *Les mathematiqués de l'hérédité.* Paris: Masson.

Margulis, S. W. 1998. Differential effects of inbreeding at juvenile and adult life-history stages in *Peromyscus polionotus. Journal of Mammalogy* 79:326–336.

Meagher, S., D. J. Penn, and W. K. Potts. 2000. Male–male competition magnifies inbreeding depression in wild house mice. *Proceedings of the National Academy of Science, USA* 97:3324–3329.

Meek, J. 2002. Scientists pledge to clone extinct Tasmanian tiger. *The Guardian.* Retrieved from http://www.guardian.co.uk/world/2002/may/29/animalwelfare.highereducation.

Meffert, L. M. 1999. How speciation experiments relate to conservation biology: The assumption of captive breeding strategies—that founder events reduce genetic variation—may not always be correct. *BioScience* 49:701–711.

Miller, P. S. 1994. Is inbreeding depression more severe in a stressful environment? *Zoo Biology* 13:195–208.

Miller, P. S. and R. C. Lacy. 1999. *VORTEX version 8 users manual. A stochastic simulation of the simulation process.* Apple Valley, Minn.: IUCN/SSC Conservation Breeding Specialist Group.

Montgomery, M. E., J. D. Ballou, R. K. Nurthen, P. R. England, D. A. Briscoe, and R. Frankham. 1997. Minimizing kinship in captive breeding programs. *Zoo Biology* 16:377–389.

Morin, P. A. and O. A. Ryder. 1991. Founder contribution and pedigree inference in a captive breeding colony of lion-tailed macaques, using mitochondrial DNA and DNA fingerprint analysis. *Zoo Biology* 10:341–352.

Myers, N. and A. H. Knoll. 2001. The biotic crisis and the future of evolution. *PNAS* 98:5389–5392.

Rabb, G. B. 1994. The changing roles of zoological parks in conserving biological diversity. *American Zoologist* 34:159–164.

Ralls, K. and J. Ballou. 1983. Extinction: Lessons from zoos. In *Genetics and conservation. A reference for managing wild animal and plant populations,* ed. C. M. Schonewald-Cox, S. M. Chambers, B. MacBryde, and W. L. Thomas, 164–184. Menlo Park, Cal.: Benjamin/Cummings.

Ralls, K., J. D. Ballou, and A. R. Templeton. 1988. Estimates of lethal equivalents and the cost of inbreeding in mammals. *Conservation Biology* 2:185–193.

Ralls, K., K. Brugger, and J. Ballou. 1979. Inbreeding and juvenile mortality in small populations of ungulates. *Science* 206:1101–1103.

Reinartz, G. E. 1977. *Patterns of genetic variation in the bonobo (*Pan paniscus*)*. Milwaukee: University of Wisconsin.

Robertson, A. 1960. A theory of limits in artificial selection. *Proceedings of the Royal Society of London—Series B: Biological Sciences* 153:234–249.

Ryan, K. K., R. C. Lacy, and S. W. Margulis. 2002. Effects of inbreeding and kinship on components of reproductive success. In *Reproduction and integrated conservation science,* ed. B. Holt, 82–96. London: Zoological Society of London.

Saccheri, I., M. Kuussaari, M. Kankare, P. Vikman, W. Fortelius, and I. Hanski. 1998. Inbreeding and extinction in a butterfly metapopulation. *Nature* 392:491–494.

Santiago, E. and A. Caballero. 2000. Application of reproductive technologies to the conservation of genetic resources. *Conservation Biology* 14:1831–1836.

Smith, D. 2002. Jurassic Park techniques may bring back thylacine. *Sydney Morning Herald.* Retrieved from http://www.smh.com.au/articles/2002/05/28/1022569773166.html.

Snyder, N. F. R., S. Derrickson, S. R. Beissinger, J. W. Wiley, T. B. Smith, W. D. Toone, et al. 1996. Limitations of captive breeding in endangered species recovery. *Conservation Biology* 10:338–348.

Soulé, M., M. Gilpin, W. Conway, and T. Foose. 1986. The millenium ark: How long a voyage, how many staterooms, how many passengers? *Zoo Biology* 5:101–113.

Stanley-Price, M. R. 1989. *Animal reintroductions. The Arabian oryx in Oman.* Cambridge: Cambridge University Press.

Taylor, A. C., W. D. Greville, and W. B. Sherwin. 2002. Consequences of inbreeding on population fitness. In *Reproduction and integrated conservation science,* ed. D. E. Wildt and B. Holt. Cambridge: Cambridge University Press.

Templeton, A. R., R. J. Robertson, J Brisson, and J. Strasburg. 2001. Disrupting evolutionary processes: The effect of habitat fragmentation on collared lizards in the Missouri Ozarks. *PNAS* 98:5426–5392.

Vrijenhoek, R. C. and P. L. Leberg. 1991. Let's not throw out the baby with the bathwater: A comment on management for MHC diversity in captive populations. *Conservation Biology* 5:252–254.

Willis, K. 1993. Use of animals with unknown ancestries in scientifically managed breeding programs. *Zoo Biology* 12:161–172.

——. 2001. Unpedigreed populations and worst-case scenarios. *Zoo Biology* 20:305–314.

Wilson, A. C. and M. R. Stanley-Price. 1994. Reintroduction as a reason for captive breeding. In *Creative conservation: Interactive management of wild and captive animals,* ed. P. J. S. Olney, G. M. Mace, and A. T. C. Feistne, 243–264. London: Chapman and Hall.

Woodworth, L. M., M. E. Montgomery, D. A. Briscoe, and R. Frankham. 2002. Rapid genetic deterioration in captive populations: Causes and conservation implications. *Conservation Genetics* 3:277–288.

Wright, S. 1931. Evolution in Mendelian populations. *Genetics* 16:97–159.

——. 1948. On the role of directed and random changes in gene frequency in the genetics of populations. *Evolution* 2:279–294.

——. 1977. *Evolution and the genetics of populations.* Volume 3. *Experimental results and evolutionary deductions*. Chicago: University of Chicago Press.

Stephen R. Palumbi

The Emerging Theme of Ocean Neighborhoods in Marine Conservation

> Though it lash the shallows that line the beach,
> Afar from the great sea-deeps,
> There is never a storm whose might can reach
> Where the vast leviathan sleeps.
>
> John Boyle O'Reilly, *Prelude to the Amber Whale*

The Importance of Scale

The oceans evoke a sense of mystery and vastness that pervades our literature and colors our perception of the nature of marine life. Deep beneath the ocean surface roils a dangerous ecology that is portrayed to be more exotic and more predatory than the mostly tamed ecosystems on land. With effort and technology we can breach the surface and dive down to observe, catalogue and understand marine life, but many crucial aspects of it remain hidden or obscure. In this obscurity lies much of the charm of the sea, but in it also lies a great deal of danger because this mysterious and vast ocean seems to be too wide and unknown for us to protect, too powerful for us to restore.

Yet protection and restoration are now needed because the ocean has been threatened by so many large-scale anthropogenic problems that we are now able to damage even the largest of the earth's seas. Just a thousand years ago, human efforts could scarcely imperil a lake or a river. But industrial agriculture and modern technology have allowed us to steadily increase the size of the body of water we could significantly degrade. Areas affected are first ponds and lakes; then small seas such as the Aral Sea, nearly drained for agricultural irrigation; then large estuaries such as the Chesapeake Bay and inland seas such as the Black Sea. Finally affected are oceanic areas such as the

Gulf of Mexico, which has erupted with diseases such as red tides and dead zones, and the great expanses of the Atlantic, Pacific, and Southern oceans.

Given the global nature of ocean threats, human capacity to save the seas appears weak. Small-scale conservation efforts often appear hopeless when one considers the vastness of the oceans. And it often seems as if small-scale efforts will be diluted by the constant movement and swirl of ocean currents. If this is true—if our conservation efforts can be successful only if they are on scales of whole oceans—then the task of launching these efforts is extremely daunting. However, if it is possible to begin conservation efforts on a smaller scale and still provide local benefits, then a pathway to ocean protection opens up, a pathway that begins with small efforts and incrementally protects more and more of the sea.

In recent years, marine science has developed ways of looking at the geographic scale of marine populations and the distances across which they respond as a single ecological unit to outside environmental influences. This scale defines the size of ocean neighborhoods, the areas in which marine populations tend to live out their lives, reproduce, and foster new generations. Like human neighborhoods, these ocean neighborhoods cannot be expected to have rigid walls and rigid limits of dispersal. But we can try to understand the basic scale of these neighborhoods and use this information in marine restoration and conservation efforts. Whether these neighborhoods exist on scales of kilometers or thousands of kilometers remains unknown for even the most important species. But recent advances in experimental ecology, satellite technology, chemical tagging, and population genetics are beginning to reveal times and places in which the size of ocean neighborhoods is surprisingly small.

Ocean Eye in the Sky

Flitting above Earth's atmosphere are some of the most important oceanographic devices ever built: satellites that observe the surface of the ocean and use the light it reflects to deduce volumes about the condition of the sea. Among the welter of data that these satellites digest every day, some of the simplest show how different some parts of the ocean are from others. Particularly along coastlines, the temperature of the ocean and its productivity often are dramatically different than they are just a few kilometers offshore. This is because the water that hugs the coast collects heat from the sun and nutrients from the land to turn into a marine soup filled with abundant life. Single-celled plants bloom along these coasts and reveal themselves to the satellites through the color of their chlorophyll. Zooplankton gorge on these small plants and themselves are eaten by myriad coastal fish and invertebrates. The result is that the coastal oceans are dramatically different from the waters of the open sea. These coastal expanses are usually considered different ocean neighborhoods by virtue of the different environment of the coast and the fact that the coastal water stays near the coast long enough to be substantially altered. Ocean movements are not always so rapid that different water masses are always well mixed. Instead, oceans can have habitat differences like those caused by different elevations or soil types or rainfall on land.

Other ocean eyes have been designed to monitor the pattern of currents and temperatures over smaller scales (Harms and Winant, 1998). These also show that the waters hugging the coasts are very different from those in the middle of the ocean. But they also show that coastal currents are complicated with eddies, gyres, backflows, and countercurrents (Palumbi et al., 2003). In particular, gyres and eddies create circular current patterns that can trap water and generate different conditions for the growth of marine populations (Menge et al., 1997). Such eddies are particularly common downstream of promontories or shallow banks that deflect currents offshore (Largier, 2003). For example, off the coast of Oregon, a shallow bank called the Heceta Bank rises up from the sea bed to deflect the southern-flowing California current (figure 6.1). In the lee of the bank, a circular gyre often forms that pulls bottom water up from the depths. This bottom water is nutrient rich and highly productive, making the Heceta Bank one of the most productive fishing areas in Oregon (Menge et al., 1997).

The conclusion from these satellite results is that the oceans have habitats that differ as much as meadows, forests, and deserts do on land. The central oceans, with their deep waters and poor access to nutrients, are the deserts of the sea. The breadbaskets of the sea are the coastal zones, whose warm, rich waters hug the shoreline and support abundant marine life.

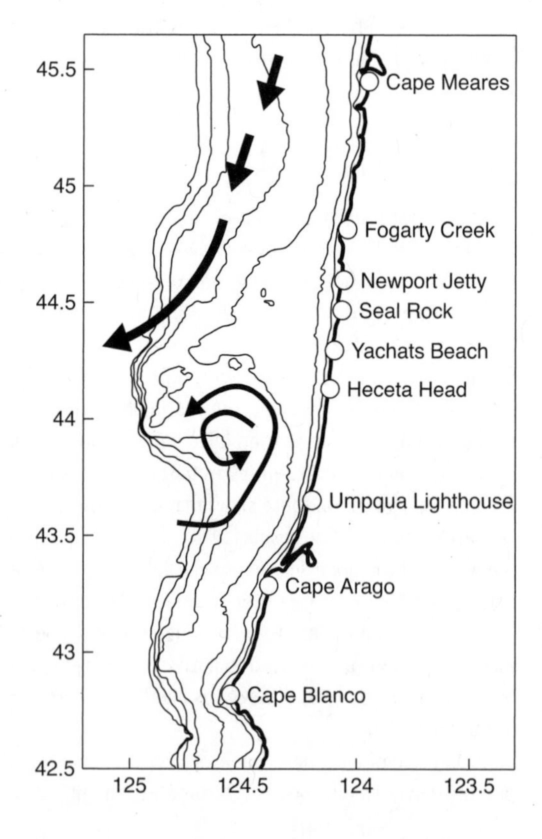

Figure 6.1 Offshore eddies near Heceta Head, Oregon trap nutrients, phytoplankton, and larvae, generating a different marine environment than the waters outside the gyre. The coastline is drawn with a thick line, and offshore bathymetry is represented by thin lines. The southern-flowing California current (thick arrows) is deflected to the west by the shallow Heceta Bank, generating a region of circular currents south of the bank. Figure redrawn from Sotka and Palumbi, in review.

The Equation of Development and Dispersal

For the species that exist along the coast, living off the warm, nutrient-laden waters at the edge of a wide, less productive open ocean, what are the fates of eggs and larvae? These species often live as adults in a small section of their species' home range, such as a single submerged rocky reef or estuarine bay. They typically have small eggs and larvae that can be carried miles away by churning ocean currents (Shanks et al., 2003). Most commercially important marine species have a stage in their life history in which an egg develops into a feeding, swimming larva that grows for days, weeks, or even months (Shanks et al., 2003) and then settles to the bottom to begin the juvenile phase of life. Laboratory studies have shown how long this process takes and have revealed that for ocean life the developmental periods differ widely. Some larvae spend as little as fifteen minutes in the water before settling. Others spend months drifting and feeding before reaching the size necessary for juvenile form (figure 6.2; see Shanks et al., 2003, for a review).

These planktonic phases can result in potential for long-distance dispersal, and in fact larvae sometimes settle in regions far from their origins. Open ocean plankton tows often contain the larvae of coastal species (Scheltema, 1986). Furthermore, the Gulf Stream brings the larvae of Caribbean fish to Cape Cod Bay each year, and El Niño events in the

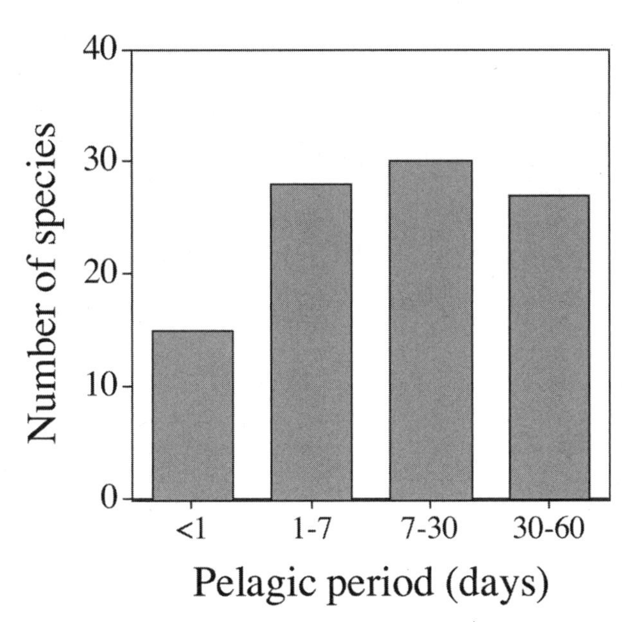

Figure 6.2 Number of invertebrates from the West Coast of the United States that have larvae with various pelagic periods. Most commercial species have larvae in the 7- to 60-day categories, whereas most species with less than 1-day pelagic periods are colonial tunicates, ascidians, or gastropods with young that hatch directly from benthic egg cases. Data from Shanks et al. (2003).

Pacific have deposited Pacific tropical species onto cold Galápagos reefs thousands of kilometers from their populations of origin (Richmond, 1990). Such long-distance dispersal connects populations across large distances genetically (Palumbi, 1992) and has given rise to the conventional wisdom that few marine larvae remain close to their parents (Scheltema, 1988; Strathmann, 1990).

But data to corroborate this notion are difficult to obtain. Because larvae are so small—each one smaller than the period after this sentence—they cannot be tagged the way giant pelagic fish can be (Block et al., 1998, 2001). As a result, marine scientists have typically estimated dispersal potential from a simple equation for potential dispersal distance (D) to the product of developmental period (P) and ocean current speed (C): $D = P \times C$.

The first part of this equation is the duration of larval life and is derived from laboratory studies of development or from the measurement of daily growth rings in the fish otoliths (ear bones) or invertebrate statoliths (calcareous particles used in balance and orientation). The second part of the equation is the average speed of currents that carry larvae from place to place. When multiplied together, these two values give an estimate of dispersal range for a species in a particular ocean setting. Such averages are typically quite large; for example, if the average speed of the California current is 20 centimeters per second (Largier, 2003), then larvae will move about 0.8 kilometers per hour, or about 20 kilometers a day. Over two weeks, this will transport a larva 280 kilometers from its origin. Faster current speeds are common; for example, the Makassar Current in Indonesia clocks in at about 1 meter per second (Barber et al., 2000).

One initial source of confirmation for these large dispersal distances was buoys that transmit their locations to satellites as they move along with ambient ocean currents. Such buoys have been found to move hundreds of kilometers over a two- to four-week span, the planktonic period of many larvae (figure 6.3), which suggested that the simple mathematics (distance = speed × duration) described actually reflects dispersal in the sea.

However, recent work has emphasized that this simple calculus is based on assumptions that are largely incorrect, or at best of limited usefulness. First, average current speeds are not good descriptions of the behavior of flowing ocean water. Instead, real currents tend to be highly variable, with regions of faster movement, regions of stalled water, and many regions where backwashes and eddies move water opposite to the prevailing direction (Largier, 2003). In addition, whereas the buoy's movement is affected only by the ocean currents, larval swimming behavior can move larvae between currents that are opposite in direction and can substantially change their dispersal trajectory (Cowen et al., 2000). As a result, the movement patterns of larvae are likely to be only poorly modeled by simple equations, and other means of determining their fates are needed.

DNA Tags for Larvae

Although larvae cannot be tagged with satellite transponders, they carry with them a natural tag in the form of their DNA that can be decoded to reveal patterns of dispersal. Reading the data in these DNA tags has become much easier because of the automation

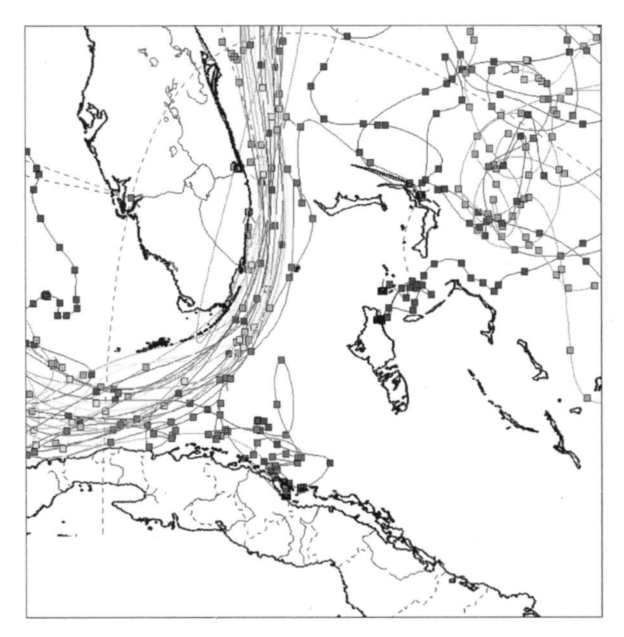

Figure 6.3 Tracks of open ocean drifters moving along the Florida current and into the Bahamas. Few tracks connect the Florida Keys and the Bahamas, despite long-distance movement from week to week. Dots represent positions at weekly intervals.

invented for the Human Genome Project. This automation allows us to assay the tags in hundreds or thousands of individuals rapidly and accurately.

Once tags are identified for these individuals, generally collected from many localities across a species' range, they can be compared to build a picture of the distribution of DNA variants in space and time. At this stage, detecting patterns in the data becomes a serious challenge, and new analytical tools have been developed to detect patterns of small and large dispersal from place to place.

One way to think about this problem is to consider a list of DNA tags to be similar to family names listed in a metropolitan phone book. Clusters of family names that occur in one city but not another can suggest patterns of low dispersal. By contrast, if dispersal is large, then family names become scattered across the landscape, and clumps of similar names are less common. For example, you will find clusters of Palumbis in Baltimore and Pittsburgh but not in most other U.S. cities, reflecting my family's recent arrival in the United States and low tendency to move. Because the DNA tags are inherited like family names, we can detect dispersal distances by examining lists of DNA tags for clumps of identical sequences within populations. Clusters of identical DNA sequences that occur only in one locality are very informative about patterns of organism movement from place to place. With thousands of data points, we can afford to look for geographic patterns of moderately rare DNA sequences, a technique of rare allele analysis proposed in the 1980s for use with allozyme data (Slaktin and Barton, 1989).

Genetic Signals of Population Ecology

Evidence is mounting that long-distance dispersal in many marine species is much more limited than previously thought. Retention of larvae around oceanic islands (Swearer et al., 1999) or in estuaries (Gaines and Bertness, 1992) shows that marine populations may provide their own recruits over scales of tens to hundreds of kilometers. Such populations receive larval input from elsewhere but at a lower level than the production of larvae that stay at home. Recently, we have shown through genetic comparisons that populations of the West Coast barnacle *Balanus glandula* show striking but previously hidden patterns of genetic differentiation (Palumbi et al., 2003). Although *B. glandula* is one of the best-studied marine species in the world and an ecological foundation species that figures prominently in coastal research, these genetic results overturn long-held assumptions of long-distance movement of barnacle larvae along the West Coast.

The results show two striking patterns. First, there is a steep genetic cline from Monterey Bay to Cape Mendocino in California at two genetic loci, mitochondrial cytochrome oxidase I and nuclear elongation factor 1 alpha (Sotka et al., 2004). Barnacles south of Monterey are part of a southern population genetically distinct from the populations north of Cape Mendocino (figure 6.4). Natural selection in a highly dispersive population can generate such strong genetic signatures, but this is unlikely because we see parallel results from two independent loci. Instead, it may be that this species has been divided into northern and southern populations—perhaps by historic separation of populations in ice age refugia—and that these populations are now intermingling along the California coast. The patterns of genetic change suggest slow intermixing between these two populations, because of either low dispersal distances of only 10 to 20 kilometers or strong selection against long-distance migrants. In either case, ecological connections of these populations within the area of the cline are shown by these data to be very low. Despite a three- to four-week planktonic larval period and the potential to drift hundreds of kilometers in the swift California current, B. glandula populations between intertidal sites in different West Coast states are largely separate.

The second genetic pattern is a distinct genetic signature of barnacle populations along the Oregon coast, where they are washed by a strong oceanic gyre formed in the lee of Heceta Head. The Heceta Gyre is a well-known area of high productivity and good fishing, but it has never before been considered as a boundary of separate marine populations. Our genetic results show that populations north and south of the gyre are different from those within it (Sotka and Palumbi, 2006). This gyre is the source of many oceanic differences in productivity and growth rate. That it also creates genetic boundaries for barnacles shows that populations within the gyre form a separate ocean neighborhood demographically distinct from populations outside. Even over the small scale of the gyre, with a diameter of only 60 to 80 kilometers, these genetic tools have revealed ecologically closed populations.

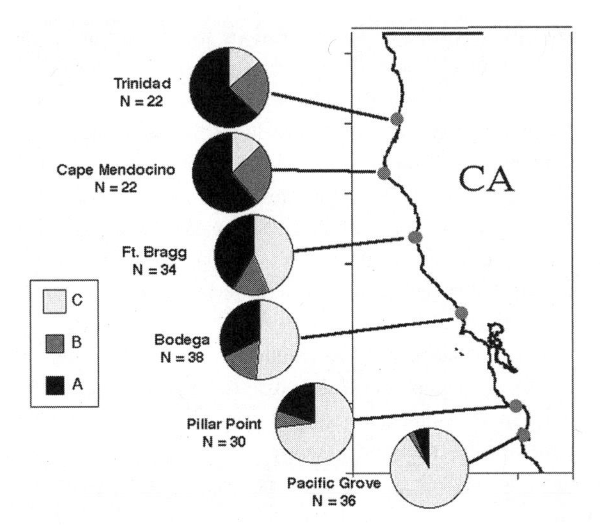

Figure 6.4 Cline along the California coast in the frequency of DNA sequence variants of a section of the mitochondrial cytochrome oxidase I gene of the intertidal barnacle *Balanus glandula*. Populations between Monterey and Cape Mendocino (ca. 400 kilometers) show a shift in the frequency of 3 mitochondrial haplotype clades (A, B, and C). Type C haplotypes dominate southern populations, whereas types A and B dominate in the north. (Redrawn from Sotka, Wares, Grosberg, and Palumbi, submitted manuscript). Such steep clines indicate either low dispersal or strong selection, both of which reduce the rate at which barnacle larvae, which spend 2 to 4 weeks in the plankton, successfully recruit into distant populations.

Chaos and Genetic Structure

Genetic investigations of marine populations have a strange skeleton in the closet. In many detailed studies, adjacent populations have shown slight but significant genetic differences, whereas more distant populations have been genetically similar (Johnson et al., 1993). This has been called chaotic patchiness, reflecting the patchy but unpredictable nature of genetic differences. Theoretically, small differences over small geographic scales should translate into larger differences over larger scales (e.g., isolation by distance [Bohonak, 1999; Palumbi, 2003]), but in chaotic patchiness, this pattern is not observed.

Three potential explanations of chaotic patchiness have been proposed, all of which have conservation implications. First, Slatkin (1993) has shown that when populations in a stepping stone model have not had time to reach genetic equilibrium after a perturbation, genetic differences can rise quickly over small distances yet remain the same over large spatial scales. Recent glacial perturbations of temperate ranges probably have affected the genetic structure of many coastal species, and the resulting slow return to genetic equilibrium may explain some cases of genetic patchiness. In these cases, the slight but significant differences between adjacent populations are the best indication of dispersal and should be taken as indications of the scale of genetic neighborhoods.

Second, settlement of larval clouds might perturb genetic structure. Suppose the offspring of small number of spawning adults travel in the plankton as a group, settling down in one place. The genetic makeup of such a cloud is likely to depend on which parents happen to reproduce, and the allele frequencies of the settlers are likely to be different from those of other settlers and local adults. This scenario will generate a pattern similar to chaotic patchiness—even if dispersal is high—as long as larval clouds hang together. The mechanisms that could maintain larvae in a single cloud are unclear but might include mesoscale eddies (Largier, 2003). The ability of this mechanism to create the observed patterns requires that each larval cloud be genetically distinct, either because of selection (Edmands et al., 1996) or because the alleles possessed by a random subset of parents are likely to be unique. Some evidence for both selection (Flowers et al., 2002) and limited parentage of larvae in clouds shows that these mechanisms are possible. However, other work suggests that larval settlers are highly diverse.

A third mechanism to produce small-scale shifts in genetic structure is the impact of coastal gyres on patterns of larvae dispersal, such as the pattern we have observed near Heceta Head, Oregon (Sotka and Palumbi, 2006). Although this is only a single example, if other species are similarly affected by this gyre and if other gyres along the complex Pacific coast also affect genetic structure, then this might explain some observations considered to be genetically patchy.

Marine Reserves: An Experiment on Small Ocean Neighborhoods

A very different type of evidence about the scale of coastal populations comes from the study of marine reserves. These regions in which no resource extraction is allowed have been the focus of a great deal of review over the past decade (Agardy, 1994; Castilla, 1996; Creese and Cole, 1995; Roberts et al., 1995; Roberts and Polunin, 1991; Ruckelshaus and Hayes, 1998; Russ and Alcala, 1996), in part because they represent an ecosystem management tool that can help manage entire marine communities rather than manage single species one at a time (Lubchenco et al., 2003; Palumbi, 2003).

Marine reserves are rather small: The average size among more than 100 reserves reviewed by Halpern (2003) was about 0.1 square kilometer. Yet when overexploitation of marine species within reserves is halted, the result is a blossoming of these local populations. On average, biomass of exploited species doubled within reserves (Palumbi, 2003, based on data in Halpern, 2003). Species that were previously overfished showed larger average sizes and higher densities within reserves than just outside their borders (Halpern, 2003), and in some cases ecosystems within reserves shifted dramatically (Babcock et al., 1999).

These data show that reserves have a dramatic effect within their borders, even on populations of mobile fish that are capable of moving away from protected areas. These results also show how the composition of a local marine ecosystem responds to local changes in exploitation. Although there are many pelagic species that may benefit less from local protection, there are hundreds of species for which ecosystem shifts inside reserves generate dramatic changes.

The best-documented effects of reserves are these changes within their borders, but there is increasing evidence that the large populations within reserves create a spillover effect that enhances marine ecosystems just outside reserve boundaries (Roberts et al., 2001b). In Florida, trophy fish of some species are most commonly caught adjacent to the Merrit Island Estuarine Reserve, which surrounds Cape Canaveral (Roberts et al., 2001b). And fishers all over the world move their fishing efforts to the waters next to local reserves in order to increase their catches (McClanahan and Kaunda-Arara, 1996). Sometimes this positioning is so precise that the reserve borders are clearly demarcated by fishing traps or anchored boats.

Some evidence of larval or egg spillover also exists, but this facet of the impact of reserves on populations is the poorest studied of all. Limpets that are much more abundant in protected areas in the Canary Islands show higher settlement inside reserves but also show higher settlement within about 1 kilometer of the reserve borders (figure 6.5, Branch). In addition, higher scallop densities just outside a trawling refuge on Georges Bank of Cape Cod suggest a spillover from inside the protected areas (Murawski et al., 2000). Tracking of coral larvae through capture of settlers at various distances from reefs (Sammarco and Andrews, 1988) also shows that movements of these larvae can be local.

Spillover across reserve borders over small spatial scales—by either movement of adults or dispersal of larvae—shows that ocean neighborhoods may sometimes be on the same 1- to 10-kilometer scale as current reserves. To date, such conjectures about neighborhood sizes based on reserve data derive from very few examples, but future scrutiny of

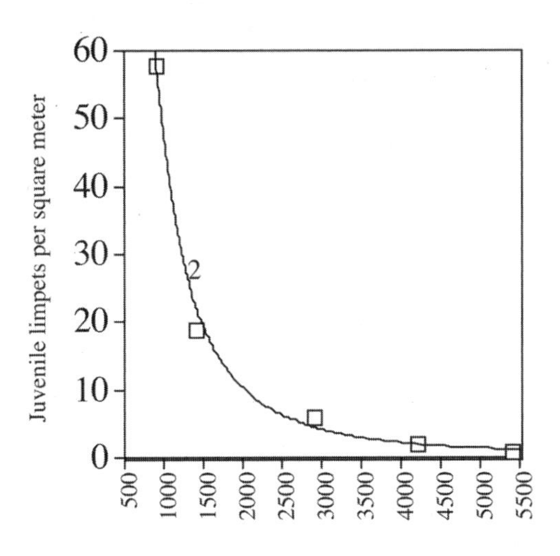

Figure 6.5 Settlement of limpets declines with increasing distance from the borders of a marine reserve in Tenerife, Canary Islands. Redrawn from Hockey and Branch (1994).

the ecosystem impact of reserves may help resolve these issues. In particular, if evidence from data sources as distinct as genetic studies and the ecology of reserves provides similar answers about neighborhood size, then confidence in the implications might increase.

The Implications of Neighborhood Size

If marine neighborhoods are large, and populations are generally open over large spatial scales, then offspring are unlikely to settle in the same place as their parents, generating two important implications for marine conservation biology. The first implication is that larvae produced in one small area are expected to disperse into other areas. If adults move after spawning, this geographic mixing from one generation to the next may be even more pronounced. In cases in which this is true, protection of adults over small areas does not necessarily lead to a stable self-seeding population, as is usually assumed in terrestrial parks, because offspring exit the park.

The second implication of large neighborhoods and small reserves is that the larvae that settle in a local area probably came from somewhere else. This requires that reproductive populations be maintained outside the boundaries of small protected areas; otherwise there will be no larval settlement in the protected area, and the ecosystem will degrade. Either additional protected areas must be maintained within dispersal distance of one another, or exploitation outside protected areas must be low enough to allow reproduction to occur.

The importance of both these implications hinges on the relative size of population neighborhoods and the scale of conservation efforts. When the neighborhood is large compared with protected areas, then these implications strongly apply (Botsford et al., 2003). If neighborhoods are too small, then there will be little spillover to the outside, and populations wiped out by some catastrophe or by overexploitation will be replenished only slowly from elsewhere.

If neighborhoods and protected area sizes are about the same size, then local adults can produce local larvae, and these protected populations can be self-seeding in the future (Botsford et al., 2001). In addition, because neighborhood edges are not impermeable barriers, there will be enough spillover across the edges of a protected area to enhance the populations outside. In these cases, self-seeding populations can have a positive impact on surrounding ecosystems and potentially enhance overall marine ecosystem stability (Botsford et al., 2003).

If marine neighborhoods are generally very large, then the best conservation efforts will require very large protected areas. These areas, perhaps the size of U.S. national parks, will be difficult to designate and will need to be few in number. By contrast, if marine neighborhoods can be small, then even small, local conservation efforts could have a local benefit. A dense network of small protected areas might be an effective management tool and might resemble a system of county parks instead of a system of large national parks. The differences between these two visions of marine protection are dramatic, but the choice between them hinges on a better picture of neighborhood size.

Diversity as the Last Word

Marine ecosystems will not have a single neighborhood size, because they are made up of species with very different dispersal traits. Species with different tendencies to move probably use the same coastline in different ways. In the face of such diversity, the challenge is to define the range in neighborhood sizes used by the wide variety of species that inhabit typical marine ecosystems. This range is likely to be large for every marine community because they all consist of species with very different generation times, dispersal mechanisms, swimming abilities, longevities, and modes of parental care (Palumbi and Hedgecock, 2003). As an example, for the Pacific shorelines of North America, some colonial invertebrates are annuals and disperse for fifteen minutes; kelp disperse as spores for a day but might raft in storms for miles; some rockfish live for a century after dispersing as larvae for months; otters give birth to young that often remain close by for their whole lives; and squid return to coastlines to spawn each year from migrations that take them into unknown parts of the open ocean. Diversity in life strategy translates into diversity in neighborhood size, and this determines management strategies.

This diversity in life strategies does not mean that we must define the neighborhoods for every species. Instead we could begin with a practical list of the range of neighborhoods needed by the species in a marine community. We also need a quantitative framework for understanding how different ocean management strategies will benefit or endanger species with those different neighborhood needs. Although this seems like a daunting task, we currently possess a huge amount of basic life history information about thousands of species (see Shanks et al., 2003, for an example). In addition, we have a suite of new technology tools—satellite tracking (Block et al., 2001), microchemistry (Swearer et al., 1999), high-throughput genetics (Palumbi, 2003), and detailed monitoring of marine reserves (Roberts et al., 2001a)—that are poised to help put this life history information into perspective. Using this information to help chart neighborhood sizes—through continued mapping of dispersal, migration, and habitat use strategies—is a key goal for the future.

References

Agardy, M. T. 1994. Advances in marine conservation: The role of marine protected areas. *Trends in Ecology and Evolution* 9:267–270.

Babcock, R. C., S. Kelly, N. T. Shears, J. W. Walker, and T. Willis. 1999. Changes in community structure in temperate marine reserves. *Marine Ecology Progress Series* 189:125–134.

Barber, P. H., S. R. Palumbi, M. V. Erdmann, and M. K. Moosa. 2000. Biogeography: A marine Wallace's line? *Nature* 406:692–693.

Block, B. A., H. Dewar, S. B. Blackwell, and E. D. Williams. 2001. Migratory movements, depth preferences, and thermal biology of Atlantic bluefin tuna. *Science* 293:1310–1314.

Block, B. A., H. Dewar, C. Farwell, and E. D. Prince. 1998. A new satellite technology for tracking the movements of Atlantic bluefin tuna. *Proceedings of the National Academy of Sciences USA* 95:9384–9389.

Bohonak, A. J. 1999. Dispersal, gene flow, and population structure. *Quarterly Review of Biology* 74:21–45.

Botsford, L. W., F. Micheli, and A. Hastings. 2001. Dependence of sustainability on configuration of marine reserves and larval dispersal distances. *Ecology Letters* 4:144–150.

Botsford, L. W., F. Micheli, and A. Hastings. 2003. Principles for the design of marine reserves. *Ecological Applications* 3:S25–31.

Castilla, J. C. 1996. The future Chilean marine park and preserves network and the concepts of conservation, preservation and management according to the national legislation. *Revista Chilena de Historia Natural* 69:253–270.

Cowen, R. K., K. M. M. Lwiza, S. Sponaugle, C. B. Paris, and D. B. Olson. 2000. Connectivity of marine populations: Open or closed? *Science* 287:857–859.

Creese, R. G. and R. G. Cole. 1995. Marine conservation in New Zealand. *Pacific Conservation Biology* 2:55–63.

Edmands, S., P. E. Moberg, and R. S. Burton. 1996. Allozyme and mitochondrial DNA evidence of population subdivision in the purple sea urchin *Strongylocentrotus purpuratus*. *Marine Biology* 126:443–450.

Flowers, J. M., S. C. Schroeter, and R. S. Burton. 2002. The recruitment sweepstakes has many winners: Genetic evidence from the sea urchin *Strongylocentrotus purpuratus*. *Evolution* 56:1445–1453.

Gaines, S. D. and M. D. Bertness. 1992. Dispersal of juveniles and variable recruitment in sessile marine species. *Nature* 360:579–580.

Halpern, B. 2003. The impact of marine reserves: Do reserves work and does reserve size matter? *Ecological Applications* 13:S117–137.

Harms, S. and C. D. Winant. 1998. Characteristic patterns of the circulation in the Santa Barbara Channel. *Journal of Geophysical Research* 103:3041–3065.

Hockey, P. A. R. and G. M. Branch. 1994. Conserving marine biodiversity on the African coast: Implications of a terrestrial perspective. *Aquatic Conservation: Marine and Freshwater Ecosystems* 4:345–362.

Johnson, M. S., K. Holborn, and R. Black. 1993. Fine scale patchiness and genetic heterogeneity of recruits of the corallivorous gastropod *Drupella cornus*. *Marine Biology* 117:91–96.

Largier, J. 2003. Considerations in estimating larval dispersal distances from oceanographic data. *Ecological Applications* 13:S71–S89.

Lubchenco, J., S. R. Palumbi, S. D. Gaines, and S. Andelman. 2003. Plugging a hole in the ocean: The emerging science of marine reserves. *Ecological Applications* 13:S3–S7.

McClanahan, T. R. and B. Kaunda-Arara. 1996. Fishery recovery in a coral-reef marine park and its effect on the adjacent fishery. *Conservation Biology* 10:1187–1199.

Menge, B. A., B. A. Daley, P. A. Wheeler, E. Dahlhoff, E. Sanford, and P. T. Strub. 1997. Benthic–pelagic links and rocky intertidal communities: Bottom-up effects on top-down control? *Proceedings of the National Academy of Sciences USA* 94:14530–14535.

Murawski, S. A. R. Brown, H. L. Lai, P. J. Rago, and L. Hendrickson. 2000. Large scale closed areas as a fishery management tool in temperate marine systems: The Geirges Bank experience. *Bulletin of Marine Science* 66:775–798.

Palumbi, S. R. 1992. Marine speciation on a small planet. *Trends in Ecology and Evolution* 7:114–118.

——. 2003. Population genetics, demographic connectivity and the design of marine reserves. *Ecological Applications* 13:S146–158.

Palumbi, S. R., S. D. Gaines, H. Leslie, and R. R. Warner. 2003. A new wave: High tech tools held design effective marine research. *Frontiers in Ecology and the Environment* 1:1–12.

Palumbi, S. R. and D. Hedgecock. 2003. The life of the sea. In *Marine conservation biology*, ed. E. Norse and L. Crowder. Washington, D.C.: Island Press.

Richmond, R. H. 1990. The effects of the El Niño/Southern Oscillation on the dispersal of corals and other marine organisms. In *Global ecological consequences of the 1982–83 El Niño–Southern Oscillation,* ed. P. W. Glynn, 127–140. Amsterdam: Elsevier.

Roberts, C., W. J. Ballentine, C. D. Buxton, P. Dayton, L. B. Crowder, W. Milon, et al. 1995. *Review of the use of marine fishery reserves in the US southeastern Atlantic.* NOAA technical memorandum NMFS-SEFSC-376.

Roberts, C., B. Halpern, S. R. Palumbi, and R. R. Warner. 2001a. Designing marine reserve networks: Why small isolated protected areas are not enough. *Conservation Biology in Practice* 2(3):10–17.

Roberts, C. M., J. A. Bohnsack, F. Gell, J. P. Hawkins, and R. Goodridge. 2001b. Effects of marine reserves on adjacent fisheries. *Science* 294:1920–1923.

Roberts, C. M. and N. V. C. Polunin. 1991. Are marine reserves effective in management of reef fisheries? *Reviews in Fish Biology and Fisheries* 1:65–91.

Ruckelshaus, M. H. and C. G. Hayes. 1998. Conservation and management of species in the sea. In *Conservation biology for the coming decade,* ed. P. Fiedler and P. Kareiva, 28–37. London: Chapman and Hall.

Russ, G. R. and A. C. Alcala. 1996. Marine reserves: Rates and patterns of recovery and decline of large predatory fish. *Ecological Applications* 6:947–961.

Sammarco, P. W. and J. C. Andrews. 1988. Localized dispersal and recruitment in Great Barrier Reef corals: The Helix experiment. *Science* 239:1422–1424.

Scheltema, R. S. 1986. Long distance dispersal by planktonic larvae of shoal-water benthic invertebrates among central Pacific islands. *Bulletin of Marine Science* 39:241–256.

——. 1988. Initial evidence for the transport of teleplanic larvae of benthic invertebrates across the east Pacific barrier. *Biological Bulletin* 174:145–152.

Shanks, A. L., B. A. Grantham, and M. A. Carr. 2003. Propagule dispersal distance and the size and spacing of marine reserves. *Ecological Applications* 13:S159–169.

Slatkin, M. 1993. Isolation by distance in equilibrium and non-equilibrium populations. *Evolution* 47:264–279.

Slatkin, M. and N. Barton. 1989. A comparison of three indirect methods for estimating average levels of gene flow. *Evolution* 4:1349–1368.

Sotka, E. E. and S. R. Palumbi. 2006. The use of genetic clines to estimate dispersal distances of marine larvae. *Ecology* 87:1094–1103.

Sotka, E. E., J. P. Wares, R. K. Grosberg, and S. R. Palumbi. 2004. Strong genetic clines and geographic variation in gene flow in the rocky intertidal barnacle *Balanus glandula*. *Molecular Ecology* 13:2143–2156.

Strathmann, R. R. 1990. Why life histories evolve differently in the sea. *American Naturalist* 30:197–207.

Swearer, S. E., J. E. Caselle, D. W. Lea, and R. R. Warner. 1999. Larval retention and recruitment in an island population of a coral reef fish. *Nature* 402:799–802.

Paul Z. Goldstein

7 Genetic Data and the Interpretation of Restoration Priorities of the *Cicindela dorsalis* Say Complex (Coleoptera: Carabidae)

The Components of Conservation Genetics Revisited

Known historically to have occurred "in great swarms" on coastal beaches from Massachusetts to Chesapeake Bay (Leng, 1902:161), the northeastern beach tiger beetle *Cicindela d. dorsalis*, one of four named subspecies, all but disappeared from its recorded range in the latter half of the twentieth century. Since the recognition of its decline in the early 1970s (Stamatov, 1972), the animal has received attention from biologists and conservationists as an important flagship species for issues in invertebrate conservation, beach stewardship, and, most recently, conservation genetics. Knisley et al. (1987) illustrated the precipitous decline of collection records of this insect, which appears to have accelerated in the 1930s and 1940s, and considered it extirpated from New England. In 1989, Simmons rediscovered a relict population on Martha's Vineyard Island, Massachusetts (Nothnagle and Simmons, 1990), and in 1991 *C. d. dorsalis* was listed as threatened under the federal Endangered Species Act. This prompted the recovery planning process, including the drafting of reintroduction plans rangewide, as federally mandated by the Endangered Species Act (U.S. Fish & Wildlife Service, 1994).

C. dorsalis presents an important range of dilemmas for conservation efforts. These may be grouped into issues stemming directly from wildlife conservation concerns and those derived from the theoretical rationale for making conservation decisions and distinguishing management units based on genetic data. Numerous opportunities for reintroduction programs in the southern part of the beetle's range arose from high population numbers (i.e., a healthy stock population) and a significant number of potential reintroduction sites. The New England populations, suspected of behavioral and genetic differentiation, have proven more problematic. The 1994 rediscovery of a small mainland Massachusetts population and its continued decline since have illustrated the fragility of small *C. dorsalis* populations and called into question the feasibility of using New England stock for reintroduction purposes. Meanwhile, several intense storms between 1991 and 1996 caused dramatic fluctuations in the core New England population on Martha's Vineyard. It is against this backdrop that that I review the qualitative relevance of available genetic information to the restoration of northeastern beach tiger beetles.

The Natural History, Decline, and Possible Recovery of *Cicindela d. dorsalis*

Declines of *C. d. dorsalis* have been variously attributed to intensive human use and development of shoreline habitats, primarily barrier beaches. Although intensive development along the eastern seaboard and the construction of jetties, revetments, and other shoreline stabilization structures have played a tremendous role in fragmenting and obliterating populations of shoreline plants and animals, human recreational activities, particularly off-road vehicle (ORV) use on otherwise pristine beaches, appear to have prevented reestablishment of *C. d. dorsalis* colonies and may have played a role in their local extirpation (Stamatov, 1972). Tiger beetles as a group are especially sensitive to vehicle traffic (Nagano, 1980) because their larval burrows cannot withstand perpetual substrate disturbance, and northeastern beach tiger beetles appear to be especially vulnerable because they belong to a group of *Cicindela* species with larval development times extending to nearly two years. It has also been observed that although they are wary enough to avoid pedestrian traffic, adult northeastern beach tiger beetles are vulnerable to ORVs.

Unfortunately, most regulation of ORV activity has centered on the management and nesting needs of migratory shorebirds, most notably the federally threatened piping plover (*Charadrius melodus*). Because these animals are nonpermanent residents, restrictions on ORVs necessitated by wildlife protection laws where plover nests occur are generally relaxed once chicks are fledged, leaving the remaining resident beach biota less protected from ORV traffic for most of the year. The proponents of ORV recreation are particularly vociferous in southern New England and have at times successfully lobbied to relax enforcement of the endangered species legislation by pitting the interests of recreation against those of endangered species. Some arguments, incongruously, have centered on incorrect notions of ecosystem management that downplay species concerns.

From the standpoint of coastal wildlife protection, the northeastern beach tiger beetle illustrates a number of life history characteristics of beach-dwelling organisms. Barrier beaches are physically harsh and deceptively complex natural areas visited by a variety of meteorological disturbances and geophysical forces. Intense exposure to sun and seawater makes for a forbidding environment. Larval tiger beetles are ambush predators that survive these conditions, including inundation by high ocean tides, by subsisting on a diet of amphipods and other invertebrates associated with beach wrack. Adult tiger beetles, which are cursorial predators, also appear resilient to high salinity and often concentrate near the incoming tide, where they may be seen getting rolled by the edges of incoming waves.

C. dorsalis also appears to be well adapted for the kind of natural disturbances typically visited on barrier beaches in the form of coastal storms and hurricanes. The two-year life cycle is such that *C. dorsalis* is among numerous summer-active species whose flight period is restricted to mid-summer and whose larval development takes nearly two years, especially near the northern limit of the species' range. Although a certain amount of variability with respect to life history timing is inevitable, the implication for individual beetle populations is that there are two primary cohorts of adult beetles. Nothnagle et al. (1993) observed differential mortality of larvae such that the landfall of Hurricane Bob on Martha's Vineyard in August 1991 was estimated to have taken more than 90 percent of first instar larvae (the 1993 adult cohort, offspring of the 1991 cohort) and roughly 50 percent of the yearling 1992 cohort's larvae. With its two-year life cycle and differential survivability among larval instars, populations of *C. dorsalis* appear able to withstand singleton storm events, at least when the larval reserve is able to avoid storms in consecutive years. Clearly this (and any) strategy is less viable as population fragmentation is coupled with increased hurricane frequencies.

The cycles and tremendous flux of erosional and accretionary sands in such areas are easily interrupted by human stabilization measures, whether initiated to protect real estate or to create recreational opportunities via "permanent" ORV tracks. *C. dorsalis* larvae appear to use dune blowouts preferentially (P. Nothnagle, personal communication, 2001), suggesting that the life history of this animal not only is adapted to buffering against storm activity but may actually depend on it. Although it may seem counterintuitive that a species would become adapted to an unpredictable disturbance regime, disturbance-dependent species exist throughout nature. Whether they are dependent on fire-prone pine barrens, heavily grazed savannas, or light gaps in tropical forests, as long as the expanse and heterogeneity of the habitat are such that recently disturbed areas are in close enough proximity to sources for recolonization, populations of these species persist.

In contrast to this kind of stochastic disturbance, the kind of chronic disturbance associated with the interruption of normal cycles of erosion and accretion by shoreline stabilization measures or, more obviously, by persistent ORV traffic appears to be something to which tiger beetles cannot adapt easily. Ironically, early studies in New England, some of which were commissioned explicitly to evaluate the effects of ORV activity on beach wildlife, did not take into account permanent resident species associated with the forebeach and therefore concluded that such areas were compatible with ORV use (Godfrey, 1978; Leatherman and Godfrey, 1979).

Logistical, Ecological, and Genetic Considerations for Restoration

After the 1989 rediscovery and subsequent listing of *C. d. dorsalis*, efforts to restore it to as much of its former range as possible have relied on life history, ecological, and, to some extent, genetic data. Thanks to early reintroduction efforts led by Knisley in Chesapeake Bay and by Nothnagle in New England, a variety of logistical and biological obstacles to the beetle's recovery were overcome, in particular with respect to the actual transplanting of the animals. One fact that emerged from the early reintroduction efforts was that although reintroductions of adult beetles failed as a result of their tendency to overdisperse, adults that emerged in situ from transplanted larvae were able to find each other and reproduce despite what one would have assumed to be higher pre-reproductive mortality among transplanted larvae than among adults.

Although logistics and feasibility are always central to the conservation decision-making process, a more important question arose early on as to whether to use stock from the remnant New England populations or the larger Chesapeake Bay populations for the purposes of reintroducing *C. dorsalis* to New England proper. The supposition that the northern remnant ocean-fronted populations and the more southern bay-fronted populations might be adapted to different intensities and cycles of winter wave action, coupled with observations that larvae in these two areas appear to be behaviorally distinct (northern larvae appear to retreat further into the dunes during periods of high-energy winter waves [Nothnagle and Simmons, 1990]), led to the notion that these populations might best be managed independently.

Meanwhile, Vogler and coworkers brought much data to bear on the questions of population structure, phylogeography, and the delineation of conservation units in the *C. dorsalis* complex (Vogler and DeSalle, 1993a, 1993b; Vogler et al., 1993). Vogler's primary data set comprises samples of 420 individuals of all four named subspecies of *C. dorsalis* from thirty-two localities from Martha's Vineyard, Massachusetts, south to Florida and west to Texas. In all, seventeen mitochondrial haplotypes were evinced. Among the primary conclusions of Vogler's work was that the only diagnosably distinct (extant) population was the island population on Martha's Vineyard, Massachusetts, which unsurprisingly is the most isolated. Based on Vogler's sampling, that population is diagnosable by virtue of a single base pair at position 5016 on the cytochrome oxidase III gene; this population exhibits unique haplotypes at two other loci as well. Although two other populations exhibit unique haplotypes (the Fire Island, New York population at one locus and the Coney Island population at two loci), these populations are polymorphic and hence not diagnosably distinct. The question of retaining putative genetic integrity of the broader New England stock population, including the introduced or "experimental" populations, as distinct from their southern counterparts thus called for a revisiting of Vogler's data.

Genetic Considerations Revisited

The northeastern beach tiger beetle illustrates a number of issues surrounding the use and relevance of genetic data to conservation decision making (Goldstein and DeSalle, 2003; Vogler, 1994; Vogler and DeSalle, 1993a, 1993b, 1994; Vogler et al., 1993). Two sets of biological observations make the *C. dorsalis* complex a useful set of flagship species for exploring these issues. First, the species is distributed clinally along the Atlantic seaboard, west to Texas, with the four described subspecific entities—*dorsalis, media, saulcyi,* and *venusta*—grading into one another as adjacent segments in an essentially linear habitat.

This complex illustrates a common pitfall in the use of subspecies in systematic biological conservation endeavors and possibly in the application of phylogenetic or cladistic methods at infrasubspecific levels. Second, based on Vogler's intensive range-wide sampling, it is noteworthy that none of these subspecies appears to be monophyletic, reciprocally or otherwise, with respect to any single haplotype (figure 2 in Vogler, 1994).

Of particular interest is the contrast between species concepts ("conservation units" or "management units") as delineated by topological criteria (tree-based methods, most commonly the criterion of reciprocal monophyly interpreted from within the framework of the biological species concept [BSC]) and by patterns of character fixation (diagnosability as it is understood from the original formulation of the phylogenetic species concept [PSC]) (Cracraft, 1983, 1989; Nixon and Wheeler, 1991). Authors have articulated variants of each of these broad categories, discussing questions surrounding the separate or simultaneous analysis of molecular data sets, methods for tree construction, and the preeminence of preserving microevolutionary processes or simply patterns of character state distributions. The putative adaptive distinction was consistent with Vogler and DeSalle's (1993a) finding that the northern population was genetically distinct, albeit by only a single base pair within the available sequence data. On the other hand, Vogler's same data indicate a southern origin, and the question quickly became whether to interpret the genetic data in light of protecting the current pattern or to recreate historical processes or history (i.e., colonization from the south).

Vogler's data are of interest to the species debate because they present a reductio ad absurdum: an entity technically legitimate as a phylogenetic species but only on the basis of a single nucleotide polymorphism. As Vogler (1994) was quick to describe, such an example highlights the obvious role of fixation (to be equated with the extinction of polymorphism in this context) in achieving diagnosability, the essence of the phylogenetic species criterion (Davis and Nixon, 1992; Nixon and Wheeler, 1991). Vogler's (1994, figure 7) thoughtful example is useful because it illustrates a case in which two populations become diagnosably distinct and reciprocally monophyletic at the instant of extinction of an intermediate, polymorphic population. However, mutually exclusive character fixation is not necessarily coincident with reciprocal monophyly (Goldstein and DeSalle, 2000).

Under the PSC, whether or not some individuals within a particular phylogenetic species appear more or less closely related to each other than to nonmembers of their own is irrelevant to the reconstruction of relationships between species. Even after character fixation has occurred, not all diagnosable groups necessarily appear as monophyletic, as the *C. dorsalis* case illustrates, and the topology of sampled individuals will certainly be polyphyletic or paraphyletic with respect to other unsampled individuals in their respective population. Rather than rely on tree-based species definitions ab initio, one should conduct pooled population aggregation analyses of all available nuclear data for patterns of fixed character states before undertaking any tree-based diagnoses. It may be argued that to do otherwise is to put the cart before the horse.

The putative diagnosis of any population by evidence as meager as a single nucleotide polymorphism prompts a discussion of whether this attribute is an artifact, particularly given the essentially linear distribution of the species in question. It also invites an empirical assessment of Vogler's (1994) prescient supposition that diagnosability in this case has been achieved through the (largely anthropogenic) extirpation of geographically intermediate polymorphic populations and his parallel illustration of the relevance of such character extinction to the formation of identifiable phylogenetic lineages. By sequencing a short fragment flanking the diagnostic site in question in extractions from museum specimens, Goldstein and DeSalle (2003) demonstrated that the character fixation event did indeed occur in the twentieth century: Populations throughout the historical range of *C. d. dorsalis,* including Martha's Vineyard in the early twentieth century, were polymorphic at this site, and fixation at either end of the geographic cline appears to have accompanied the extirpation of more centrally located polymorphic populations. This conclusion emphasizes the importance of caution against overinterpreting limited data from extant populations, although it illustrates aptly the relationship between character fixation and the recognition of phylogenetic species, as first articulated by Nixon and Wheeler (1991). What remains is to synthesize all the available biological data with an eye toward restoration of *C. d. dorsalis* and to articulate a framework for understanding the powers and limitations of phylogenetic species theory in conservation practice.

The Components of Conservation Genetics Revisited

Goldstein et al. (2000) observed three common areas of confusion surrounding the interpretation of genetic data in conservation contexts: the interpretation of frequency data or distance data as diagnostic (a term used here in reference to individuals as opposed to samples), the interpretation of raw character data as necessarily adaptive or reflective of evolutionary potential, and the application of hierarchy-based concepts, terms, and analyses to nonhierarchically related groups of organisms from which inference of speciation patterns or biogeographic histories may be spurious. Although elements of Goldstein et al. (2000) have been characterized with various degrees of accuracy, the salient points of their argument have yet to be countered.

It is possible to identify several broader categories of issues surrounding the integration of phylogenetic theory and molecular evolution in conservation biology. First, much debate and confusion surround the differences between various versions of the PSC. These differences, real, imagined, or subtle, have been reviewed elsewhere (de Queiroz, 1998; Goldstein and DeSalle, 2000; Harrison, 1998), and we will not resurrect those treatments here except insofar as the distinction between tree-based and character-based species diagnoses is pertinent. Although the cladograms in Vogler's initial studies are valuable in corroborating a southern origin of the species, they may be seen as largely irrelevant to the distinction of biological management units. More relevant to that endeavor are the character state distributions themselves.

Second, as articulated in earlier articles (e.g., Goldstein and Brower, 2002; Goldstein and DeSalle, 2000; Goldstein et al., 2000), systematics and population genetics complement one another best when their respective strengths and limitations are acknowledged. The "intraspecific phylogeography" neologism of Avise (2000) has represented for many a bridge between disciplines, and although the notion of reciprocal monophyly may indeed be a valuable contribution to our understanding of biological divergence patterns at or near the species boundary, particularly from the perspective of coalescent theory, it has been my position that the paradigm of "intraspecific phylogeography" taken alone may be inappropriate for introducing phylogenetic histories to conservation endeavors, particularly when the generation of phylogeographic hypotheses is not undertaken critically.

The intended synthesis of systematics and population genetics, based on both the tone and the content of the writings of preeminent phylogeographers, is perhaps better described as an integration of systematics into population genetics, a pattern that seems to recapitulate the new synthesis by leaving the field of systematics essentially orphaned, as it was for several decades in the twentieth century. In my estimation, this intended synthesis has not been well served by methodologically uncritical phylogeographic studies. In particular, the methods by which topologies are generated are important, as is the correspondence between entities that appear to be reciprocally monophyletic and entities that can be diagnosed by suites of fixed character states. In systematic circles, debates over phylogenetic inference methods persist most particularly between proponents of maximum likelihood and those of cladistic parsimony, but phenetic methods such as neighbor joining are still commonly used in conservation studies. This appears to be because neighbor joining is a quick way to obtain single, fully resolved topologies without cumbersome, computer-intensive searching. Unfortunately, neighbor joining is highly suspect as a scientific or conservation tool because in all cases it obtains a fully resolved set of relationships, even when all the terminals are identical; and, relatedly, it is sensitive to input order shuffling, that is, the order in which the terminals appear in the matrix (Farris et al., 1996).

Third, it is most critical to distinguish between the operations of recognizing groups of individuals as species and exploring the evolution of reproductive isolation or cohesion. Proponents of the BSC generally fail to see that these two operations are divorced for the purpose of assigning names to species, one of the more basic practices in taxonomy and systematics. It must be reiterated that although adopting the PSC in no way hampers the study of reproductive isolation, evolutionary cohesion, or their origins, an insistence on a role for the BSC in taxonomy hampers the systematization of nature, at a time when the world's biota is disappearing faster than it can be described. Although phylogeography's stated hope in generating hierarchical cladograms is to uncover phylogenetic history at the finest levels, that goal is potentially compromised by interpretation of the hierarchically depicted relationships between mtDNA haplotypes as relationships between organisms that are not hierarchically related. Regardless of whether uniparentally inherited organellar genomes are de facto related hierarchically, their inferred relationships need not correspond to relationships between the individuals that bear them.

An apparent (and possibly unnecessary) source of disagreement derives from the obvious point that any study, phylogeographic or otherwise, purporting to be "intraspecific" requires that a species boundary first be delimited. A result or conclusion that is intraspecific within the paradigm of the BSC may not be intraspecific at all with respect to the PSC. Avise (2000), the founder and most active proponent of phylogeography, has explicitly tied the paradigm to the Mayrian BSC and argued that reproductive cohesion is essential to the study of both microevolution and macroevolution, that subspecies should retain an important if not preeminent role in systematic studies and correspond to evolutionary significant units in conservation programs, and that to think otherwise represents ideological entrenchment and sophistry. Proponents of the PSC (sensu Wheeler and Nixon, 1991) do not dispute that nested sets of hierarchically related organisms may exist within what have been deemed reproductively isolated biological species; indeed, they assert that reproductive isolation is a poor arbiter of relationship and should be formally divorced from the business of systematizing nature. They argue that there exist minimal units for phylogenetic analysis and that these units correspond to things we called species regardless of whether they are reproductively isolated or microevolutionarily equivalent. Under a character-based approach (as opposed to a tree-based approach) all individuals can be assigned to species or "units" without it being necessary to determine their relationships to one another. Arguments over "the" proper approach to conservation genetic data are necessarily at cross purposes, to the extent that agreement on the fundamental notion of species concept is not first achieved. Without such agreement, results of intraspecific phylogeographies may be seen as artifacts of adherence to the BSC.

At times, debates over species concepts have been misdirected or characterized by nonscientific arguments, often in the form of statements to the effect that to focus on patterns of character state distributions is at best ignorant and at worst antievolutionary. In this discussion, I have focused on what I reluctantly call a pattern-oriented approach, not as an end in itself but as a means of understanding (and ultimately preserving) the variation that is the fodder for microevolution.

If successful, efforts to restore the northeastern beach tiger beetle to more of its historical range will have relied most fundamentally on overcoming ecological and demographic obstacles. As a resident species that is particularly sensitive to environmental alteration and that exhibits an extended life cycle, *C. dorsalis* also highlights the importance of managing for beaches as dynamic ecosystems rather than piecemeal in terms of individual components of the fauna, particularly when they are migratory species or nonpermanent residents. The case of *C. dorsalis* illustrates the importance of rooting our broad-scale management practices in appropriate species-level information. The divorce of this beetle's life history requirements from the stewardship of beaches where it once occurred has presented significant obstacles to its restoration.

Acknowledgments

This chapter benefits from years of discussion with and field work in the company of Alfried Vogler, Rob DeSalle, and George Amato and eleven years of field work in the company of Tim Simmons, Susi Von Oettingen, and Phil Nothnagle. Of course, any errors are my sole responsibility. I wish to acknowledge the Field Museum's Pritzker Laboratory for Molecular Systematics and Evolution and the Lewis B. and Dorothy Cullman Program in Molecular Systematics; numerous landowners for their kind permission for continuing access to critical habitat for *C. dorsalis;* the Massachusetts Division of Fish & Wildlife and the Massachusetts Natural Heritage and Endangered Species Program for ongoing funding; the Trustees of Reservations for logistic support and coordination; and all the conservationists and friends who assisted in the field since Simmons's discovery in 1989. This chapter is dedicated to the memory of Phil Nothnagle, who passed away on June 1, 2002.

References

Avise, J. C. 2000. *Phylogeography. The history and formation of species*. Cambridge, Mass.: Harvard University Press.

Cracraft, J. 1983. Species concepts and speciation analysis. *Current Ornithology* 1:159–187.

——. 1989. Speciation and its ontology: The empirical consequences of alternative species concepts for understanding patterns and processes of differentiation. In *Speciation and its consequences,* ed. D. Otte and J. A. Endle, 28–57. Sunderland, Mass.: Sinauer.

Davis, J. I. and K. C. Nixon. 1992. Populations, genetic variation, and the delimitation of phylogenetic species. *Systematic Biology* 41:421–435.

de Queiroz, K. 1998. The general lineage concept of species, species criteria, and the process of speciation: A conceptual unification and terminological recommendations. In *Endless forms: Species and speciation,* ed. D. J. Howard and S. H. Berlocher, 57–78. New York: Oxford University Press.

Farris, J. S., V. A. Albert, M. Kallersjo, and D. Lipscomb. 1996. Parsimony jackknifing outperforms neighbor-joining. *Cladistics* 12(2):99–124.

Godfrey, P. J. 1978. Impact of off-road vehicles on coastal ecosystems. In *Proceedings of the symposium on technical, environmental, socioeconomic and regulatory aspects of coastal zone planning and management*. March 14–16, 1978, San Francisco.

Goldstein, P. Z. and A. V. Z. Brower. 2002. Molecular systematics and the origin of species: New syntheses or methodological introgressions? In *Molecular systematics and evolution: Theory and practice,* ed. R. DeSalle, G. Giribet, and W. Wheeler, 147–161. Basel, Switzerland: Birkhauser/Verlag.

Goldstein, P. Z. and R. DeSalle. 2000. Phylogenetic species, nested hierarchies, and character fixation. *Cladistics* 16:364–384.

Goldstein, P. Z. and R. DeSalle. 2003. Calibrating phylogenetic species formation in a threatened insect using DNA from historical specimens. *Molecular Ecology* 12:1993–1998.

Goldstein, P. Z., R. DeSalle, G. Amato, and A. P. Vogler. 2000. Conservation genetics at the species boundary. *Conservation Biology* 14:120–131.

Harrison, R. G. 1998. Linking evolutionary pattern and process: The relevance of species concepts for the study of speciation. In *Endless forms, species and speciation,* ed. D. J. Howard and S. H. Berlocher, 19–31. New York: Oxford University Press.

Knisley, C. B., J. I. Luebke, and D. R. Beatty. 1987. Natural history and population decline of the coastal tiger beetle *Cicindela dorsalis dorsalis* Say (Coleoptera: Cicindelidae). *Virginia Journal of Science* 38(4):293–303.

Leatherman, S. P. and P. J. Godfrey. 1979. *The impact of off-road vehicles on coastal ecosystems on Cape Cod national seashore: An overview*. Report no. 34. Amherst: Park Service Cooperative Research Unit, University of Massachusetts.

Leng, C. W. 1902. Revision of the Cicindelidae of Borel America. *Transactions of the American Entomological Society* 28:93–186.

Nagano, C. D. 1980. Population status of the tiger beetles of the genus *Cicindela* (Coleoptera: Cicindelidae) inhabiting the marine shoreline of southern California. *Atala* 8(2):33–42.

Nixon, K. C. and Q. D. Wheeler. 1991. An amplification of the phylogenetic species concept. *Cladistics* 6:211–223.

Nothnagle, P. and T. Simmons. 1990. *Ecology of the northeastern beach tiger beetle* Cicindela d. dorsalis *in southeastern Massachusetts*. Unpublished report to Massachusetts Natural Heritage and Endangered Species Program.

Nothnagle, P., T. Simmons, and B. Knicely. 1993. Cicindela dorsalis: *Populations and habitat conditions on Martha's Vineyard 1990–1992.* Westborough: Massachusetts Natural Heritage and Endangered Species Program.

Stamatov, J. 1972. *Cicindela dorsalis* endangered on North Atlantic Coast. *Cicindela* 4:78.

U.S. Fish & Wildlife Service. 1994. *Northeastern beach tiger beetle* Cicindela d. dorsalis *(Say) recovery plan.* Hadley, Mass.: Author.

Vogler, A. P. 1994. Extinction and the formation of phylogenetic lineages: Diagnosing units of conservation management in the tiger beetle *Cicindela dorsalis*. In *Molecular ecology and evolution: Approaches and applications,* ed. B. Schierwater, B. Streit, G. P. Wagner, and R. DeSalle, 261–274. Basel, Switzerland: Birkhauser Verlag.

Vogler, A. P. and R. DeSalle. 1993a. Mitochondrial DNA evolution and the application of the phylogenetics species concept in the *Cicindela dorsalis* complex (Coleoptera: Cicindelidae). In *Carabid beetles: Ecology and evolution,* ed. K. Desender, 69–75. Dordrecht, the Netherlands: Kluwer.

Vogler, A. P. and R. DeSalle. 1993b. Phylogeographic patterns in coastal North American tiger beetles (*Cicindela dorsalis* Say) inferred from mitochondrial DNA sequences. *Evolution* 47(4):1192–1202.

Vogler, A. P. and R. DeSalle. 1994. Diagnosing units of conservation management. *Conservation Biology* 8:354–363.

Vogler, A. P., R. DeSalle, T. Assmann, C. B. Knisley, and T. D. Schultz. 1993. Molecular population genetics of the endangered tiger beetle *Cicindela dorsalis* (Coleoptera: Cicindelidae). *Annals of the Entomological Society of America* 86:142–152.

James P. Gibbs

8

Range Collapse, Population Loss, and the Erosion of Global Genetic Resources

> Many of the criticisms directed at the "doomsayers" look even sillier when one realizes that a major component of the decline of biodiversity is the loss of genetically distinct populations. Much of the public discussion of extinctions is concentrated on species for historical reasons, but extirpation of populations is the dominant element of the extinction crisis.
>
> P. R. Ehrlich (1991:175)

Much of the current scientific and public concern over the extinction crisis centers on loss of species, and rightly so: Extinction of each species corresponds to the loss of all genetic variation that uniquely characterized it. This "pruning of the tree of life" is occurring at an alarming rate, with global extinction rates estimated at 1,000 to 10,000 times higher than at any time in the last 65 million years (Lawton and May, 1995). Because so much of aggregate genetic diversity is lodged in the branches rather than the twigs of the tree of life, the tree of life may be able to survive vigorous pruning at the species level (Nee and May, 1997). Nevertheless, the tree's twigs and leaves (species and populations, respectively) are critical to delivering ecological services and also supplying the genetic diversity crucial to development of agricultural crops and pharmaceuticals (Ehrlich and Ehrlich, 1992).

The scale of population extinction is truly alarming. Consider that there are an estimated 220 populations per species, or some 1.1 to 6.6 billion genetically distinct populations globally (Hughes et al., 1997). Rates of tropical forest loss suggest rates of local population extinction on the rate of 1,800 populations per hour in such areas alone (Hughes et al., 1997). This rate is 432 times greater than that of species loss in the same regions (cf., 100 species per day, Lawton and May, 1995).

Substantial genetic differentiation between geographically separated populations characterizes most species (Ehrlich and Raven, 1969). This nonrandom distribution of variation reflects the underlying genetic architecture of a population, which arises over time through the interacting evolutionary forces of gene flow, drift, mutation, and selection (Avise, 1994). Adequately understanding patterns of intraspecific genetic diversity within and between populations is a fundamental problem in the fields of ecological genetics and evolutionary biology (Lewontin, 1974). It is also increasingly of interest to conservation biologists, restoration ecologists, and land managers who must identify, conserve, restore, and manage species of conservation concern (Falk and Holsinger, 1991; Ledig, 1988; Templeton, 1988). That locally differentiated populations and the genetic diversity they embody represent a legitimate component of biological diversity is recognized under several important conservation laws, including the U.S. Endangered Species Act, which accommodates distinct population segments.

Recent research has shed new light on patterns of population loss and species range collapse. Populations on the periphery of a species' geographic range tend to occupy less suitable habitats and exhibit lower and more variable densities; thus it is often assumed that they are more vulnerable to loss. In fact, the reverse appears to be more common (Channell and Lomolino, 2000; Lomolino and Channell, 1995). At least for vertebrates, most remnant populations tend to persist in the periphery of a species' historical geographic range, and thus species ranges tend to collapse outward rather than inward. This is unfortunate from a conservation genetics perspective because genetic drift tends to lower genetic variability in peripheral and isolated populations (Cassel and Tammaru, 2003; Edenhamn et al., 2000; Jones et al., 2001; Lammi et al., 1999) because the remaining populations may be those most lacking in genetic diversity in the first place.

Surprisingly scant attention has been paid to the potential significance that incremental loss of locally differentiated populations represents for global genetic resources. In this chapter I attempted to integrate data on the magnitude of genetic differentiation between species versus that between populations within species and data on the extent of population loss within species. The goal was to estimate the contribution that loss of local populations makes to aggregate erosion of global genetic resources in addition to that caused by outright loss of species. The rationale for the inquiry was to examine the merits of the current focus of the conservation community on rescuing species while issues of population loss go unheeded, thus potentially risking a quiet but serious hemorrhaging of genetic diversity at the population level.

A First Approximation

The genetic distances between species are about fifteen times as large as those between populations (Avise and Aquadro, 1982; Ward et al., 1992), which might at first suggest that range collapse and loss of genetic diversity are a nonissue in the context of global genetic resources. Yet the dramatically higher rates of extinction of local populations than of species—about 420 times as high (see Hughes et al., 1997; Lawton and May, 1995)—make the issue a compelling one. To shed some light on the matter, we need data on the relative magnitude of genetic differences between species and within populations in combination with estimates of the numbers of extinct species and of those still extant but having undergone severe range contractions. With these data, aggregate losses of genetic diversity caused by outright extinctions and those caused by population extinctions within extant species can be estimated.

There is no simple way to tackle the problem. Major challenges include the spatially complex population genetic structure that characterizes most species, coupled with the paucity of comprehensive surveys of genetic variants across almost any species. Furthermore, the bewildering variety of classes of genetic markers used to assess genetic structure, each with its own idiosyncrasies, complicates efforts to make straightforward comparisons between the studies using them. Nevertheless, an immense body of published work on allozyme variation in populations has been conveniently summarized in several key publications. These permit generalizations about the relative magnitude of genetic differences between populations within species (hereafter "within-species variation") and of those between species (hereafter "between-species variation"). Nei's (1975) genetic distance, *D*, which estimates the magnitude of nucleotide substitutions separating taxa, has frequently been summarized for both within- and between-species comparisons of allozyme variation and provides a useful metric for this analysis.

Two key publications have compiled many hundreds of allozyme studies to summarize the magnitude of genetic variation within and between species. Ward et al. (1992) summarized variation in forty-seven proteins in 648 vertebrates and 370 invertebrates to estimate the genetic distance and hence the amount of population subdivision between animal taxa (values for vertebrates are summarized in table 8.1). Avise and Aquadro (1982) summarized several thousand pairwise genetic distances between vertebrate species and found genetic distances one to three orders of magnitude higher between species than within species (see also table 8.1).

Because Ward et al. (1992) and Avise and Aquadro (1982) used the same currency, Nei's *D*, their data permit a comparison of the loss of genetic diversity associated with population and species losses. However, these measures are valid only if estimates of the number of species and populations lost during the recent extinction crisis are considered. Accordingly, two such data sets were used. For the United States, a database developed by The Nature Conservancy for its Natural Heritage Programs was used (Master et al., 2000) to determine the number of species that have become extinct (i.e., presumed extinct [GX], not located despite intensive searches; and possibly extinct [GH], or of historical occurrence with still some hope of rediscovery) or imperiled and reduced to so few populations that most genetic diversity that resided within them has since been lost (i.e., critically imperiled [G1], with typically five or fewer occurrences or 1,000 or fewer individuals). A second database on global species status, developed by the International Union for Conservation of Nature (e.g., Mace and Stuart, 1994), was used to determine the number of species that have become extinct (i.e., extinct [Ex], when there is no reasonable doubt that the last individual has died; or extinct in the wild [EW], when it is known only to survive in cultivation, in captivity, or as a naturalized population well outside the past range)

Table 8.1 Relative contribution to erosion of global genetic diversity of loss of populations within extant species versus outright species extinctions in birds, mammals, amphibians, reptiles, and fishes. Loss fraction represents the estimated amount of genetic diversity lost via extinction of local populations in extant species as a fraction of that lost to outright extinctions.

			The Nature Conservancy list, United States					International Union for the Conservation of Nature list, global				
			Number of species		Aggregate loss of genetic diversity			Number of species		Aggregate loss of genetic diversity		
Taxon	D_{within} [a]	$D_{between}$ [b]	Species extinct	Species critically imperiled	Species extinct	Species critically imperiled	Loss fraction	Species extinct	Species critically endangered	Species extinct	Species critically endangered	Loss fraction
Mammals	0.031	0.30	1	9	0.3	0.3	45%	87	180	28.8	5.6	16%
Birds	0.007	0.10	26	27	2.8	0.2	6	131	182	14.0	1.3	8
Reptiles	0.048	0.51	0	8	0.0	0.4	100	22	56	12.3	2.7	18
Amphibians	0.061	1.12	6	33	7.1	2.0	22	5	25	5.9	1.5	21
Fish	0.018	0.36	21	93	7.9	1.7	17	92	156	34.8	2.8	8

[a]Average minimum genetic distance between subpopulations (Ward et al., 1992).

[b]Average minimum genetic distance between species within genera, unweighted by number of pairwise species comparisons per genus (Avise and Aquadro, 1982).

and those reduced to so few populations that little within-species variation remained (i.e., critically endangered [CR], when it is facing an extremely high risk of extinction in the wild in the immediate future).

Databases on genetic distances between and within species were integrated with those on species extinction and imperilment to estimate the fraction of genetic diversity lost to date that can be attributed to loss of species outright and loss of populations within extant species (table 8.1). Calculations were made for five classes of vertebrates—mammals, birds, amphibians, reptiles, and fishes—because these were the only taxa in common in all databases. The cumulative genetic diversity lost that can be attributed to loss of species was estimated as sum of (1) the product of the number of extinct taxa and the average genetic distance between them and (2) the product of the number of extinct taxa and the average genetic diversity between populations within them. Thus, this estimate of the loss of genetic diversity incorporated the component unique to each species and that unique to the populations within them. The cumulative loss of genetic diversity that can be attributed strictly to loss of populations within extant species was estimated as the product of the number of imperiled species and the average genetic diversity between populations within them. Both metrics can be interpreted as the accumulated nucleotide substitutions lost to the two types of extinctions.

Together, these analyses (table 8.1) suggest that loss of populations accounts for a significant component of erosion of global vertebrate genetic diversity that has occurred to date as a result of the recent extinction crisis. At the global level, aggregate losses of genetic diversity in vertebrates due to population loss alone amount to 8 to 21 percent of that due to outright species extinctions. Weighted by species number, the global average across taxa attributable to population losses alone was about 12 percent. In the United States, taxon-specific values ranged from 6 to 100 percent, with the weighted average at 11 percent. Thus, genetic resources lost to population extinctions in extant species may represent about 10 percent of that lost to outright species extinctions. Another way to portray the results is that, on average, the aggregate effects of every set of ten imperiled species account for about as much of a threat to global genetic diversity as does every single species that goes extinct, at least for the vertebrates examined.

These estimates are subject to many uncertainties and assumptions. The first assumption is that the genetic divergence metrics represent unique and exclusive components of genetic diversity. Yet in reality, some amount of between-population differences may also be manifested in between-species differences. Moreover, we assumed that imperiled species have lost almost all their genetic diversity. In many cases this will be true, yet where remnant populations are composed of once widely dispersed individuals concentrated into a single or a few remaining populations, a large fraction of the original variation may still remain. The use only of imperiled species may actually be a conservative approach, insofar as species listed under categories of less imperilment (vulnerable, endangered, special concern) may also have suffered dramatic population losses. Finally, the estimates are certainly tenuous because the species status lists used are continually in flux, and they are experiencing a distinct downward trend of more species entering in the more vulnerable classes as the extinction crisis proceeds.

Conclusions

Despite the uncertainties of this analysis, if the current distribution of species in classes of endangerment were maintained, or even if many of the endangered species listed recovered, the amount of genetic diversity already lost from them through the disappearance of local, distinct populations is potentially substantial. Such losses have lasting consequences for both the evolutionary potential of those species and the economic welfare of humanity (Ehrlich and Ehrlich, 1992). Some refocusing of conservation attention on preventing population losses appears to be warranted to protect global genetic resources.

Acknowledgments

I am grateful to organizers of the original symposium for inviting me to speak on the topic, Eleanor Sterling for prodding me to produce a manuscript, Lisle Gibbs for orienting me to some important citations, and Mark Lomolino for enlightening me on issues of range collapse.

References

Avise, J. C. 1994. *Molecular markers, natural history and evolution.* New York: Chapman & Hall.

Avise, J. C. and C. F. Aquadro. 1982. A comparative summary of genetic distances in the vertebrates. *Evolutionary Biology* 15:151–185.

Cassel, A. and T. Tammaru. 2003. Allozyme variability in central, peripheral and isolated populations of the scarce heath (*Coenonympha hero:* Lepidoptera, Nymphalidae): Implications for conservation. *Conservation Genetics* 4:83–93.

Channell, R. and M. V. Lomolino. 2000. Dynamic biogeography and conservation of endangered species. *Nature* 403:84–86.

Edenhamn, P., M. Höggren, and A. Carlson. 2000. Genetic diversity and fitness in peripheral and central populations of the European tree frog *Hyla arborea. Hereditas* 133:115–122.

Ehrlich, P. R. 1991. Letter: Population diversity and the future of ecosystems. *Science* 254:175.

Ehrlich, P. R. and A. H. Ehrlich. 1992. The value of biodiversity. *Ambio* 21:219–226.

Ehrlich, P. R. and P. H. Raven. 1969. Differentiation of populations. *Science* 165:1228–1232.

Falk, D. A. and K. E. Holsinger, eds. 1991. *Genetics and conservation of rare plants*. New York: Oxford University Press.

Hughes, J. B., G. C. Daily, and P. R. Ehrlich. 1997. Population diversity: Its extent and extinction. *Science* 278:689–692.

Jones, B., C. Gliddon, and J. E. G. Good. 2001. The conservation of variation in geographically peripheral populations: *Lloydia serotina* (Liliaceae) in Britain. *Biological Conservation* 101:147–156.

Lammi, A., P. Siikamäki, and K. Mustajärv. 1999. Genetic diversity, population size, and fitness in central and peripheral populations of a rare plant *Lychnis viscaria*. *Conservation Biology* 5:1069–1078.

Lawton, J. H. and R. M. May, eds. 1995. *Extinction rates.* New York: Oxford University Press.

Ledig, F. T. 1988. The conservation of diversity in forest trees. *BioScience* 38:471–479.

Lewontin, R. C. 1974. *The genetic basis of evolutionary change*. New York: Columbia University Press.

Lomolino, M. V. and R. Channell. 1995. Splendid isolation: Patterns of range collapse in endangered mammals. *Journal of Mammalogy* 76:335–347.

Mace, G. M. and S. N. Stuart. 1994. Draft IUCN Red List categories, version 2.2. *Species* 21–22:13–24.

Master, L. L., B. A. Stein, L. S. Kutner, and G. A. Hammerson. 2000. Vanishing assets: Conservation status of U.S. species. In *Precious heritage: The status of biodiversity in the United States*, ed. B. A. Stein, L. S. Kutner, and J. S. Adams, 93–118. New York: Oxford University Press.

Nee, S. and R. M. May. 1997. Extinction and the loss of evolutionary history. *Science* 278: 692–694.

Nei, M. 1975. *Molecular population genetics and evolution*. New York: Elsevier.

Templeton, A. R. 1988. *The genetic consequences of habitat fragmentation*. Paper presented at the 35th Annual Systematics Symposium, October 8, Missouri Botanical Garden, St. Louis.

Ward, R. D., D. O. F. Skibinski, and M. Woodwark. 1992. Protein heterozygosity, protein structure, and taxonomic differentiation. In *Evolutionary biology*, ed. M. K. Hect, B. Wallace, and R. J. Macintyre, 73–159. New York: Plenum.

Part III

Saving Genetic Resources

How we conserve endangered species depends greatly on how we preserve their genetic resources. This section contains five chapters that examine the role of collections in conservation biology. The first chapter, by Robert Hanner, Angélique Corthals, and Rob DeSalle, describes the role of classic museum collections and not-so-classic frozen tissue collections in museums. By far, the world's museums and herbaria hold the greatest diversity of life of any kind of collection, and the potential for obtaining genetic resources from these collections is great. The authors suggest that the road to the future in museum and herbarium collections wanting to preserve genetic resources will involve more careful preservation of tissues in liquid nitrogen repositories. The second chapter, by Leona Chemnick, Marlys Houck, and Oliver Ryder, describes the Frozen Zoo collection at the San Diego Zoo. This collection houses frozen tissue cell lines from a broad range of vertebrates. The Frozen Zoo is a state-of-the-art collection of more than 6,000 specimens that can provide materials for studying genome organization, genome evolution, and the molecular basis for adaptation. The third chapter, by Vitaly Volobouev, describes the importance of cryopreservation in helping with the current biodiversity crisis. Volobouev clearly articulates the utility of cryopreservation in modern biodiversity studies as a source of RNA, DNA, and proteins that can in turn detail our genetic understanding of organismal diversity in a systematic context. The next chapter, by Deborah Rogers, Calvin Qualset, Patrick McGuire, and Oliver Ryder, describes the silent biodiversity crisis. This crisis concerns the loss of biorepositories and biological research collections. This chapter concisely describes the crisis in no uncertain terms and points to the urgency of stopping the loss of this important part of our biodiversity legacy. We end this part with a chapter by Cathi Lehn, Rebeccah Bryning, Rob DeSalle, and Richard Cahoon titled "Who Owns the Ark?" This chapter discusses archiving and material transfer issues. Together the chapters in this part will provide the reader with some tools to understand the silent biodiversity crisis and how scientists are approaching the storage and utility of biological and genetic resources in collections.

Robert Hanner, Angélique Corthals, and Rob DeSalle

9 Biodiversity, Conservation, and Genetic Resources in Modern Museum and Herbarium Collections

The biodiversity that exists today is the result of 3.5 billion years of evolution. Its preservation is of critical importance for many reasons, including the intrinsic value of functional ecosystems, species, and, ultimately, genes. The conservation of biological diversity is one of the most critical problems of our time. Unfortunately, our window of opportunity for sampling and studying this diversity is closing as habitats and the organisms they contain are lost to extinction at an ever-increasing rate. Much work lies ahead in simply documenting the existence of the world's remaining unknown species, as perhaps less than 10 percent of all species diversity is yet described. Modern uses of biodiversity involve the many varieties of plants, animals, and microorganisms now being used worldwide for farming and myriad other human activities. For example, access to genetic resources is of critical importance for agriculture and food production, conservation, economic development, environmental monitoring, epidemiology, forensic studies, gene pool preservation, molecular pathology, retrospective healthcare studies, and more. Genetic resource thinking and collections in museums and herbaria have come of age because of the convergence of needs and approaches in many areas of modern organismal, conservation, systematic, and genomic biology.

Converging Issues in Archiving and Storage of Biodiversity

The roles and approaches of modern museums and herbaria with regard to genetic resources have changed dramatically over the past decade as a result of the use of more genomic approaches to specimen analysis. This change has happened on two intersecting fronts. The first comes from personnel at the collection-based institutions themselves, who have realized the vast importance of their materials to molecular and genomic studies (Higuchi et al., 1984; Thomas et al., 1990). Over the past decade, almost all larger museums and herbaria have also started molecular systematics programs, and these have expanded the vision of collection-based institutions to include the realm of storage of materials for museum scientists who work in this area. The ability to extract viable DNA from existing and long-standing specimens in collections has extended the utility of the collections to the realm of genetic and molecular analysis. Prendini et al. (2001) and references therein summarize the various sources of tissues that have been stored in classic collections in ethanol, formalin, and dried conditions. All of these sources of tissues have yielded viable DNA for examination, with some of the methods of preservation being better than others. So-called paleo DNA studies (Greenwood, 2001) are also possible, and so even these kinds of specimens can be very useful in certain cases. Finally, we point out that museums and herbaria are already adapted to the genomic revolution because the methods of preservation for older genetic analysis techniques, such as allozyme analysis, implemented the storage of high-quality tissues (Dessauer and Haffner, 1984).

The second front comes from the opposite direction and involves the realization by molecular systematists, molecular biologists, and, more recently, genomicists that the materials they work with are in dire need of archiving and storage (Ruedas et al., 2000). The need for well-characterized genetic material has not escaped the attention of individual researchers who have informally acquired large holdings of frozen tissues from a variety of living organisms in the course of their studies. Such materials are known to have value beyond the purpose of their original collection. However, it is important to note that although such collections constitute a valuable biological resource, they are largely untapped because their existence is not well known. For many research collections maintained in institutions not specifically charged with archival functions, development of a system that provides long-term access to the material has been problematic. Moreover, the materials in such collections often lack proper documentation and are often stored in an inappropriate manner from the standpoint of long-term preservation and archiving of genetic resources.

The Twin Problems of Preservation and Archiving

Institutions at the forefront of biodiversity collecting, such as herbaria and museums, maintain coverage of most of the world's known biodiversity and are therefore of critical importance to the future of humanity as well as to the future of the species represented by the collections in these institutions. However, these collections consist primarily of materials that are unavailable for, or of limited utility to, modern lines of genomic and genetic research because traditionally collected and preserved specimens lack many levels of biological information that are relevant to modern science. These historic specimens were originally collected under a paradigm of scientific inquiry concerned primarily with the preservation of the morphological integrity of the collected specimen rather than with preserving a broad range of biochemical constituents therein. Biological materials change and deteriorate with time. In order to preserve biological materials for later study or use, some means of halting these processes must be used that will not fundamentally alter the nature of the material. The permanently denatured proteins and nucleic acids of traditional formalin-fixed museum specimens are useless for studies at the molecular level.

Degradation of materials in museums and herbaria is an ongoing problem; collections continue to be at risk of deterioration and loss due to a lack of understanding of the issues associated with storage temperature. Critical preservation temperature for biomolecules should be the temperature below which no physical or chemical changes will occur. This temperature is called the glass transition temperature (Tg), which is approximately –132°C. Simply keeping the specimens frozen is not sufficient for long-term preservation. At temperatures above Tg, all storage achieves is the delay of the process of degradation (Franks, 1985). Many current facilities archive biomaterials in mechanical freezers above critical preservation temperature, and backup freezer space is often nonexistent. Given the possibility of a mechanical freezer failure and the increasing frequency of blackouts and brownouts in major cities, it is obvious that many collections are in jeopardy. Ideally, the failure mode of the system used to archive biomaterials should allow sufficient time for corrective action in the event of failure. If we are preserving our samples for as yet unidentified uses, it is incumbent on us to ensure that the maximum potential of the specimen is preserved for future scientific exploitation.

Because museum archives of tissue specimens are intended to serve in documenting biodiversity at the molecular level, supporting a broad array of comparative genetic and genomic research initiatives is now more necessary than ever. For genetic analyses, that tissues and their included polymers must be collected and maintained in a biochemically active form. Ideally, tissue samples should be collected fresh and immediately frozen or buffered so as to stabilize and preserve biomolecules for future study. Freezing remains the most reliable and versatile method of preserving tissue for long-term storage (Engstrom et al., 1999). Ultracold storage of biological tissues preserves a broad suite of biochemical characteristics (Cook et al., 1999), although protein and lipid changes can occur during frozen storage if the storage conditions are not sufficiently cold (Florian, 1990).

A parallel problem to the one just mentioned is the need to archive voucher specimens from molecular work. A growing body of sequence-based exploratory work cannot be confirmed or duplicated because the specimen used by the investigator cannot be reexamined, since no designated voucher specimen was deposited in a reference collection (Ruedas et al., 2000). It is imperative to deposit into a reference collection all specimens used in generating molecular data. The crucial significance of collections is that they make possible the interpretation of published data in scientific journals such that it can be reconfirmed in other field or laboratory studies. Without voucher specimens or cultures, it is not possible to confirm that subsequent studies, or even parallel studies elsewhere, involve the same species (Hawksworth and Mound, 1991).

Museums have the fundamental role of building and maintaining biological collections for documentation of biodiversity (Engstrom et al., 1999). Reference collections that are properly and permanently preserved are a crucial component of the information transfer system of biological diversity. This is far more important than a mere concern for nomenclature might at first imply, because nomenclatural assignments shape our fundamental perceptions of how the biological world is organized. Reference collections are essential for identification, as vouchers for the application of names and as vouchers of species used in research projects. Ethical researchers are in the habit of depositing voucher specimens of the organisms they study in natural history museums and herbaria. This provides a long-term record of their work and hedges against changes in taxonomy, which could alter the interpretation of their results (Thomas, 1994).

Some Solutions: The Ambrose Monell Cryo Collection as a Model Collection

In a time of extensive species loss, efforts to offer a comprehensive and widely accessible record of the earth's biological diversity are essential. With the digital revolution under way, natural history museums are poised at the forefront of biodiversity information access, especially concerning biomaterial collections used in modern genetic, genomic, and taxonomic studies. Many museums and herbaria have taken the requisite steps to address these twin problems (see appendix 9.1; see also Prendini et al., 2001).

Although all these facilities are adequate and admirable in their structures and approaches, we wish to focus on the workings of one collection that can serve as a paradigm for modern genetic resource banking in herbaria and museums: the Ambrose Monell Cryo Collection (AM-CC) at the American Museum of Natural History. This collection was established to help meet the demand for properly documented frozen tissue specimens needed by modern scientific researchers. Launched in May 2001, the AM-CC will house approximately 1 million frozen tissue samples representing the DNA of a wide range of species. Potentially the largest and most comprehensive initiative of its kind, the museum's frozen tissue collection will support a broad range of research and allow scientists, today and in the future, to take full advantage of advances in genomic technology.

Dynamics of the AM-CC

The AM-CC addresses an underserved niche among cryogenic biorepositories by attempting to catalogue global biogenetic diversity at the molecular level in tandem with traditional morphological voucher specimens (Corthals and DeSalle, 2005). The AM-CC maintains tissue samples containing the genomic information from a wide range of species. Unique and important biomaterials will be available to the scientific community in perpetuity. In this way, the specimens themselves form databases, incorporating a massive body of information on distribution, seasonality, and so on (Hawksworth and Mound, 1991) with the potential to derive genetic and genomic information from the cryopreserved tissues. The facility is designed to maintain samples below –150°C using an array of vapor phase liquid nitrogen freezers.

Databasing and Inventory

In its daily operations, the AM-CC tissue samples are indexed using Freezerworks, a relational database application program intended for the task of freezer inventory management (Ioannou, 2000). The program creates a record for each specimen, giving it a unique barcode ID. This technology is now being used both in museums (Monk, 1998) and in other biodiversity assessment programs (Oliver et al., 2000). Data entry in the database is made easy and reliable with the "import" feature, allowing donors to generate their own data spreadsheet and have it imported into the AM-CC database without ever having a third party (lab tech) modify anything manually. Each record contains data ranging from the collecting event (who collected it, when, where, how, and even why), to physical characteristics of the donor animal, to the position of each vial in the collection's many freezers. After data entry, the program generates a printed cryo-resistant label, which includes the unique ID number both as a barcode and as a human-readable numeric string. This feature allows lab technicians to retrieve any vial from the freezers quickly and reliably by scanning the label to retrieve its associated data and thus confirm the identity of the specimen they are attempting to retrieve. The computer database tracks each barcoded vial, noting the specimen's taxonomic identity, where the specimen was collected, and by whom, and tracks how many times the sample has been thawed and refrozen. This is critical because each thermal cycling of the specimen degrades the biomolecules, limiting the research potential of the specimen.

Online Database

Because the relational database, Freezerworks Unlimited, has not yet been made compatible with the Web, a shadow database had to be built in order to render the holdings of our collection accessible to the scientific community worldwide. In order to host the fully searchable database, the AM-CC launched its own Web site in 2002. The data available on the AM-CC Web site are a subset of data from the facility's database, Freezerworks Unlimited. This allows the AM-CC to control the amount of data published on the Web. The database is updated every three months, at which time the entire data set is exported from Freezerworks into a text file. The DataCheck software is then run to make the process faster and conform to standards of database information by diagnosing nondamage data problems, such as gremlins (nonprintable characters in data), not-a-number numbers, and bad booleans. Once checked and quality controlled, the data set is ready for the last steps of Web interface.

The AM-CC Online Database and GenBank/NCBI

The database is designed to be integrated with the National Center for Biotechnology Information (NCBI) Entrez indexing and retrieval engine. This allows AM-CC records with nucleotide sequence accession numbers to link out to corresponding pages on the NCBI GenBank and Taxonomy databases and for GenBank sequences to link out back to the AM-CC.

The core tables of the database are the specimen and taxonomy tables, and these are joined by an associative table that allows a many-to-many relationship between specimens and taxonomic names. In addition, both the specimen and taxonomy tables form a many-to-one relationship with a table for "foreign resources," which flexibly stores information about external URLs, files, and database resources. At the moment, only URLs to Web sites are stored in these tables, but in theory the design allows any type of resource location information to be stored that pertains to an address that can be associated with an institution. In the future, the taxonomic names will be joined many-to-many with a common names table. The database schema as it exists now is believed to be reasonably sound, although it is subject to further development.

E-Vouchers in the AM-CC

Monk and Baker (2001:3) define an e-voucher as "a digital representation of a specimen. An e-voucher may be ancillary to a classical voucher specimen or it may be the only representative of the specimen in the collection." The importance of e-vouchers has emerged not only with the advent of better digital imaging technology but also with new approaches to specimen collection and the redefinition of museum vouchers. Traditionally, voucher specimens deposited into museum collections were morphological vouchers, and in most cases they consisted in the cadaver of the whole animal. In the case of a frozen tissue repository, this traditional definition of a voucher becomes not only impractical but, for the majority of animals sampled, simply impossible to apply. Thus, the digital picture of the specimen sampled becomes the morphological voucher in the many cases where the remains of the animal sampled are unobtainable. To that effect, the AM-CC has linked, whenever possible, the specimen records to digital images (hosted by the museum's Digital Library server), making a complete connection between sequence data and the visual identity of the specimen examined.

Future Considerations

Modern bioinformatics initiatives ultimately will link tissue specimen collection records with bibliographic citations, competing taxonomic determinations, geospatial referencing information, and much more. The AM-CC database is capable of linking out to other resources on the Web relevant to the collection, and in the future it may make use of Web services to integrate more completely with other such information providers.

We also must address the state of current specimens in natural history collections and their role in modern genetic and genomic studies. Plans to salvage classically preserved specimens such as dried, pinned ethanol-preserved or formalin-preserved specimens should be of high priority in thinking about genetic resources at museums and herbaria. In addition, special consideration should be given to current frozen tissue collections in natural history museums and herbaria. For instance, even –80°C preserved specimens will experience some degradation over time. Consideration of such degradation even in carefully preserved genetic resource collections should be a high priority.

Freezer malfunction has occurred in many collections maintained at –80°C, and the impact on specimens that have experienced these freezer disasters is an important concern. Specimens that thaw for days may not contain the high–molecular weight DNA necessary for polymerase chain reaction, sequencing, restriction fragment length polymorphism, and a host of other research applications. In addition, dried and ethanol-preserved specimens and formalin-preserved specimens, though useful in some instances in molecular systematic and population genetic studies, also are not sufficiently preserved for genomic-level studies. Because of the rarity of some materials and because partially degraded specimens can often yield important results, priorities in existing collections must be established to determine which specimens merit salvage. According to studies conducted in the AM-CC, reptilian blood, thigh, and cardiac tissue samples are still valuable after an extended thaw. Other tissue types—liver, kidney, and small intestine—do not contain sufficient high–molecular weight DNA after prolonged incubation at room temperature. Therefore, such samples have little value in a genetic repository to document biodiversity and were subsequently discarded by the AM-CC staff after extreme degradation of these samples had been determined. However, thawed blood, thigh, and cardiac tissue samples are more informative; such tissues may contain the total genomic DNA for an extended period of time at room temperatures after freezer failure. In fact, many of these specimens have since been successfully used in PCR-based studies. Interestingly, ovarian tissues showed little or no degradation despite the thaw. Reproductive tissues are highly recommended as part of the sampling strategy because of their stability.

Conclusions and Recommendations

Major research institutions can assist in the development of frozen genetic resource collections by exchanging, donating, or otherwise making available duplicates of authoritatively named specimens used in their research. Development of such collections can be further enhanced by submission of karyotype, chromatographic profiles, electrophoretic banding pattern preparations, GenBank accession numbers, and reprints along with voucher specimens.

Technological advances and the existence of targeted scientific collections provide the basis for genomic and molecular genetic studies. Such studies are revolutionizing our understanding of comparative biology and evolution. One of the challenges to further progress in preserving genetic resources has been the need to collect samples under a modern paradigm of preservation that maintains not only the collecting locality and morphological identity of the specimens but also the integrity of the biomolecules within those collected specimens. We must endeavor to sample biodiversity with an eye toward minimal regret over the loss of biological information associated with the collections that we have made and continue to make. Special efforts are now needed to secure tissue samples from threatened species and from species that are rare and difficult to obtain. The most effective means of preserving biological materials is by freezing and subsequently storing at low temperature. Yet this presupposes the existence of facilities that are adequate for the long-term storage, preservation, and archival of such biomaterials.

References

Cook, J. A., G. H. Jarrell, A. M. Runck, and J. R. Dembosk. 1999. *The Alaska Frozen Tissue Collection and associated electronic database: A resource for marine biotechnology.* OCS Study MMS 99-0008. Fairbanks: University of Alaska.

Corthals, A. and R. DeSalle. 2005. An application of tissue and DNA banking for genomics and conservation: The Ambrose Monell Cryo Collection (AMCC). *Systematic Biology* 54(5):819–823.

Dessauer, H. C. and H. S. Haffner. 1984. *Collections of frozen tissues: Value, management, field and laboratory procedures, and directory of existing collections.* Washington, D.C.: Association of Systematics Collections.
Engstrom, M. D., R. W. Murphy, and O. Haddrath. 1999. Sampling vertebrate collections for molecular research: Practice and policies. In *Managing the modern herbarium: An interdisciplinary approach,* ed. A. Metsger and S. Byers, 315–330. Vancouver: D. Elton-Wolf.
Florian, M. L. 1990. The effects of freezing and freeze-drying on natural history specimens. *Collection Forum* 6:45–52.
Franks, F. 1985. *Biophysics and biochemistry at low temperatures.* Cambridge: Cambridge University Press.
Greenwood, A. D. 2001. Mammoth biology: Biomolecules, phylogeny, Numts, nuclear DNA, and the biology of an extinct species. *Ancient Biomolecules* 3:255–266.
Hawksworth, D. L. and L. A. Mound. 1991. Biodiversity databases: The crucial significance of collections. In *The biodiversity of microorganisms and invertebrates: Its role in sustainable agriculture,* ed. D. L. Hawksworth, 17–29. Wallingford, U.K.: CAB International.
Higuchi, R., B. Bowman, M. Freiberger, O. A. Ryder, and A. C. Wilson. 1984. DNA sequences from the quagga, an extinct member of the horse family. *Nature* 312:282–284.
Ioannou, Y. A. 2000. A frozen database. *Science* 288:1191.
Monk, R. R. 1998. Bar code use in the mammal collection at the Museum of Texas Tech University. *Museology, Museum of Texas Tech University* 8:1–8.
Monk, R. R. and R. J. Baker. 2001. e-Vouchers and the use of digital imagery in natural history collections. *Museology of the Museum of Texas Tech University* 10:1–8.
Oliver, I., A. Pik, D. Britton, J. M. Dangerfield, R. K. Colwell, and A. J. Beattie. 2000. Virtual biodiversity assessment systems. *BioScience* 50:441–450.
Prendini, L., R. Hanner, and R. DeSalle. 2001. Obtaining, storing and archiving specimens and tissue samples for use in molecular studies. In *Methods and tools in biosciences and medicine: Techniques in molecular evolution and systematics,* ed. R. DeSalle, G. Giribet, and W. Wheeler, 176–249. Basel, Switzerland: Birkhauser.
Ruedas, L. A., J. Salazar-Bravo, J. W. Dragoo, and T. L. Yates. 2000. The importance of being earnest: What, if anything, constitutes a "specimen examined?" *Molecular Phylogenetics and Evolution* 17:129–132.
Thomas, R. H. 1994. Molecules, museums and vouchers. *Trends in Ecology and Evolution* 9:413–414.
Thomas, W. K., S. Pääbo, F. X. Villablanca, and A. C. Wilson. 1990. Spatial and temporal continuity of kangaroo rat population shown by sequencing mitochondrial DNA from museum specimens. *Journal of Molecular Evolution* 31:101–112.

Appendix 9.1. Selected Frozen Tissue Collections

Academy of Natural Sciences, Philadelphia, PA, USA: http://www.acnatsci.org
Alaska Frozen Tissue Collection, University of Alaska Museum, University of Alaska at Fairbanks, Fairbanks, AK, USA: http://www.uaf.edu/museum/collections/
Ambrose Monell Collection for Molecular and Microbial Research, American Museum of Natural History, New York, NY, USA: http://research.amnh.org/amcc/
Burke Museum, University of Washington, Seattle, WA, USA: http://www.washington.edu/burkemuseum/collections/genetic/index.php
Captive Propagation Research Group, Patuxent Wildlife Research Center, U.S. Geological Survey and U.S. Fish and Wildlife Service, Laurel, MD, USA: http://www.pwrc.usgs.gov; http://patuxent.fws.gov
Center for Reproduction of Endangered Species, Zoological Society of San Diego, San Diego, CA, USA: http://cres.sandiegozoo.org/

Collection of Genetic Resources, Museum of Natural Science, Louisiana State University, Baton Rouge, LA, USA: http://www.museum.lsu.edu/LSUMNS/Museum/NatSci/tissues.html

Division of Biological Materials, Museum of Southwestern Biology, University of New Mexico, Albuquerque, NM, USA: http://www.msb.unm.edu/dgr/index.html

Field Museum of Natural History, Chicago, IL, USA: http://www.fmnh.org/research_collections

Laboratory of Molecular Systematics, National Museum of Natural History, Smithsonian Institution, Washington, DC, USA: http://lab.si.edu/frozencollections.html

The Millennium Seed Bank Project, Science and Horticulture, Royal Botanic Gardens Kew, Richmond, UK: http://www.kew.org.uk/msbp/index.htm

Museum of Natural History, University of Kansas, Lawrence, KS, USA: http://www.nhm.ku.edu/Hdocs/Collections.html

Museum of Vertebrate Zoology, University of California, Berkeley, CA, USA: http://mvz.berkeley.edu/

National Biomonitoring Specimen Bank, National Institute of Standards and Technology, Chemical Science and Technology Laboratory, Charleston, SC, USA: http://www.nist.gov/public_affairs/gallery/specimen.htm

National Plant Germplasm Network, U.S. Department of Agriculture, Beltsville, MD, USA: http://www.ars-grin.gov/npgs/

Natural Products Repository, National Cancer Institute, National Institutes of Health, Frederick, MD, USA: http://dtp.nci.nih.gov/branches/npb/repository.html

Natural Science Research Laboratory, Museum of Texas Tech University, Lubbock, TX, USA: http://www.depts.ttu.edu/museumttu/

Peabody Museum of Natural History, Yale University, New Haven, CT, USA: http://www.peabody.yale.edu/collections

Leona G. Chemnick,
Marlys L. Houck, and
Oliver A. Ryder

Banking of Genetic Resources

The Frozen Zoo at the San Diego Zoo

It is a testimony to the rapid advances in the field of genetics that the banking efforts for genetic samples envisioned more than thirty years ago largely anticipated the need for germplasm resources. In the intervening years, the importance of saving biomaterials in a form that can be propagated has not diminished, and the value of saving samples of DNA as resources for studying genome organization, genome evolution, and the molecular basis for adaptation has been seen as a worthwhile goal in and of itself (Ryder et al., 2000).

The importance of collecting and providing access to biological samples from a number of diverse animal species has been recognized for many years by the scientists at the Zoological Society of San Diego (ZSSD) Center for Reproduction of Endangered Species (CRES). To help provide a repository for this valuable material, CRES initiated its efforts to preserve biological specimens with the formation of the Frozen Zoo twenty-eight years ago, and the history of the Frozen Zoo has been provided by its founders, Arlene Kumamoto (1998; Kumamoto and Houck, 1992) and Kurt Benirschke (Benirschke and Kumamoto, 1991). Although there is as yet no recognized worldwide directory of genetic resource collections, it is believed that this endeavor to save viable cells and DNA preparations from rare, threatened, and endangered mammals, birds, and reptiles at the facilities of the ZSSD stands among the most significant efforts of its kind attempted anywhere in the world.

An important component of the CRES collection has been the development of a policy statement to clearly express the ZSSD's view on how these valuable resources are to be managed and for what purpose samples are to be provided to investigators. The Frozen Zoo policy states that the primary purpose of the collection is for conservation, with an emphasis on scientific research that is related to conservation efforts involving endangered species and their habitats. Additional policy points cover stewardship responsibilities, the protection of the resources represented by the Frozen Zoo, the ZSSD itself, and the request that any research be done mainly for noncommercial purposes, recognizing the possibility of commercial use resulting from the research. In such an event, the policy affirms that the ZSSD will apply any commercially derived benefits toward conservation of wild populations and will encourage others to do the same.

In this chapter, the accessions of the Frozen Zoo, the rationale for their collection, and the recent efforts to expand are described. The uses of this collection are too numerous and diverse to describe here. The Frozen Zoo collections have played a crucial role in studies from diverse fields, including medicine, physiology, evolution, the biology of aging, and other biomedical studies. Formative contributions in the phylogeny of birds and mammals and many highly influential studies related to the conservation of endangered species have been facilitated by access to samples in the Frozen Zoo.

Current Holdings

The Frozen Zoo currently holds viable fibroblast cell lines and tissue pieces (the living tissue collection) or nucleic acid preparations from approximately 7,000 individual mammals, birds, and reptiles. Artiodactyls, perissodactyls, primates, and carnivores are very heavily represented (figure 10.1), together making up more than 85 percent of the total collection. The emphasis of the collection in the past was on specimens from mammalian taxa, and twenty of the twenty-six mammalian orders (Wilson and Reeder, 1993) are now represented. More than ninety-five species within the primate order alone are held in the collection. Another focus was to seek significant representation in the collection of samples from Africa, Asia, and South America. More recently, this aim has been expanded in the effort to obtain samples from other regions, including San Diego's own local species.

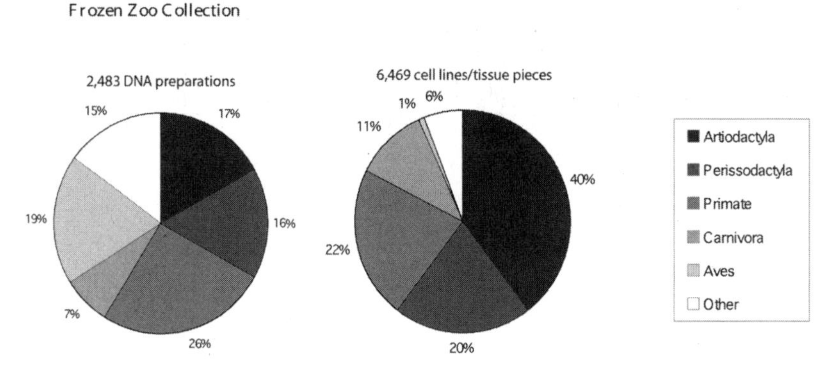

Figure 10.1 The Frozen Zoo repository contains diploid cell lines and minced tissue pieces from more than 6,400 individual species and DNA preps from almost 2,500 specimens, many of which were isolated from the cell lines. Artiodactyls, perissodactyls, primates, and carnivores are predominant in the living tissue collection, making up 94% of the total, and these same mammalian orders and the bird samples make up 85% of the DNA inventory.

Table 10.1 Species Survival Plan (SSP) and Europäische Erhaltzungszucht Programme (EEP) mammals represented in the Frozen Zoo.

Carnivora, 30/31	**Primates, 28/32**	**Cetaceans, 1/1**
Cat, Geoffroy's (0) E	Baboon (5 spp.) (8 D) S	Dolphin, bottle-nosed (1 D) E
Cat, Pallas's (14) S	Bonobo (63 D) S, E	
Cat, sand (0) S	Chimpanzee (80 D) S	**Chiroptera, 1/2**
Cat, Temmnick's golden (0) E	Colobus (89 D) S, E	Bat, Livingstone's fruit (0) E
Cheetah (64 D) S, E	Drill (23 D) S, E	Bat, Rodrigues fruit (15) S, E
Dog, African wild (16 D) S, E	Gelada (5 D) E	
Dog, bush (1 D) E	Gibbon (*Hylobates* sp.) (34 D) S, E	**Dasyuromorphia, 0/1**
Ferret, black-footed (10 D) S	Gorilla, lowland (158 D) S, E	Kowari (0) E
Fossa (10 D) E	Guenon (10 spp.) (49 D) S, E	
Jaguar (4 D) S	Langur (4 spp.) (54 D) S, E	**Diprotodontia, 1/2**
Leopard, Amur (3) E	Lemur, Alaotra gentle (0) E	Bettong, brush-tailed (0) E
Leopard, clouded (17) S, E	Lemur, black (19 D) S, E	Kangaroo, tree (18 D) S, E
Leopard, Persian (5) E	Lemur, Coquerel's mouse (0) S	
Leopard, snow (27 D) S, E	Lemur, mongoose (7 D) S, E	**Edentates, 1/1**
Leopard, Sri Lankan (0) E	Lemur, ring-tailed (63 D) S	Anteater, giant (7 D) E
Lion (26 D) S, E	Lemur, ruffed (72 D) S, E	
Margay (8) E	Loris, pygmy slow (29 D) S, E	**Perissodactyla, 10/10**
Mink, European (2 D) E	Macaque (*Macaca* sp.) (101 D) S, E	Ass, Asian wild (109 D) S, E
Ocelot (4) S	Mandrill (12 D) S	Ass, Somali wild (23 D) E
Otter, Asian small-clawed (16 D) S	Mangabey (19 D) S	Horse, Asian wild (505 D) S, E
Otter, European (2) E	Marmoset, white-fronted (4) E	Rhinoceros, Asian (17 D) S, E
Panda, giant (7 D) S	Monkey, Colombian spider (11 D) E	Rhinoceros, black (82 D) S, E
Panda, red (23 D) S, E	Monkey, De Brazza's (11 D) S	Rhinoceros, Sumatran (8 D) S
Tiger, Amur (13 D) S, E	Monkey, Goeldi's (6 D) S, E	Rhinoceros, white (52 D) S, E
Tiger, Indochinese (12 D) S	Monkey, woolly (37 D) E	Tapir, Malayan (28 D) E
Tiger, Sumatran (10 D) S, E	Orangutan (48 D) S, E	Zebra, Grevy's (50 D) S, E
Wolf, maned (11 D) S, E	Saki, white-faced (19 D) S, E	Zebra, Hartmann's mountain (130 D) S, E
Wolf, Mexican gray (11) S	Tamarin, cotton-top (7) S, E	
Wolf, red (4) S	Tamarin, emperor (6 D) E	**Proboscidae, 2/2**
Wolf, Siberian (0) E	Tamarin, golden lion (1 D) S, E	Elephant, African (3 D) S, E
Wolverine (0) E	Tamarin, gold-headed lion (0) E	Elephant, Asian (4 D) S, E
	Tamarin, pied (0) E	

The acquisition strategy for the Frozen Zoo has been influenced by the close cooperation of zoos in the formation of the North American-based SSPs and the similar European-based management EEPs, whereby breeding and management of designated species are conducted by joint effort. Many SSPs and EEPs have deemed it worthwhile to preserve the genomes of individual animals in their programs, whose histories are recorded in species studbooks. Some of these samples from now-deceased founder animals will be of immense value when retrospective studies of the captive populations are undertaken. The mammals shown in the tables are from the most recent SSP and EEP listings. The numbers in parentheses represent the number of fibroblast cell lines and tissue pieces frozen as of December 2002, and the "D" shows that DNA has been isolated. "S" indicates listing by the SSP; "E" indicates listing by the EEP.

Of major importance are animals listed as endangered and threatened, and of the approximately 500 species and subspecies in the collection, more than 200 are listed in the International Union for Conservation of Nature Red List of Threatened Species. An equally important concern from the onset was the interaction of our conservation efforts with those of the Species Survival Plan (SSP) groups under the auspices of the American Zoo and Aquarium Association (AZA). Seventy-two of the seventy-four SSP mammalian taxa are now represented in the CRES repository (table 10.1), and close cooperation with several SSP groups has enabled the continuing collection and storage of specimens from all captive-held individuals of certain targeted species.

Living Tissue Collection

The living tissue holdings of the Frozen Zoo have been carefully prepared and frozen to maintain viability. As of December 31, 2002, this part of the collection included 6,469 accessions, 86 percent of which are preserved as low-passage diploid fibroblast cell strains. The balance are kept as cryopreserved minced tissue pieces, from which early passage diploid fibroblast cell strains may be established. Although more than 480 vertebrate taxa are represented in the three classes Mammalia, Reptilia, and Aves, more than 98 percent of taxa in the cell line collection are mammals.

The first nonmammalian cells were added in 1995, and these were avian specimens for which sex determination by chromosome analysis was desired. Subsequently, forty-nine avian specimens from nine orders have been preserved. In 2000, the first reptilian cell lines were established. Although nonmammalian cell strains are an extremely small part of the current collection, we are working to increase the availability of cell lines from avian and reptilian taxa to provide similar opportunities for scientific study and conservation assessments that the collection of mammalian cell strains has afforded.

Nucleic Acid Preparations

The Frozen Zoo has also stored numerous nucleic acid preparations, mainly DNA, and as of December 31, 2002 this part of the repository held 2,483 samples. As with the cell lines, the early emphasis was primarily on the preparation of DNA from mammalian samples, with the primate species currently making up the largest percentage from a single order (figure 10.1). This emphasis reflects, and has also contributed to, the numerous primate studies conducted here at CRES and at other institutions. In addition to our efforts to incorporate extensive diversity in primate taxa into the collection as nucleic acids, emphasis has been placed on amassing significant population samples, especially from as many individuals as possible in certain primate species.

Other mammals currently represented in large numbers as DNA are the Perissodactyla, especially the Przewalski's horse, which makes up 25 percent of the total DNA samples from that order, and the Artiodactyla, with 142 of 220 taxa (Wilson and Reeder, 1993) represented. Initially, only a limited number of DNA samples from birds and reptiles were prepared for in-house studies, but as the number of cell lines has increased in these taxa, so has the number of DNA preparations. As an example of our focus on nonmammalian species, we have DNA isolated from the entire extant population of the California condor and from many of the deceased founders, and these preparations have been used extensively in long-term collaborative studies.

A wide variety of studies using the cell line or nucleic acid samples from the Frozen Zoo have been conducted over the years by ZSSD investigators and other scientists throughout the world. As of December 31, 2002 approximately 390 investigators and co-investigators at more than 210 institutions in 39 states, the District of Columbia, and 17 countries have been assisted in their work, and they have been provided with 4,174 samples from 445 different species. The results from these studies have helped other research scientists, zoo curators, and field biologists via information disseminated through scientific publications, husbandry manuals, and clinical reports.

Recent Efforts to Expand

The establishment of the Frozen Zoo was a significant step in the ZSSD goal of conserving genetic resources of endangered species for science and conservation. However, in 1999, an important decision was made to expand the role of the Frozen Zoo. It was decided that a major augmentation of the genome resources (represented mainly by frozen skin cells and their derivatives, such as DNA) would serve as a substantial step toward furthering the ZSSD goal of conservation.

Although the Frozen Zoo currently includes genetic material from approximately 500 taxa, additions to the Frozen Zoo formerly lacked a directed, strategic collection focus; samples were added on an opportunistic basis or requested for specific research projects. Several critically endangered species of great interest were underrepresented in the collection or were missing altogether. In order to identify the precise animals that could provide samples of the greatest genetic importance, CRES generated lists of desired specimens for each of the U.S. institutions that held individuals from a number of threatened species.

CRES maintained its focus on SSP mammals in the initial stage of this recent expansion effort. These species were selected because the AZA, and the conservation community in general, have made a particular commitment to all SSP species. The AZA chooses threatened and endangered species based on their ability to inspire the dedication of both professionals and the public to work toward the preservation of the species and its habitat. For these same reasons, additional mammalian species chosen by European institutions belonging to the Europäische Erhaltzungszucht Programm (EEP) or European Conservation Breeding Programs are also being targeted for inclusion, and currently seventy-six of the eighty-eight EEP taxa are also banked in the collection (table 10.1).

In September 2000, specific requests were sent to hundreds of U.S. institutions that currently hold a number of threatened species, asking for specimens when they became available on an opportunistic basis. This has resulted in a remarkable increase in samples contributed by other zoological institutions. Since the inception of the planned expansion, the percentage of samples frozen from other institutions has been more than 40 percent of the total, and the percentage probably will increase in the coming years.

As part of the effort of CRES to offer its collection for use in research and development, a formal Frozen Zoo species list was generated in time for the "Genetic Resources for the New Century" meeting in 2000 to represent the samples that were readily available. Currently, when this institution is contacted for information on how to obtain samples, the inquiring investigator can be directed to the Frozen Zoo Web site (http://cres.sandiegozoo.org/projects/gr_frozen_zoo.html), where the Frozen Zoo sample list and a frequently asked questions (FAQ) list are available. The FAQ outlines the procedure for requests to CRES, which includes asking the investigator for a formal letter describing the scope of the project and for the completion of a Research Material Agreement, and it describes potential access and handling fees and the shipping arrangements. The sample list will be updated as needed, and it is expected that a greater number of individuals per taxon and total taxa will be available. By prior arrangement, materials not on this list can be provided. Some cell lines must be thawed and expanded before shipment, and all cell lines are diploid fibroblasts that have had their karyotypes delineated. All materials available through CRES have been obtained in compliance with the Convention on International Trade in Endangered Species of Wild Fauna and Flora.

The Frozen Zoo is focusing much of its expansion effort on a joint five-year project with National Science Foundation funding. Initiated in 2001, this project is known as the Integrated Primate Biomaterials and Information Resource (IPBIR) and provides additional resources to investigators who need primate samples. In collaboration with Coriell Institute for Medical Research, the International Species and Identification System, Princeton University, and the San Diego Supercomputer Center, CRES is helping to greatly enlarge the pool of available primate resources and increase the amount of information available on each individual sample. By of the end of 2002, we had provided samples to IPBIR from 283 individuals representing forty-seven nonhuman primate species. These primate samples will be available primarily through the IPBIR Web site or, in special cases, the Frozen Zoo.

Conclusion

This unique and irreplaceable resource has become the largest of its kind, and it continues to expand with the invitation to institutions for specific specimens and the addition of avian and reptilian species to the repository. The samples now held in the Frozen Zoo at CRES serve not only as a library of genetic diversity representing taxa that are rapidly disappearing from the wild but also as the material used in research conducted in multiple arenas.

As the importance of collecting and providing access to samples of biological diversity for describing, understanding, and conserving the living world becomes more recognized and ingrained, we predict that efforts similar to the Frozen Zoo and other collections will foster a much-needed collaborative system for sharing information and samples. However, currently the collective effort is incompletely described and its global extent not fully recognized. Additionally, the value of genetic resource banks is not yet as broadly appreciated as would benefit investigators in the future (Benford, 1992).

A promising development was the global initiative sponsored by Diversitas, an international program of diversity science that promotes and synthesizes scientific research on biodiversity. As part of the International Biodiversity Observation Year (IBOY) in 2001–2002, this initiative marked a new level of worldwide recognition of the need to bank genetic material. The IBOY focus included establishing cell and tissue banks and collections of DNA, providing irreplaceable resources for the human quest to understand the biosphere, expanding medical knowledge and healthcare, and providing a broader appreciation of and tools for conserving as rich a legacy as possible of biological diversity.

Acknowledgments

Dr. Kurt Benirschke and Arlene Kumamoto initiated the Frozen Zoo at CRES and served as its first curators. Over the years we have been fortunate to have had many valuable employees, volunteers, outside consultants, and students assisting in the Frozen Zoo effort. Currently Beth Baum, Suellen Charter, Heidi Davis, Julie Fronczek, Emily Stremel, and Grace Magee help us in the lab, and Catherine Avila and Susan Hansen provide administrative support. We thank Wendy Bulger, Gary Adler, and J. Craig Potter for legal suggestions and guidance, and from the past we owe an extra thanks to Tammy Wright, Jelinda Pepper, Renee Cook, Julie Concha, Mary Jane Bundry, Peter Land, Gressa McDowell, Stacy Graham, and Steve Kingswood.

References

Benford, G. 1992. Saving the library of life. *Proceedings of the National Academy of Science* 89(N22):11098–11101.

Benirschke, K. and A. T. Kumamoto. 1991. Mammalian cytogenetics and conservation of species. *Journal of Heredity* 82:187–191.

Kumamoto, A. T. 1998, Winter. The Frozen Zoo: Extending the living animal collection. *CRES Report,* pp. 1–2.

Kumamoto, A. T. and M. Houck. 1992, June. Zoo on ice. *Zoonooz,* pp. 14–16.

Ryder, O. A., A. McLaren, S. Brenner, Y.-P. Zhang, and K. Benirschke. 2000. DNA banks for endangered species. *Science* 288:275–277.

Wilson, D. E. and D. M. Reeder, eds. 1993. *Mammal species of the world.* 2nd ed. Washington, D.C.: Smithsonian Institution Press.

Vitaly Volobouev

11

The Role of Cryopreserved Cell and Tissue Collections for the Study of Biodiversity and Its Conservation

The Rio de Janeiro Convention on Biodiversity (1992) was the first and most complete document substantiating the importance of biodiversity conservation on a global scale, presenting an integrated program on the study, protection, and use of species diversity. This initiative was taken up and developed in Systematics Agenda 2000 (1994) (hereafter SA 2000), an ambitious program of research with the aim of answering four major questions over the next twenty-five years: What are the earth's species? Where do they occur? What properties do they have? How are they related? This document made evident the fact that that we can realize no conservation or practical use of biodiversity without first discovering, describing, and classifying the world's species, and the accomplishment of these tasks will result in a series of significant benefits to science and society.

The feasibility of this program is related first and foremost to the use of new methods of molecular biology collectively known as molecular systematics (sensu Hillis and Moritz, 1990), which include the analysis of nucleic acids, proteins, and chromosomes. When correctly used, these methods permit the solution of various problems of systematics and evolutionary biology (Hillis and Moritz, 1990). In particular, they make possible the establishment of a series of highly specific genetic characters for every species. Being necessary and sufficient for unambiguous identification and characterization, these genetic characters make it possible to determine species' distribution limits and to establish phylogenetic relationships between species. As a result, they are invaluable in meeting the challenge of SA 2000.

Most of these techniques were developed and intensively applied to systematics in the past few decades. Their increasing contribution to our knowledge of species diversity and relationships is remarkable. However, if we draw up a summary of what has been done during this period and compare it with what remains to be done, the amount of data already obtained appears surprisingly modest.

Besides the chronic lack of appropriate funding for systematic research (discussed in detail elsewhere; see Ehrlich and Wilson, 1991; Morowitz, 1991; Raven and Wilson, 1992; SA 2000, 1992), there is another important reason, which may or may not be directly related to the level of funding available. This is the scarcity of available material to carry out a thorough genetic characterization of a species by means of molecular systematic approaches.

There are two different aspects to this problem. The first consists of the well-known difficulty experienced in obtaining any material at all from a particular taxon for immediate study. This is due to several factors, some biological (e.g., rarity or geographic remoteness of a species), some political (e.g., the inaccessibility of a region), and some human related (e.g., lack of competent personnel in situ or absence of effective cooperation). The implication of this lack of readily available information is a chronic incompleteness of comparable genetic and phylogenetic information about related taxa. It is rare to be able to locate genera for which the same kind of genetic data are available for all species, and the higher the taxon's rank, the worse the situation.

The second aspect of the availability problem is related to storage techniques, which limit the kind of material available for the taxon under study. Storage techniques such as alcohol-preserved or directly frozen tissues, specimens from museum collections, or fossil remains may be completely inappropriate for certain types of analysis.

Therefore, the nonavailability of appropriate material for a full range of molecular systematic analyses may be a serious obstacle to the achievement of the SA 2000 program within the scheduled time period.

The problem may be effectively solved by the appropriate development and use of materials stored in other types of biological collections, especially cryopreserved cell and tissue collections (CCTCs) (see chapter 9, this volume). These collections consist of nontransformed somatic cells (usually primary fibroblast cultures) and samples of tissue fragments of a few cubic millimeters each, both supplemented with cryoprotectors and stored in liquid nitrogen below at least –132°C. For details on cryopreservation of animal cells, see Doyle and Morris (1991) and Simione (1992).

CCTCs Are the Sole Source of Living Biological Materials

The well-developed methods of cryobiology permit indefinite storage of cells and tissue samples in liquid nitrogen without alteration of their vital characteristics. Unfrozen, the cells retain their capacity for growth and reproduction and thereby preserve and transmit their genetic information. They can be multiplied at any time until the necessary amount is obtained. In the occasional case of loss of a cell culture due to contamination or degradation resulting from the numerous rounds of freezing and thawing, cryopreserved tissue fragments can be used again to start a new culture. It follows that sets of a few cell or tissue samples derived from biopsies 0.1–0.3 cm^3 in size can prevent the loss of the genetic information of the taxa concerned and serve at any time as overall sources of living biological material for an essentially limitless range of research projects in the fields of molecular systematics and evolutionary biology, and in biotechnology as well.

Capabilities of CCTCs

CCTC = DNA + RNA + Protein Banks = Genome Banks

The 100 million cells easily obtainable after the third division cycle of a primary cell culture represent a sufficient quantity of material to perform the various analyses of both nuclear and organellar DNAs, even without amplification by means of the polymerase chain reaction technique (the use of which reduces the quantity of cells needed by at least three orders of magnitude). When necessary, in addition to DNA analyses, this material may be shared and successfully used for isozyme electrophoresis. For comparison, the weight of 10×10^7 cells is many times greater than that of an individual louse or mosquito, as routinely used for electrophoretic analysis (see Pasteur et al., 1988), which means that CCTCs may be considered at the same time as collections of both nucleic acids (DNA and RNA) and proteins.

Only CCTCs Provide Access to Chromosomal Analysis and Its Various Applications

Among the different types of existing collections, CCTCs are the sole source of material for chromosome analysis, without which numerous new animal species would not have been discovered, and which still remains an invaluable means of discerning sibling species. More importantly, in the case of rapid chromosome evolution, when the development of effective postzygotic isolation is not accompanied by significant measurable genetic divergence at either the gene or the DNA level, karyotypes are the sole genetic marker permitting the identification of new species (King, 1993; Ortells, 1995; Taylor, 2000; Vié et al. 1996). Consequently, the unambiguous identification of new "chromosomal" species among those remaining to be discovered may be impossible, whether or not highly sensitive DNA sequencing is applied.

Furthermore, chromosomal analysis remains an irreplaceable tool in the study of the mechanisms of animal speciation (Britton-Davidian et al., 2000; King, 1993; Searle and Wojcik, 1998; Taylor, 2000). In mice and shrews, which have become the models of choice for studying the initial steps of the speciation process, including formation of chromosome races, their internal structure, limits of distribution, hybrid zones, and consequences of interbreeding may be precisely documented only by means of chromosome banding analysis (Britton-Davidian et al., 2000; Searle, 1993; Searle and Wojcik, 1998). Where chromosomal analysis is really indispensable is in gene mapping by fluorescent in situ hybridization (FISH). FISH always requires precise chromosome maps, elaborated chromosome nomenclatures, and high-quality chromosome preparations. Recent developments in interspecies karyotype comparison by means of FISH, using flow sorting and generation of chromosome-specific DNA libraries as hybridization probes, have opened up new avenues to a rapid genetic mapping of any vertebrate species. Once again, the availability of chromosome banding data, or materials to perform such an analysis when necessary, is essential to the success of various fundamental and applied research projects requiring genetic maps. In addition, the precise identification of chromosome rearrangements, which became possible thanks to interspecies karyotype comparisons by high-resolution banding and FISH, turned out to be extremely useful for the reconstruction of phylogenetic relationships and pathways of genomic evolution (Carter et al., 1990; Grafodatsky et al., 2000; O'Brien and Stanyon, 1999; Yang et al., 2000a, 2000b).

CCTCs Open New Perspectives for Conservation Strategies

Considering the growing technical progress in cloning animals by nuclear transfer from a cultured differentiated somatic cell (Campbell et al., 1996; Cohen, 1997; Polejaeva et al., 2000), it is realistic to expect that cloning will become routine in the near future. As a result, CCTC may change both ex situ and in situ conservation strategies. For this reason it would be wise to begin the immediate cryopreservation of material derived from as many specimens as possible of rare and endangered species, especially those for which, at present, the gamete and embryo cryopreservation techniques do not work well.

CCTCs Do Not Harm Sampled Animals

CCTCs are the only type of collection that allows cytogenetic analyses without killing the animals, the importance of which is evident in a variety of research projects requiring the knowledge of karyotype constitution and, of course, in the case of rare and endangered species.

Although there are many other applications of CCTC, from environmental monitoring to various breeding and conservation programs and biotechnology, the aim of this chapter is to show that only this type of biological collection possesses all the virtues capable of accelerating the accomplishment of SA 2000 goals.

Before discussing a number of practical measures to be undertaken in order to realize the potential of CCTCs for describing species diversity, let us consider some frequent objections to CCTCs by comparison with alcohol-preserved and directly frozen tissue collections.

Revising Supposed Weaknesses of CCTCs

A Limited Life Duration of Cell Lines and a Risk of Possible Transformation

Because of the genetically determined number of divisions any cell can undergo, it is evident that nonimmortalized cells cannot be cultured ad infinitum. Fortunately this is not necessary because the quantity of cells obtained after a third cycle of divisions of primary culture is generally sufficient for the application of various techniques of molecular systematics and thereby elucidates a complex of highly species-specific genetic characters. In contrast, although it provides stability and unlimited capacity for division, immortalization of the cells leads to genetic transformation often accompanied by undesirable chromosomal rearrangements and thus is inappropriate for the purposes of genetic characterization.

Does CCTC Reflect True Genetic Variability of the Species?

The objection often made by population geneticists and conservation biologists is that CCTC does not represent the true genetic variability of species. The establishment and maintenance of cell cultures are time-consuming and expensive, and except in some research projects CCTCs generally contain cell samples from a limited number of specimens. However, cryopreservation of tissue samples is not expensive, and the quantity of these samples depends exclusively on the availability of a biopsy from the taxa of interest, a problem common to all the types of collections. Taking into account that the main goal of CCTC is to accelerate identification and characterization of unknown species diversity following a strategy developed later in this chapter, this objection is not directly relevant to the problems discussed here. Nevertheless, keeping in mind the extreme importance of availability of representative samples for the study of population genetics, our laboratory strategy is to preserve tissue samples from as large a number of specimens of rare, endangered, and genetically unstudied species as possible.

Difficulties Related to a Planned Replenishment of CCTCs

The most complex problem facing biomaterial banking is planning CCTCs with the aim of rapidly identifying species that are still to be discovered and described. Where and what do we need to collect and study first? One can compare these difficulties to that of collecting material derived from wild endangered species. Access to and collection of material from this category of species is difficult for the reasons specified earlier, but especially because of the existing draconian international legislation concerning the transfer of material (see Ryder et al., 2000 for details). Paradoxically, one would expect that access to these species would constitute a lesser problem because most of them have been already identified, some are well-studied genetically, and often there are one or more people investigating these taxa.

In reality, the major part of the problem concerns a huge category of genetically unstudied taxa: those to which no technique of molecular systematics has ever been applied. Even among mammals, the best-studied animal group, this category comprises no less than 60 percent of the species recognized (sensu Wilson and Reeder, 1993), not to mention those remaining to be discovered and described. The problem is compounded by the fact that some taxa regarded as full species by some are considered subspecies by others, and the attention paid to these two taxonomic divisions is highly unequal, with a general disregard for the latter. For example, in the genus *Acomys,* scientists' estimates range from fourteen full species with fifty-two subspecies to thirty-eight full species.

These studies involving only part of the intraspecific taxa identified within the genus do not imply that all or most subspecies necessarily represent full species. As well exemplified by *Acomys,* the taxonomic status—species or subspecies—of all forms previously identified using morphological analysis alone cannot be considered definitive until they have undergone reexamination by means of molecular systematic techniques.

Although the recent spectacular discoveries in eastern Asian tropical forests of the Saola (Dung et al., 1993) and the giant muntjac (Schaller and Vrba, 1996), two large artiodactyl species, morphologically very distinct from their relatives, have demonstrated to the public the incompleteness of our knowledge of species diversity, it seems that most species yet to be "discovered" are hidden among taxa that show little or no morphological distinctness and that do not necessarily occur in parts of the world difficult to access. This idea is well supported by the recent discoveries, by means of molecular systematic techniques, of three new species within the well-studied vespertilionid bats in Europe (Barratt et al., 1997; Keifer and Veith, 2001; von Helversen et al., 2001). If we add to these examples all the growing series of new sibling species detected recently by a molecular systematic approach, particularly among African murid rodents (Dobigny et al., 2002; Volobouev et al., 2002a, 2002b), it becomes clear that intraspecies taxa are indisputably the best candidates for detection of unknown species diversity. Such widespread confusion at lower taxonomic levels reflects the evident insufficiency of morphological criteria. Nevertheless, we can and should build on the foundation built by previous taxonomists.

Toward a Rapid Identification of Unknown Species Diversity

If, as a starting point, we take from this huge list of taxa those that have not yet been studied by means of molecular systematic techniques, let us consider how CCTCs may contribute to the study of unexplored species diversity and to the rapid identification of those among them that are endangered.

Working Strategy of CCTCs

As our years of experience show, a pilot study of three to five specimens per taxon, including a detailed description of their karyotype and a nucleotide sequence of at least one gene, may be sufficient to characterize a taxon and give a good idea of its phylogenetic affinities, especially when comparable genetic information is available for related taxa. Given that even one genetic marker for a species may permit its discrimination from other related species, the comprehensive genetic analysis of a single specimen is by no means a waste; it might even be sufficient for the identification of a new species. Clearly, when such information is available it will determine the next research strategy, comprising an appropriate spatial and quantitative sampling, and the choice of the most suitable genetic markers, in order to place the new form into a biogeographic and phylogenetic context.

Therefore, the principal mission of CCTCs should consist of the continuous cryopreservation and parallel molecular systematic investigation of each previously described but as yet genetically unstudied taxa. Taking into account continuously growing sequencing and chromosomal databases as the first step, the sequencing of one gene and a detailed karyotypic analysis applied to a limited number of specimens will be sufficient to characterize a taxon and to establish its phylogenetic affinities.

Although these two genetic approaches (especially sequencing) are extremely useful in the lab, they turn out to be difficult to apply in field conditions. When a species, once identified, shows no clear morphological characters distinguishing it from its relatives living in sympatry, or even in syntopy, the next step consists of searching for other genetic markers that are easily detectable in field conditions after analysis of saliva, urine, feces, hair, and buccal, anal, and vaginal smears (Neet and Hausser, 1991). For this purpose many more additional molecular, biochemical, and cytological analyses must be carried out on previously studied specimens for which material is available. This series of species-specific genetic markers will act as a kind of genetic identity card for the species, accompanying the holotype and specimens from the type series and allowing a rapid understanding of species distribution patterns, population densities, and other attributes. It follows that CCTCs supported by appropriate facilities and skilled personnel may become a kind of factory, producing species-specific genetic markers, the essential tool for species identification and characterization. Support for the existing CCTCs and creation of new ones (initially concentrating in hotspots of biodiversity) may significantly transform the present research strategy by reducing the efforts of the scientific community to reach the goals of SA 2000.

Realization of CCTC Potential

Although a number of important CCTCs already exist all over the world (partly identified by Dessauer and Hafner, 1984; Perez-Suarez, 1995), until now they have developed in complete isolation from one another. Moreover, many probably remain unrecognized by the conservation geneticist, and the existence of some of them is threatened by a chronic lack of financial support at both national and international levels. Our laboratory CCTC (available shortly as a Web site), which is one of the largest in the world, comprises material from some thousand specimens from about 400 mammalian species belonging to 16 orders, 62 families, and 176 genera (classification after Wilson and Reeder, 1993), more than 50 bird species (mostly Falconiformes), and about 30 species of reptiles and amphibians. The lack of appropriate funding results in the very slow completion of the collection and limited scientific output.

In order to realize the potential offered by CCTCs on a global scale, it has become urgently necessary to organize the existing collections into an infrastructure similar, for example, to that of the World Federation for Culture Collections (WFCC), comprising more than 345 microbial collections, detailed information about which is available at the Mondial Data Centre of the WFCC (Takishima et al., 1989).

I am convinced that the organization of existing CCTCs into a network would be an excellent opportunity to combine all the different types of genetic resource collections. The creation of associated databases would open enormous possibilities for a coordinated international collaboration, resulting in an immediate and rapid acceleration of world species identification, their thorough genetic characterizations, and the organization of this information into predictable classifications.

The three main tasks of such a network or federation should consist of preparation of a taxonomic directory of existing collections, including the precise geographic origin, quantity of available materials per specimen per taxon down to the subspecific rank, and their conservation status; creation of a database of established genetic characteristics of the taxa with special reference to their genetic markers; and an outline of an international collaboration strategy in order to proceed to a planned collection and study of genetically unstudied taxa. If the former two tasks appear feasible in the immediate future, the latter is much more complex and cannot be accomplished in the absence of effective cooperation at the international level.

Creation of a Genome Database on a Global Scale

In 1992–1994, as a first step toward the realization of such a project, a group of European scientists worked on a feasibility study for the creation of a European network of wild vertebrate species CCTCs. In the framework of a concerted action contract, modestly financed by the European Commission (Commission of European Communities), we have identified eighteen CCTCs and seventy-four collections of alcohol-preserved and directly frozen tissues and have prepared a report on the information gathered about the kind of materials stored, their use, taxonomic content, number of specimens and species per class, and biogeographic regions covered (Perez-Suarez, 1995). Moreover this final report specified the issues in need of urgent consideration by the European Commission with regard to further development of genetic resource collections in Europe and worldwide. Unfortunately, the request for further funding to continue and accomplish the project was rejected.

Recently another group of scientists proposed setting up a Web-based network to register DNA banks for endangered species (Ryder et al., 2000). This important publication was followed by two international meetings: "Genetic Resources for the New Century," organized by the Zoological Society of San Diego in 2000 (Loder, 2000), and "Conservation Genetics in the Age of Genomics," organized by the American Museum of Natural History and the Wildlife Conservation Society in 2001, which gathered numerous researchers to discuss crucial conservation issues related to genome resources. These events encourage us to believe that the time is right to unite the separate and complementary efforts across the world concerning the different kinds of genome preservation and to work together for a more rapid identification and study of vanishing species diversity.

Acknowledgments

I would like to thank my colleagues involved in the European Commission contract, which resulted in preparation of the survey of European genetic banks holding cryopreserved cells and tissues, namely G. Perez-Suarez, F. Palacios, L. Granjon, S. Salvidio, and A. Lattes. I would also like to thank C. Denys, L. Granjon, G. Pasteur, and B. M. N. Wallace for critical reading and suggestions on an earlier draft of the manuscript.

References

Barratt, E. M., R. Deaville, T. M. Burland, M. W. Bruford, G. Jones, P. A. Racey, et al. 1997. DNA answers the call of pipistrelle bat species. *Nature* 387:138.

Britton-Davidian, J., J. Catalan, M. da Graça Ramalhinho, G. Ganem, J. C. Auffray, R. Capela, et al. 2000. Rapid chromosomal evolution in island mice. *Nature* 403:158.

Campbell, K. H. S., J. McWhir, W. A. Ritchie, and I. Wilmut. 1996. Sheep cloned by nuclear transfer from a cultured cell line. *Nature* 380:64–67.

Carter, N. P., M. E. Ferguson-Smith, N. A. Affara, H. Briggs, and M. A. Ferguson-Smith. 1990. Study of X chromosome abnormality in XX males using bivariate flow karyotype analysis and flow sorted dot blots. *Cytometry* 11:202–207.

Cohen, J. 1997. Can cloning help save beleaguered species? *Science* 276:1329–1330.

Dessauer, H. C. and M. S. Hafner, eds. 1984. *Collection of frozen tissues: Value, management, field and laboratory procedures, and directory of existing collections*. Lawrence: Association of Systematics Collections, University of Kansas Press.

Dobigny, G., V. Aniskin, and V. Volobouev. 2002. Explosive chromosome evolution and speciation in the gerbil genus *Taterillus* (Rodentia, Gerbillidae): A case of two new cryptic species. *Cytogenetics and Genome Research* 96:117–124.

Doyle, A. and C. B. Morris. 1991. Maintenance of animal cells. In *Maintenance of Microorganisms*, 227–241. New York: Academic Press Limited.

Dung, V. V. et al. 1993. A new species of living bovid from Vietnam. *Nature* 363:443–445.

Ehrlich, P. R. and E. O. Wilson. 1991. Biodiversity studies: Science and policy. *Science* 253:758–762.

Graphodatsky, A., F. Yang, P. O'Brien, N. Serdukova, B. Milne, V. Trifonov, et al. 2000. A comparative chromosome map of the arctic fox, red fox and dog defined by chromosome painting and high resolution G-banding. *Chromosome Research* 8:253–263.

Hillis, D. M. and C. Moritz. 1990. An overview of applications of molecular systematics. In *Molecular systematics (1990)*, ed. D. M. Hillis and C. Moritz, 588. Sunderland, Mass.: Sinauer.

Keifer, A. and M. Veith. 2001. A new species of long-eared bat from Europe (Chiroptera: Vespertilionidae). *Myotis* 39:5–16.

King, M. 1993. *Species evolution: The role of chromosome change.* Cambridge: Cambridge University Press.

Loder, N. 2000. Gene bank to offer family album of mammals. *Nature* 405:265.

Morowitz, H. J. 1991. Balancing species preservation and economic considerations. *Science* 253:752–754.

Neet, C. R. and J. Hausser. 1991. Biochemical analysis and determination of living individuals of the Alpine karyotypic races and species of the *Sorex araneus* group. *Mémoires de la Société Vaudoise des Sciences Naturelles* 19:97–106.

O'Brien, S. J. and R. Stanyon. 1999. Phylogenomics: Ancestral primate viewed. *Nature* 402:365–366.

Ortells, M. O. 1995. Phylogenetic analysis of G-banded karyotypes among the South American subterranean rodents of the genus *Ctenomys* (Caviomorpha: Octodontidae), with special reference to chromosomal evolution and speciation. *Biological Journal of the Linnean Society* 54:43–70.

Pasteur, N., G. Pasteur, F. Bonhomme, J. Catalan, and J. Britton-Davidian. 1988. *Practical isozyme genetics*. Chichester, U.K. and New York: Horwood/Wiley.

Perez-Suarez, G. 1995. *Survey of European genetics banks holding animal cryopreserved collections.* Typescript available on request.

Polejaeva, I. A., S. H. Chen, T. D. Vaught, R. L. Page, J. Mullins, S. Ball, et al. 2000. Cloned pigs produced by nuclear transfer from adult somatic cells. *Nature* 407:86–90.

Raven, P. H. and E. O. Wilson. 1992. A fifty-year plan for biodiversity studies. *Science* 258:1099–1100.

Ryder, O. A., A. McLaren, S. Brenner, Y.-P. Zhang, and K. Benirschke. 2000. DNA banks for endangered species. *Science* 228:275–277.

Schaller, G. B. and E. S. Vrba. 1996. Description of the giant muntjac (*Megamuntiacus vuquangensis*) in Laos. *Journal of Mammalogy* 77:675–683.

Searle, J. B. 1993. Chromosomal hybrid zones in eutherian mammals. In *Hybrid zones and evolutionary process,* ed. R. G. Harrison, 309–353. Oxford: Oxford University Press.

Searle, J. B. and J. M. Wojcik. 1998. Chromosomal evolution: The case of *Sorex araneus*. In *Evolution of shrews,* ed. J. M. Wojcik and M. Wolsan, 219–268. Bialowieza: Mammal Research Institute, Polish Academy of Sciences.

Simione, F. P. Jr. 1992. Key issues related to the genetic stability and preservation of cells and cell banks. *Journal of Parenteral Science & Technology* 46:226–232.

Systematics Agenda 2000: Charting the Biosphere. 1992. Technical report. Produced by Systematics Agenda 2000. A Consortium of the American Society of Plant Taxonomists, the Society of Systematic Biologists, and the Willi Henning Society, in cooperation with the Association of Systematics Collections, pp. 1–34.

Takishima, Y., T. Shimuira, Y. Udagawa, and H. Sugawara. 1989. *Guide to World Data Centre on Microorganisms with a list of culture collections in the world.* Saitama, Japan: World Data Centre on Microorganisms.

Taylor, P. 2000. Patterns of chromosomal variation in southern African rodents. *Journal of Mammalogy* 81:317–331.

Vié, J. C., V. Volobouev, J. L. Patton, and L. Granjon. 1996. A new species of *Isothrix* (Rodentia: Echimyidae) from French Guiana. *Mammalia* 60:393–406.

Volobouev, V. T., V. M. Aniskin, E. Lecompte, and J.-F. Ducro. 2002a. Chromosomal characterization of *Arvicanthis* species (Rodentia, Murinae) from western and central Africa: Implications for taxonomy. *Cytogenetics and Genome Research* 96:250–260.

Volobouev, V. T. et al. 2002b. Patterns of karyotype evolution in complexes of sibling species within three genera of African murid rodents inferred from the comparison of cytogenetic and molecular data. *Cytogenetics and Genome Research* 96:261–275.

von Helversen, O., K.-G. Heller, F. Mayer, A. Nemeth, M. Volleth, and P. Gombkötö. 2001. Cryptic mammalian species: A new species of whiskered bat (*Myotis alcathoe* n.sp.) in Europe. *Naturwissenschaften* 88:217–223.

Wilson, D. E. and D. A. M. Reeder. 1993. *Mammal species of the world: A taxonomic and geographic reference.* 2nd ed. Washington, D.C.: Smithsonian Institution.

Yang, F., B. S. Milne, C. Schelling, G. Dolf, J. Schläpfer, M. Switonski, et al. 2000a. Chromosome identification and assignment of DNA clones in the dog using a red fox and dog comparative map. *Chromosome Research* 8:93–100.

Yang, F., P. C. M. O'Brian, and M. A. Ferguson-Smith. 2000b. Comparative chromosome map of the laboratory mouse and Chinese hamster defined by reciprocal chromosome painting. *Chromosome Research* 8:219–227.

Deborah L. Rogers,
Calvin O. Qualset,
Patrick E. McGuire,
and Oliver A. Ryder

The Silent Biodiversity Crisis: Loss of Genetic Resource Collections

For many centuries biological materials have been collected to meet human needs—originally for sustenance and utility but later to satisfy scientific curiosity. Later collections of living plants, animals, and microbes were classified, curated, conserved, and shared in parallel with public and private collections of human artifacts and natural history materials. Biological resource collections may be viewed as genetic resource collections when they encompass genetically definable material. Genetic resource collections provide a basis for contributing to the conservation of species and the diversity within species. It is our goal to provide an understanding of genetic resource collections and their place in the broader context of biodiversity conservation. We call attention to the fact that massive losses in genetic resources are quietly occurring, and this threatens the integrity of biological diversity and the ability to restore losses in nature, to advance scientific discoveries, and to use genetic variability in producing new, useful products through genetic manipulations.

When entire species go extinct or hover on the edge of extinction, this may catch the attention of media, kick laws into effect, or spur conservation advocates into action. However, when populations of a species, entire genetic resource collections, or genetic information is lost, the losses may not be noticed or valued, hence a silent biodiversity crisis. These cumulative and continuing losses of biological diversity can dramatically affect the adaptive potential and long-term survival of species and limit the advancement of biological research education and the development of new agricultural commodities and healthcare products. Continuing losses at the species level are spectacular and evocative, whereas the loss of genes—the ultimate component of biological diversity—is less obvious and dramatic but of equal concern.

The purpose of this chapter is to elucidate the nature and types of genetic resource collections, to note their traditional and emerging uses and values, to call attention to a representative array of conditions that contribute to this silent biodiversity crisis, and to offer some suggestions that can direct attention and support to these resources.

What Are Genetic Resources?

Genetic resources include any representation of genetic diversity, from natural occurrences to highly domesticated and managed collections. In the most inclusive sense, genetic diversity embraces not only variation in the DNA of individual organisms but also the variation in whole genomes and the diversity in expression of genes in individuals and populations that account for intraspecies and interspecies diversity. Genetic resources encompass all aspects of genetic diversity, not just the subset that is immediately or potentially identifiable and useful to our species. Genetic diversity is the sum total of all allelic differences within a species; although a collection of genetic resources can span this range of genetic diversity, it more commonly focuses on a subset defined by some implied use or purpose (e.g., cultivated genotypes, defined genetic stocks, or characterized mutations). Collections can consist of small or redundant (e.g., cloned) samples of a species' genetic diversity or any larger proportion, including complete representation of a species (e.g., an endangered animal that has gone extinct in the wild and all that remains are the individuals held in zoos). Genetic resource collections include the natural collections of living species in their natural habitats if confined and managed as nature reserves or in situ conservation, sensu Soulé (1991). We focus our attention on genetic resources amassed and held ex situ, that is, outside their natural habitats in the case of wild species, and assemblies of domesticated species and special derivatives obtained through genetic research (genetic stocks). These resources may be living or nonliving and still have value as a representation of genetic diversity. Typically, genetic resource collections include species, populations, genotypes or individuals, tissues, genes, DNA or RNA sequences cloned into bacterial plasmids, organelles, viruses, and information such as published DNA sequences (table 12.1). Genetic diversity is measured by allelic differences in genes between individuals and summarized as a statistical parameter (Nei, 1973) or by inferences from morphological, developmental, or physiological measurements. It is highly dynamic. Genetic diversity is conserved over time and space but can be altered by evolutionary forces. Finally, genetic diversity is shaped in response to environmental fluctuations and by interactions with other species.

Genetic diversity is central to biological diversity at all levels and scales (Ledig, 1993). Soulé (1991) presented a comprehensive biodiversity preservation scheme framed as a biospatial hierarchy that considers protection of whole ecosystems, communities, species, and populations, down to cryobanked biomaterials. The loss of biological diversity can be measured in extinctions—the loss of entire species. Therefore, much attention is paid to and policy based on extinctions. Yet the loss of genetically distinct populations within species may be an even more prevalent and urgent problem (Ceballos and Ehrlich, 2001; Daily and Ehrlich, 1995; Ehrlich, 1988; Hobbs and Mooney, 1998; Ledig, 1993; chapter 8, this volume). These largely unseen or unrecorded population extinctions have been called the secret extinctions (Ledig, 1993).

Table 12.1 The spectrum of genetic resources and their associated modes of conservation.

Type of genetic resource	Mode of conservation
Living	
Biota living in functioning ecosystems including the genetic resources within ecosystems, communities, species, or populations	In situ genetic reserves
Biota living in domestic, off-site, or highly managed situations, often but not always represented as individuals	Zoological parks, botanical gardens, seed orchards, farm collections
Reproductive propagules such as seeds, vegetative plant parts, and spores	Seed banks, cold storage, or cryopreservation banks
Gametes, such as pollen, sperm, or ova	Cold storage or cryopreservation banks
Tissues and cells	Tissue cultures (in vitro culture)
Genes	Cloned into bacterial vectors
DNA sequences	Cloned into bacterial vectors (gene banks) or deposited in sequence databases
Nonliving	
Nonliving collections, including nonviable seeds; nonliving individuals, tissues, cells; genetic sequence information	Natural history museums, herbaria
Ancient (a)DNA	Fossilized remains, material embedded (in situ) in lake sediments, peat bogs, glaciers, permafrost, etc.
Genetic sequence information	Genetic sequence databases

Types, Uses, and Importance of Ex Situ Genetic Resource Collections

Ex situ genetic resource collections are vastly diverse, reflecting the variety in forms and levels of life. In fact, the accessibility of genetic material and hence the range of these collections has expanded rapidly over the last fifteen years with the advent of technologies for removing and using extremely small amounts of DNA. Genetic resource collections include species, various levels of genetic variation (e.g., genomes, cells, tissues, individuals), individuals in populations, and both dead and living specimens, nonliving preserved tissues (e.g., blood, skin, hair), cloned genes and genomes in genomic or cDNA libraries, microarrays, various types of molecular entities (e.g., expressed sequence tags, bacterial artifical chromosomes, yeast artifical chromosomes, restriction fragment length polymorphisms), databases of nucleotide sequences, and genetic maps. Genetic resource collections are managed institutionally by several types of organizations, including governments, nonprofit organizations, educational institutions, private consortia, and cooperatives (table 12.2). Genetic resource collections often are in the public domain; government departments and affiliates and universities account for many of the sites where collections have been initiated, maintained, and distributed. However, collaborative structures and nonprofit organizations fill an important niche that complements or otherwise provides ex situ conservation in areas not served or underserved by the more traditional government and academic institutions. For example, international agricultural research centers

Table 12.2 Examples of organizations that maintain genetic resource collections.

Organization	Roles
International and multinational	
United Nations Food and Agriculture Organization Consultative Group on International Agricultural Research international agricultural research centers	Service to crop-specific mandates; also involve wild related species; hold collections in trust for the United Nations Food and Agriculture Organization.
International Plant Genetic Resources Institute	Focus is on facilitation of national collections and policy for collections.
Nordic Gene Bank	Usually related to crops and wild relatives.
National (public)	
National collections, such as U.S. National Plant Germplasm System, U.S. Department of Agriculture Plant Materials Service, Indian National Bureau of Plant Genetic Resources	Usually related to crops and wild relatives; some involve natural biodiversity, usually as part of federal ministries or departments, often as research institutes and related to natural plant biodiversity for revegetation and restoration.
Consortia, such as U.S. Center for Plant Conservation	Center for Plant Conservation addresses native threatened plant biodiversity.
Domestic regional, private, or quasiprivate	
Academic institutions, such as University of California	Research and teaching collections.
Private collections, such as plant and animal breeders	Cultivars, breeds, and breeding lines.
Nonprofit organizations' collections, such as SeedSavers, American Type Culture Collection, Kew Royal Botanic Gardens, American Livestock Breeds Conservancy	Wild species, crop varieties, genetic variants.

of the Consultative Group on International Agricultural Research hold more than 600,000 accessions of crop plant genetic resources (Fuccillo, 1997). These collections are held in trust for the United Nations Food and Agriculture Organization. Another international consortium is the Nordic Gene Bank, a collaborative organization of Nordic countries. Through its regional networks, it coordinates all activities concerning the conservation, use, and information of its genetic resources, which include about 30,000 seed accessions. Its mission is the conservation and provision of plant genetic resources in the service of world food production and sustainable use of wild habitats (Hulden et al., 1998). The American Type Culture Collection (ATCC) (http://www.atcc.org/) is a nonprofit corporation but with a greater scope in mission and types of collections. ATCC, established in 1925 by a consortium of scientific societies, provides a wide range of biological materials (including bacteria, viruses, fungi, protozoa, algae, plant tissue cultures, plant seeds, and various other cultures and bioproducts), technical services, and educational programs primarily for the purpose of advancing scientific knowledge.

Although there is a range of types of genetic resource collections, they represent the purposes of the collector for specific uses or research objectives. Most recent collections represent the rapidly evolving genomic sciences field, leading to specialized genome resource banks (Wildt et al., 1997). A comprehensive listing of ex situ collections of biological diversity has not been undertaken, but several examples will illustrate the scale and scope of certain collections worldwide. It is estimated that there are 6,500 natural science (i.e., preserved) collections worldwide, with more than 2.4 billion specimens. These holdings are expected to increase by 500 percent by the mid–twenty-first century (Vartorella,

1999). There are more than 2,000 entries in the global zoo directory, a networking resource maintained by the Conservation Breeding Specialist Group of the International Union for the Conservation of Nature (CBSG, 2000). This list includes aquariums, falconries, safari parks, zoos, bird gardens, butterfly farms, wildlife centers, wildlife breeding facilities, marine sanctuaries, vivaria, and other zoological collections. Botanic gardens and arboreta worldwide number at least 1,300 (McNeely et al., 1990). A world directory of collections of microorganism cultures lists more than 481 collections in fifty-one countries (Roblin, 1997). The collections vary widely in their size and diversity. One collection, the Petergof Genetic Collection of microorganisms in St. Petersburg, Russia, contains some 1,000 genetically marked strains of yeasts (*Saccharomyces cerevisiae* and *Pichia methanolica*) and about 600 strains of algae (*Chlamydomonas reinhardtii, Chlorella vulgaris,* and *Scenedesmus obliquus*) (Samsonova et al., 1994).

Genetic resource collections have varied uses. As populations disappear in situ, with their attendant genetic diversity, ex situ collections are increasingly important as a means of holding some of that diversity in perpetuity or until it can be used effectively for restoration or reintroduction (Nicholson, 1994). This may be particularly important in some developing countries with rapidly expanding populations that may also be accompanied by large and continuing losses of natural forest. In Colombia, for example, a recent estimate of forest loss is 660,000 to 880,000 hectares per year, primarily to agriculture, grazing, and mining. The implied associated loss of species—combined with poverty, an unstable government structure, and the chaos caused by crime and guerrilla activities that make in situ conservation challenging—suggests that ex situ conservation is an essential strategy to supplement in situ conservation to help prevent further extinctions of species (Atehortua, 1992).

Restoration

There is general consensus that biological collections should be carefully planned for their potential genetic contributions to restoration (Hutchins, 1995). However, genetic resource collections cannot and should not replace in situ conservation. Natural evolution ceases in ex situ collections, and in fact, genetic changes that may occur in the collections do not represent changes that would have occurred in nature. In addition, the ex situ sample may be an inadequate representation of the genetic diversity that occurs in nature. Insufficient understanding of how to sample populations appropriately to capture the range of genetic variation, from neutral alleles to quantitative variation, means that many, perhaps most, collections would not yield an effective restoration population with long-term adaptive potential (Hamilton, 1994). The existence of ex situ collections should not be construed as license to disrupt or destroy natural populations of the same taxa; however, ex situ collections may end up being the last resort for conserving genetic diversity when natural diversity is destroyed. For restoration purposes, genetic resource collections should be assembled according to geographic and genetic criteria. In forestry, the importance of local adaptation has long been recognized. For many commercial forest tree species, seed is collected on a population or provenance basis and planted back on its original site after harvesting or catastrophic events such as fire. However, for most species representative genetic diversity collections are not maintained, and so a collection may not be undertaken until after much of the naturally occurring genetic diversity has already been lost.

Research

A recently published review of crop genetic research revealed a significant use of conserved genetic material in such research (Dudnik et al., 2001). Research was four times more likely to involve plant material from genetic resource collections than from natural field collections. The objectives of the research included assessments of genetic diversity among accessions, inheritance of particular traits of interest, taxonomic and phylogenetic analyses, and cytological research. Genetic resource collections also provide for unanticipated research goals. Such was the case with an accession of the thermophilic eubacterium *Thermus aquaticus*, which was collected from a hot spring in Yellowstone National Park and deposited in the American Type Culture Collection as accession number 25105 (Brock and Freeze, 1969). The availability of accessions of this organism in the ATCC repository facilitated K. B. Mullis's search for a DNA polymerase that was functional at high temperatures, providing greater control of the polymerase chain reaction (PCR), used for synthesizing defined DNA sequences (Mullis et al., 1986). His discovery of the *Taq* DNA polymerase and subsequent refinement of PCR have revolutionized molecular biology and genomic research.

Conservation

The use of biological collections for support and even refugia for genetic diversity in natural populations is a recent phenomenon, at least for some types of collections. For example, botanic gardens have historical roots in other pursuits, including pleasure gardens for royalty and elites, the contemplative and herbal gardens of the various religious faiths, and the introduction and preservation of food and medicinal plant stocks (Nicholson, 1994; Smith, 1985). Genetic resource collections have yielded accessions for agricultural use as new crop species and sources of desirable genes for breeding. A case in point is the U.S. National Plant Germplasm System, managed by the U.S. Department of Agriculture. Its historic objectives focused on introducing and maintaining plant species and varieties with agricultural or other commercial value. More recently, the mission has expanded to include preservation of wild species, particularly those that may be at risk. Genetic resource collections can contribute more generally to conservation by providing opportunities for public education, scientific research, the development of relevant technologies and products, professional training and technology transfer, ecotourism, and fundraising to support in situ conservation (Hutchins, 1995; Hutchins et al., 2000).

New Uses, New Technology

The range of genetic resource collections has been profoundly extended since the mid-1980s by new techniques and technologies that allow exploration of previously inaccessible DNA. For example, Higuchi et al. (1984) provided the first demonstration that clonable DNA sequence information can be recovered from the remains of an extinct horse species. In 1985, the advent of PCR techniques (Saiki et al., 1985) allowed amplification of extremely small DNA samples. This technique and others that improve DNA extraction and sequencing continue to increase the value of preserved collections as sources of genetic information, which are all applicable to the professional interests of taxonomists, systematists, geneticists, and conservation biologists. In many museums, preservation methods that do not result in DNA degradation are being substituted for older methods. Although preserved collections are not ordinarily amenable to genetic analysis, and museum collections vary in the degree to which intraspecific genetic diversity is represented, there is much scientific value in museum collections as genetic resources. For example, museum collections of extinct species or populations can be useful for phylogenetic inferences at the DNA level. In other cases, the use of DNA from museum specimens of extant species can add a temporal aspect to genetic studies (Thomas et al., 1990). Similarly, genetic resource collections can provide a baseline or reference group for genetic monitoring, such as assessing genetic impacts from pollution or some other catastrophic event.

New techniques also expand the usable range of genetic resource collections to natural collections or fossil archives such as those entrapped in ocean floors, glacial ice, lake sediments, peat bogs, or buried trunks. For example, using PCR technology, samples of ancient fungi trapped in glacial ice have recently been studied, facilitating fungal phylogenetic and other studies (Ma et al., 2000). A large-scale paleogenetic project in western Europe (FOSSILVA) mobilized the collective expertise of paleobotanists and geneticists to analyze ancient DNA (aDNA) as a tool in better understanding the relationship of genetic response to climatic disturbance. This emerging field of paleogenetics has great potential in the prediction of biospheric responses to future environmental changes and stimulates links between traditional biological collections (e.g., herbaria samples) and sources of aDNA (Dumolin-Lapègue et al., 1999).

Disaster Mitigation

Just as location-appropriate genetic resources are needed for restoration of locally adapted populations, so are local landraces and traditional varieties critical to agriculture, particularly in less developed countries, where there may not be access to sufficient expertise or capacity, more reliance on good genetic–environment interactions, and strong cultural preferences for certain food varieties. Genetic resource collections can serve as valuable reservoirs during politically unstable or environmentally disastrous periods. For example, traditional rice varieties of Cambodia collected and preserved in the rice gene bank of the International Rice Research Institute in the early 1970s were used to replace critical seed stocks lost during dislocations in that country after 1975 (IRRI, 1982).

The Silent Biodiversity Crisis

In general, when losses of naturally occurring species, populations, or sometimes even individuals occur, the loss may involve public recognition, comment, or action. Indeed, the term *biodiversity crisis* is commonly used to refer to the overall phenomenon of pending and realized extinction of taxa (e.g., Eldredge, 1992). However, similar losses of ex situ reserves may go largely unnoticed. Even curators in these situations may not be aware of the losses for some time if they are living collections, such as seed, that have lost viability. Mirroring Ledig's (1993) use of the term *secret extinctions* to refer to the loss of populations in nature, we use the phrase *silent biodiversity crisis* to describe the quiet loss of ex situ genetic resource collections. In particular, if these collections contain genetic diversity that is no longer present in situ, the loss of the collections means irreplaceable loss of specific alleles.

Partial or complete losses of genetic resource collections can occur through a large number of mechanisms. Although occasionally there is a discrete and obvious cause for the loss—such as a fire or natural catastrophic event—more often a series of related events that lead to the loss, such that the proximate cause is not the only problem. For example, insufficient funding usually results in insufficient staffing and facilities to adequately maintain collections. The process of developing and maintaining genetic resource collections involves at least six steps: acquisition or collection, documentation, characterization, evaluation, distribution, and maintenance. If there is a financial crisis for a collection, the fallback priority will probably be to concentrate on maintenance, simply hanging on to what was there. If this continues long enough and involves loss of staff, there will be an accompanying loss of institutional memory of what the full services and roles of the collection had been or could be. The result is ultimate loss of accessions through attrition. For example, many plant genetic resource collections—even in the United States—have been described as seed morgues (Goodman, 1990).

Here we illustrate the silent biodiversity crisis by describing a number of predominant problems.

Insufficient, Reduced, or Fluctuating Funding

Chronically insufficient or reduced funding is a leading culprit in the silent biodiversity crisis. Insufficient budgets directly lead to insufficient infrastructures to maintain collections properly and to overtaxed or undertrained staff. Because ex situ genetic resource collections often have long-term objectives, even a one-year budget deficit can be disastrous. For example, a recent study of the status of avian genetic resources in the United States and Canada determined that some 238 avian genetic stocks have been reported as lost or eliminated in the last fifteen years, and the actual number of lost stocks is probably much higher (Pisenti et al., 1999). Most notable among these lost genetic resources was the discontinuation of more than thirty genetic stocks—inbred, specialty, and historically commercial stocks—once maintained by the Canadian government (Agriculture and Agrifood Canada) that were dispersed or eliminated in 1997 because of lack of funding. One-half of these stocks had no duplicates elsewhere.

Insufficient funding is often the basis for more proximate causes of loss of genetic resource collections. Lack of proper storage resulted in the loss, by 1974, of more than half of a valuable collection of sorghum accessions that had been assembled in the 1960s by the Rockefeller Foundation in the Indian Agricultural Research Institute (Prasada Rao and Mengesha, 1988). Similarly, the maize collections in Latin America were described as being in a critical state by the early 1990s, with some unique accessions lost and a large number of others endangered, largely the result of two interacting factors: insufficient funding and lack of reliable storage facilities (Listman, 1992; Salhuana et al., 1991).

The N. I. Vavilov Research Institute of Plant Industry (VIR) in St. Petersburg, Russia, site of the world's oldest and second-largest plant gene bank, is facing political and financial pressures that could have devastating effects for this collection of more than 300,000 plant genetic resource accessions. The Russian government would like to acquire the VIR facilities in St. Petersburg for other uses, but relocation of VIR to new facilities would cost more than $70 million (Webster, 2003). In any case, modernization of facilities is badly needed (VIR ICG, 1998). Even if budgets are stable, increasing responsibilities will add financial pressure. For example, natural history collection holdings worldwide have experienced 300 percent growth in accessions over the last seventy years (Vartorella, 1999), but budgets have not increased proportionally.

Undervaluation of Collections

Genetic resource collections are particularly vulnerable to losses in funding because of their concentration in the public sector and their silent or underrepresented value in the marketplace. Thus, cutbacks or downsizing in public agencies (e.g., U.S. Department of Agriculture) or institutions (e.g., universities) that are disproportionately responsible for or involved with genetic resource collections will have much more severe impacts on collections than if their maintenance were represented by greater stakeholder diversity. Typically, and in part because of this historical public financing or subsidization of genetic resource collections, the true value of these collections has not been calculated and is not internalized in research, restoration, or educational activities. Therefore, when budgets are cut there is little recognition of the value of losses when the collections are put at risk and little motivation or precedent to find alternative funding mechanisms. This failure to internalize the cost of taxonomic services increases the drain on already scarce taxonomic resources (Hoagland, 1996).

Accidents and Natural Disasters

The most obvious losses from genetic resource collections result from human error or accidents, such as electrical outages, equipment failures, or other infrastructure weaknesses. Seeds, cells, and tissues often have critical storage temperature needs, particularly when in cryopreservation. Ex situ genetic reserves are not necessarily any safer from natural disasters than are in situ reserves. For example, in October 1999 a cyclone in eastern India caused extensive damage to the Central Rice Research Institute in Cuttack and its germplasm stocks and destroyed much of the collection of rare and exotic plants at the Regional Plant Resource Center in Bhubaneshwar (Bagla, 1999).

War

Like cultural heritage, genetic resource collections are negatively affected by war, civil strife, and low-intensity conflict. A chronicle of the effects of war on crop genetic resources by Richards and Ruivenkamp (1997) provides vivid case studies from the upper West African coastal zone from Senegal to Liberia. Losses of genetic resource collections are part of the collateral damage of war, although the term is applied more routinely to the loss of civilian life, continuing impacts on human health in the war region, and environmental damage (e.g., Salvage, 2002). During the recent war in the former Yugoslavia, genetic resource collections were lost through both direct destruction and abandonment (Anonymous, 1992). In countries where genetic resource collections are often held informally in villages rather than, or in addition to, national gene banks, serious losses of local seeds can occur when villages are abandoned during civil wars. Nothing beats the bravery and commitment of the Russian scientists at the Vavilov Institute of Plant Industry in St. Petersburg, eight of whom died of starvation in 1942 defending the country's genetic heritage while denying themselves the nourishment that resided in the seed banks just a few meters away (Alexanyan and Krivchenko, 1991).

Abandonment

Abandoned or unfunded collections are sufficiently routine in the experience of natural history museums that they have been awarded their own designation: orphaned collections. Perhaps the greatest losses occur in U.S. universities, where faculty retirements have left collections abandoned. Surprisingly, most universities do not have clear policies regarding ownership or relocation of collections (Gardner et al., 1993). Rescue of these abandoned collections presents problems to the rescuing institution because the growth is usually unanticipated and involves a rescue and reconditioning effort that is more expensive in staff time and effort than any other type of acquisition (California Academy of Sciences, 2000).

Crime

Vandalism is an additional cause of losses of genetic resource collections. In the latter case, the losses often are collateral damage in the fight against genetically modified organisms (GMOs). Specifically, genetic resource collections that are growing in neat rows in a research environment may be mistaken for GMOs. Field crops at the University of California at Davis were destroyed in four separate incidents during a three-month period in 1999. Most of these were conventional genetic resource collections that had not been genetically modified through recombinant DNA technology. The attacks were attributed to activists opposed to the study of genetically modified crops (Vanderhoef, 1999). According to one report, between November 1998 and May 2001 there were more than forty attacks in North America on genetic resource collections that were genetically modified or assumed to be (Genetix Alert News, 2001).

Inadequate Distribution Policies

An insipid form of direct loss of genetic resource collections is unintentional depletion or exhaustion of reserves through overuse or inadequate policies governing their use. Some genetic resource collections may be reproducible, and some uses (e.g., some educational uses) are not consumptive. However, consumptive uses such as research or restoration activities may compromise the collection. Although this may serve the objectives for which the collection was made, it is nevertheless important to have policies in place to determine priorities where there are competing uses for the collections and to ensure that experimental designs make judicious use of valuable and limited genetic resource collections.

Insufficient Taxonomic Expertise

In addition to direct losses of ex situ collections, several accompanying trends exacerbate this silent crisis. One such trend is a worldwide shortage of taxonomists available to characterize and inventory the world's biological diversity. Recent estimates suggest that there are only about 5,000 experienced taxonomists worldwide (ETI, 2000). This trend is sufficiently ubiquitous and serious that it has been recognized by numerous international organizations and given the title "the taxonomic impediment" by the International Union of Biological Sciences/Diversitas. Lack of taxonomic expertise has other effects, such as a reduction in biodiversity-related research (Hoagland, 1996). This worldwide shortage of taxonomists is expected to worsen in the near future because of an aging taxonomic workforce, a decline in students being trained in taxonomy, and a decline in the number of paid positions that support basic taxonomic work (Hoagland, 1996). When academic taxonomy positions decline, for example, there is a corresponding decline in the number of new students trained in this discipline, in the teaching of this discipline among graduate students in other disciplines, and in the profile and support for collections under taxonomists' care.

International Impediments

Finally, social, legal, and biotechnological developments in the last decade have driven an environment of declining access to genetic resource collections. Implementation of Article 15 of the Convention on Biological Diversity (CBD), which assigns use and control over biodiversity to individual countries and their states, has spawned many related communications to the CBD, and several countries have developed legal regimes to regulate access to genetic resources (Grajal, 1999). Ideally, regulation of access to genetic resources could protect their conservation value and encourage recognition of the market value of some resources, but the reality is that many such laws result in arbitrary moratoriums on export or collection permits for national or international museums or herbaria and in a general slowdown in biodiversity access permits. This, in turn, can greatly impede the collection, inventory, research, and monitoring of biodiversity (Grajal, 1999). As a result, existing genetic resource collections may not be replenished or used in accordance with their highest conservation value. Ironically, it is biodiversity hotspots, where biodiversity is being lost at a great rate, that are suffering from the CBD-induced national policies.

Facilitating the Appropriate Use of Genetic Resource Collections

The profile, use, value, support, and security of ex situ genetic resource collections are highly interrelated. For example, when genetic resource collections are used frequently in teaching, research, and conservation applications, their profile, value, and long-term security would normally increase. Similarly, by increasing public awareness of the existence, significance, and value of genetic resource collections, we may improve support and contribute toward institutionalizing their maintenance. By documenting and publicizing the availability of these collections, we contribute to their use; by documenting the use of the collections in research, education, and other services, we nurture a conservation ethic and increase public understanding of and support for these resources and the processes necessary for their maintenance.

Communication

Directories and communication efforts that increase the awareness of the availability of such collections should facilitate their use and hence their value and security. Although there is no single authoritative source for investigating locations of genetic resources, electronic mailing lists are increasingly used as an effective, free, rapid, and international means of communicating both requests and offers regarding genetic resource collections. For example, the Evolution Directory (evoldir@evol.biology.mcmaster.ca) posts notices for researchers who want to locate or share their genetic resource collections.

Acknowledgment in Publications

Another inexpensive and effective way to communicate availability of collections is in related research papers. Some journals require that DNA sequences published in the journal be deposited in GenBank or another suitable database and the accession number for each sequence be included in the manuscript. In a review of fifteen journals that accept genetic studies, we found that two-thirds maintained this requirement. However, regardless of publication policy, researchers could enhance ex situ collections by depositing samples of their new research collections into an appropriate repository and voluntarily entering the DNA, RNA, or protein sequences that result from their research into an appropriate databank and providing the accession information in their publication. As an informal check on this kind of researcher initiative, we examined thirty research papers from) the year 2000 in four journals that carry research findings on taxonomic, population genetic, and genetic conservation studies. These studies spanned a broad range of taxa and used samples from natural populations (twenty-four), ex situ collections (four), or both (two). We found that in most cases (twenty-two of twenty-six), new samples from wild populations were not deposited in repositories, or at least the authors did not report doing so. In studies where DNA was sequenced (nine), sequence registration information was provided in all but one case, showing researcher initiative in the journals without this requirement in their publication policy.

Policy

Critical to the reversal of the silent biodiversity crisis are the development and implementation of policies to manage collections effectively and to guide their wise use. Appropriate policies at the level of the agency, organization, or institution directly responsible for specific genetic resource collections can do much to reduce their vulnerability. Specifically, policies that address monitoring requirements to track viability of living materials, outline appropriate technical guidelines for maintaining collections, allow incorporation of improved management techniques and technologies as they are realized, and designate staff responsible for successful management of collections can overcome many common pitfalls. If appropriately implemented, policies can outlive the watch of any particular person or administration and provide the continuity needed for long-term ex situ conservation. And finally, it is essential that policies address the need for a responsible means of propagating, distributing, or otherwise using the genetic resource collections. Policies not only must take into account the biology of the species involved and whether various uses are consumptive (e.g., direct use of seeds) but also must set out a framework for priorities in the case of limited resources. Policies can appropriately divert collections to critical conservation efforts, avoid wasteful use such as research with inappropriate or experimental protocols, and remove the pressure from the individual to make such decisions about competing uses.

Opportunities to Mitigate the Silent Biodiversity Crisis

Value Should Be Recognized and Budgeted

Scientists can provide direct support to the security and availability of genetic resource collections by recognizing the contributions to their research made by ex situ collections. This recognition is best expressed in both of the common scientific currencies. First, we need to document this value and use it effectively to maintain or increase support for collections. For example, in research papers where genetic resource collections (ex situ or in situ) have been used, it is both appropriate and useful to provide recognition for the genetic resource collections and the host agency or organization in the acknowledgment section and express appreciation for their availability and access. Second, the value—or *some* value—for the use of genetic resource collections should be included in the budget in research proposals. Even if there is no expected or direct cost for the use of collections, some benevolent funding or in-kind services should be provided to recognize the value of these resources to the research activity and to make more apparent the true value of these silent or subsidized resources.

Biodiversity assessments should include genetic resource collections. These assessments have been, and continue to be, undertaken at international, national, regional, and local scales by various parties including governments and nongovernment agencies. It is important to include ex situ conservation reserves in these biodiversity reviews or assessments, both to assess their status and to form appropriate functional links between ex situ and in situ conservation, research, and educational activities. In particular, the inclusion of ex situ reserves in biodiversity assessments should support conservation efforts by revealing possible sources of genetic resources for mitigation or restoration efforts. It should also be recognized that ex situ collections may lie well outside the geographic scope of a particular species' range or biodiversity assessment project, and these reserves should be included, too.

Better Characterization of Collections Can Increase Efficiency of Maintenance and Use

Appropriate policy reviews, economic analyses, and scientific research can contribute to more economically and efficiently maintained collections. Genetic characterization of genetic resource collections can help to identify redundant accessions, allowing their removal and more efficient genetic resource management. This has been done effectively, for example, in a collection of flax (*Linum usitatissimum*) of the Centre for Genetic Resources in the Netherlands (van Treuren et al., 2001) and for a *Brassica oleracea* germplasm collection (van Hintum et al., 1996). Some grantmaking foundations and agencies—such as the U.S. National Science Foundation (NSF) and the National Institutes of Health—provide grant opportunities for genetic resource collections. For example, the NSF offers a "Support of Living Stock Collections" grants program that supports repositories of research organisms, genetic stocks, and seeds, as well as cell lines and DNA clones that are associated with the whole organisms in that collection, resources that are considered essential for national or international scientific research in the biological sciences (http://www.nsf.gov/). However, this program is not seen by NSF as a solution to long-term funding for collections, and a criterion for awards is that the taxon has a large user base or is recognized as a model research organism. Expanding this program to include a broader array of organisms, or developing a new grant program to provide support for collecting and maintaining genetic resources that are important for conservation purposes, would be a valuable service.

Funding Organizations' Focus on Collections

Grantmakers can also support genetic resource collections by revising their application guidelines to include specific policies for obtaining and maintaining genetic material. Specifically, budget items could be requested that show costs of collecting materials de novo or obtaining materials from existing collections and costs related to long-term maintenance of new genetic resource collections. Similar to the publication policy of some journals as regards the registration of DNA sequence information, grantmakers could require compensation for use of existing collections and donation of new collections to an appropriate (if feasible and desirable) repository along with some curatorial funding.

Endowments

At the national level, a national endowment for genetic resources that recognizes their inherent value and societal responsibility for their conservation has been proposed (Qualset, 1990). A similar recommendation for the support of basic and applied research in conservation biology was proposed in 1978 by a group of international biologists at a conference in San Diego, California (Douglas, 1978). Like all biodiversity, genetic diversity is part of any nation's heritage alongside its art and culture (Maxted, 1997). Endowment funds can be created to support specific collections. This tactic was used to support tomato genetic stocks and wild relatives at the University of California at Davis. First, an assessment of value was completed and recommendations were made for sustaining this collection (Tomato Genetics Stock Center Task Force, 1988). Now an endowment fund provides substantial annual support to the Charles M. Rick Tomato Genetics Resource Center. Similar assessments and proposals have been made for avian genetic stocks (Pisenti et al., 1999) and citrus genetic stocks (Kahn et al., 2001). If donation collections are to be accepted that will not be sustained by user fees, the donor is expected to endow the collection for perpetual care.

Enhanced Communications Between Users and Holders

Worldwide, many organizations support the collection, maintenance, documentation, research, and use of genetic resource collections (see table 12.2). However, the piecemeal nature of the geographic range, taxonomic coverage, and scope of mission of these organizations leaves a broad array of species not represented by genetic resource collections and collections without adequate support. Institutional homes for collections are essential because the time frame and funding requirements for collections exceed that which can be provided by individuals. Ideally, an accessible electronic clearinghouse would connect researchers with appropriate institutional homes for their samples, including molecular gene banks where preserved biological material could be maintained for future DNA recovery. Some resources provide services complementary to this goal, including the Biodiversity and Biological Collections Web Server, which maintains a collection of databases (http://biodiversity.uno.edu/), and the All Species Foundation (http://www.all-species.org/), which aims to build a searchable inventory of all species through foundation-based and collaborative activities. The proposed electronic clearinghouse for genetic resource collections should be developed in coordination with such existing resources to minimize redundancy, ensure maximum impact, and advertise its availability.

Networking and Collaboration

Finally, we encourage the organization of networks of conservation geneticists, taxonomists, systematists, and others as a means of locating specific expertise and specific collections. In addition to the connectivity and forums provided by professional societies, focused efforts to make rapid and international connections for the support and use of genetic resource collections are recommended. The Expert Center for Taxonomic Identification (ETI, 2000), for example, is a not-for-profit organization dedicated to improving on a global scale the quantity, quality, and accessibility of taxonomic information. The "Taxasphere" concept proposed by Hoagland (1996) is more ambitious in scope and proposed a clearinghouse—making use of such resources as provided by ETI—for requests for taxonomic services, for funding opportunities, and for generally harnessing expertise and other resources toward research and services in support of biodiversity inventory and management worldwide. Positive steps can be taken at every level of activity—from that of the individual professional to institutional, professional society, and international organizations—to add value, recognize significance, and provide security for ex situ genetic resource collections, giving voice to the silent biodiversity crisis and averting further loss.

Summary

Genetic diversity is central to all other levels of biological diversity. However, losses of genetic diversity are much less apparent and well publicized than losses of diversity at the species level. In particular, many ex situ genetic collections—although very important to species conservation, research, education, and commercial pursuits—are undervalued, undersupported, and under threat. Losses of these genetic collections happen quickly and quietly, leading to a silent biodiversity crisis. Genetic collections are being lost worldwide from myriad causes, including insufficient funding and infrastructure support, a deficit of trained taxonomists and curators, natural disasters, accidents, wars and civil strife, and vandalism. These proximate causes of loss are generally magnified by a historical and widespread undervaluing, often related to the traditional public-subsidized systems for genetic collections, such that their value has not been fully recognized or incorporated into research, commercial, and conservation funding. Some examples of efforts to facilitate appropriate use of genetic resource collections include improving communication between users and curators and developing improved policy on the consumptive use of collections. Recommendations for mitigation of the silent biodiversity crisis include developing appropriate policies to secure the collections in perpetuity, normalizing the costs associated with acquiring and maintaining genetic collections, expanding biological assessments to include ex situ genetic resources, initiating and expanding grant programs to support genetic collections, and increasing the networking between taxonomists, geneticists, and others with interest in genetic resources.

References

Alexanyan, S. M. and V. I. Krivchenko. 1991. Vavilov Institute scientists heroically preserve world plant genetic resources collections during World War II siege of Leningrad. *Diversity* 7(4):10–13.

Anonymous. 1992. Genetic resources: Another victim of war between former Yugoslav states. *Diversity* 8(3):12.

Atehortua, L. 1992. Floristic diversity, botanical exploration and the establishment of a germplasm bank for conservation in the Neotropic: The Colombian view. In *Conservation of plant genes: DNA banking and in vitro biotechnology*, ed. R. P. Adams and J. E. Adams, 37–44. New York: Academic Press.

Bagla, P. 1999. Cyclone wrecks rice, botanical centers. *Science* 286:454.

Brock, T. D. and H. Freeze. 1969. *Thermus aquaticus* gen. n. and sp., a nonsporulating extreme thermophile. *Journal of Bacteriology* 98:289–297.

California Academy of Sciences. 2000. http://www.calacademy.org/.

CBSG (Conservation Breeding Specialist Group). 2000. Global zoo directory. http://www.cbsg.org.

Ceballos, G. and P. R. Ehrlich. 2001. Population extinction: A critical issue. In *The red book: The extinction crisis face to face*, ed. P. Robles Gil, R. Pérez Gil, and A. Bolívar, 86, 89. Mexico City, Mexico: CEMEX.

Daily, G. C. and P. R. Ehrlich. 1995. Population extinction and the biodiversity crisis. In *Biodiversity conservation*, ed. C. A. Perrins, 45–55. Dordrecht, The Netherlands: Kluwer.

Douglas, J. 1978. Biologists urge US endowment for conservation. *Nature* 275:82–83.

Dudnik, N. S., I. Thormann, and T. Hodgkin. 2001. The extent of use of plant genetic resources in research: A literature survey. *Crop Science* 41:6–10.

Dumolin-Lapègue, S., M.-H. Pemonge, L. Gielly, P. Taberlet, and R. Petit. 1999. Amplification of oak DNA from ancient and modern wood. *Molecular Ecology* 8:2137–2140.

Ehrlich, P. R. 1988. The loss of diversity: Causes and consequences. In *Biodiversity*, ed. E. O. Wilson, 21–27. Washington, D.C.: National Academy Press.

Eldredge, N. 1992. Where the twain meet: Causal intersections between the genealogical and ecological realms. In *Systematics, ecology, and the biodiversity crisis*, ed. N. Eldredge, 1–14. New York: Columbia University Press.

ETI. 2000. Expert Center for Taxonomic Identification. http://www.eti.uva.nl/.

Fuccillo, D., ed. 1997. *Biodiversity in trust: Conservation and use of plant genetic resources in CGIAR centers*. Cambridge: Cambridge University Press.

Gardner, S. L., P. E. McGuire, and L. S. Kimsey, eds. 1993. *Management of university biological collections: A framework for policy and practice.* Report no. 11. Davis: University of California Genetic Resources Conservation Program.

Genetix Alert News. 2001. Activists destroy GE crops at research facility in Brentwood, CA. *Genetix Alert News Release,* May 17.

Goodman, M. M. 1990. What genetic and germplasm stocks are worth conserving? In *Genetic resources at risk: Scientific issues, technologies, and funding policies*, ed. P. E. McGuire and C. O. Qualset, 1–9. Report no. 5. Davis: University of California Genetic Resources Conservation Program.

Grajal, A. 1999. Biodiversity and the nation state: Regulating access to genetic resources limits biodiversity research in developing countries. *Conservation Biology* 13:6–10.

Hamilton, M. B. 1994. Ex situ conservation of wild plant species: Time to reassess the genetic assumptions and implications of seed banks. *Conservation Biology* 8:39–49.

Higuchi, R., B. Bowman, M. Freiberger, O. A. Ryder, and A. C. Wilson. 1984. DNA sequences from the quagga, an extinct member of the horse family. *Nature* 312:282–284.

Hoagland, K. E. 1996. The taxonomic impediment and the Convention on Biodiversity. *Association of Systematics Collections Newsletter* 24:61–62, 66–67.

Hobbs, R. J. and H. A. Mooney. 1998. Broadening the extinction debate: Population deletions and additions in California and Western Australia. *Conservation Biology* 12:271–283.

Hulden, M., B. Lund, G. B. Poulsen, E. Thorn, and J. Weibull. 1998. The Nordic commitment: Regional and international collaboration on plant genetic resources. *Plant Varieties and Seeds* 11:1–13.

Hutchins, M. 1995. Strategic collection planning: Theory and practice. *Zoo Biology* 14:2–22.

Hutchins, M. et al. 2000. Priority-setting for ex situ conservation. *Conservation Biology* 11:593.

IRRI. 1982. IRRI germplasm bank returns traditional rice seeds to Kampuchea. *IRRI Reporter* 3:4.

Kahn, T. L., R. R. Krueger, D. J. Gumpf, M. L. Roose, M. L. Arpaia, T. A. Batkin, et al. 2001. *Citrus genetic resources in California: Analysis and recommendations for long-term conservation*. Report no. 22. Davis: University of California Division of Agriculture and Natural Resources, Genetic Resources Conservation Program.

Ledig, F. T. 1993. Secret extinctions: The loss of genetic diversity in forest ecosystems. In *Our living legacy: Proceedings of a symposium on biological diversity*, ed. M. A. Fenger, E. H. Miller, J. A. Hohnson, and E. J. R. Williams, 127–140. Victoria, BC: Royal British Columbia Museum.

Listman, G. M. 1992. Regenerating endangered Latin American maize germplasm: The USAID/CIMMYT cooperative project. *Diversity* 8(2):14.

Ma, L. J., S. O. Rogers, C. Catranis, and W. T. Starmer. 2000. Detection and characterization of ancient fungi entrapped in glacial ice. *Mycologia* 92:286–295.

Maxted, N. 1997. Complementary conservation strategies. In *Plant genetic conservation*, ed. N. Maxted, B. V. Ford-Lloyd, and J. G. Hawkes, 15–39. London: Chapman & Hall.

McNeely, J. A., K. R. Miller, W. V. Reid, R. A. Mittermeier, and T. B. Werner. 1990. *Conserving the world's biological diversity*. Gland, Switzerland: IUCN.

Mullis, K. B., F. A. Faloona, S. Scharf, R. K. Saiki, G. Horn, and H. A. Erlich. 1986. Specific enzymatic amplification of DNA in vitro: The polymerase chain reaction. *Cold Spring Harbor Symposium on Quantitative Biology* 51:263–274.

Nei, M. 1973. Analysis of gene diversity in subdivided populations. *Proceedings of the National Academy of Sciences USA* 70:3321–3323.

Nicholson, R. 1994. Hedging a bet against extinction: The use of shears in ex situ conservation. *Biodiversity and Conservation* 3:628–631.

Pisenti, J. M., M. E. Delany, R. L. Taylor Jr., U. K. Abbott, H. Abplanalp, J. A. Arthur, et al. 1999. *Avian genetic resources at risk: An assessment and proposal for conservation of genetic stocks in the USA and Canada*. Report no. 20. Davis: University of California Genetic Resources Conservation Program.

Prasada Rao, K. E. and M. H. Mengesha. 1988. Sorghum genetic resources: Synthesis of available diversity and its utilization. In *Plant genetic resources: Indian perspective*, ed. R. S. Paroda, R. K. Arora, and K. P. S. Chandel, 159–169. New Delhi, India: National Bureau of Plant Genetic Resources.

Qualset, C. O. 1990. Conservation of genetic resources: A proposal for sharing the responsibility. In *Genetic resources at risk: Scientific issues, technologies, and funding policies*, ed. P. E. McGuire and C. O. Qualset, 35–42. Report no. 5. Davis: University of California Genetic Resources Conservation Program.

Richards, P. and G. Ruivenkamp. 1997. *Seeds and survival: Crop genetic resources in war and reconstruction in Africa*. Rome: International Plant Genetic Resources Institute.

Roblin, R. O. 1997. Resources for biodiversity in living collections and the challenges of assessing microbial biodiversity. In *Biodiversity II: Understanding and protecting our biological resources,* ed. M. L. Reaka-Kudla, D. E. Wilson, and E. O. Wilson, 467–474. Washington, D.C.: Joseph Henry Press.

Saiki, R. K., S. Scharf, F. Faloona, K. B. Mullis, G. T. Horn, and H. A. Erlich. 1985. Enzymatic amplification of β-globin genomic sequences and restriction site analysis for diagnosis of sickle cell anemia. *Science* 230:1350–1354.

Salhuana, W., Q. Jones, and R. Sevilla. 1991. The Latin American Maize Project: Model for rescue and use of irreplaceable germplasm. *Diversity* 7(1, 2):40–42.

Salvage, J. 2002. *Collateral damage: The health and environmental costs of war on Iraq*. London: Medact.

Samsonova, M. G., V. M. Andrianova, T. N. Borshchevskaia, and A. S. Chunaev. 1994. The Petergof Genetic Collections of Microorganisms. *Genetika* 30:1129.

Smith, N. J. H. 1985. *Botanic gardens and germplasm conservation*. Harold L. Lyon Arboretum Lecture no. 14. Honolulu: University of Hawaii Press.

Soulé, M. 1991. Conservation: Tactics for a constant crisis. *Science* 253:744–750.

Thomas, W. K., S. Pääbo, F. X. Villablanca, and A. C. Wilson. 1990. Spatial and temporal continuity of kangaroo rat populations shown by sequencing mitochondrial DNA from museum specimens. *Journal of Molecular Evolution* 31:101–112.

Tomato Genetics Stock Center Task Force. 1988. *Evaluation of the University of California Tomato Genetics Stock Center: Recommendations for its long-term management, funding, and facilities*. Report no. 2. Davis: University of California Genetic Resources Conservation Program.

Vanderhoef, L. N. 1999. *An open letter to the campus community*. Research Resources no. 246. Davis: University of California.

van Hintum, T. J. L. , I. W. Boukema, and D. L. Visser. 1996. Reduction of duplication in a *Brassica oleracea* germplasm collection. *Genetic Research and Crop Evolution* 43: 343–349.

van Treuren, R. et al. 2001. Marker-assisted rationalisation of genetic resource collections: A case study in flax using AFLPs. *Theoretical and Applied Genetics* 103:144–152.

Vartorella, W. F. 1999. Evolution, predator traps, and money pits: Re-thinking collection. *Society for the Preservation of Natural History Collections Newsletter* 13(2):1, 9, 13.

VIR ICG. 1998. *The N. I. Vavilov Research Institute of Plant Industry: A review and recommendations for restructuring.* Report by VIR International Consultative Group. Alnarp, Sweden: Nordic Gene Bank.

Webster, P. 2003. Prestigious plant institute in jeopardy. *Science* 299:641.

Wildt, D. E., W. F. Rall, J. K. Critser, S. L. Monfort, and U. S. Seal. 1997. Genome resource banks: Living collections for biodiversity conservation. *BioScience* 47:689–698.

Cathi Lehn, Rebecah Bryning,
Rob DeSalle, and Richard Cahoon

13

Who Owns the Ark, and Why Does It Matter?

Animal diversity from around the globe can be found in institutional collections accredited by the Association of Zoos and Aquariums (AZA). As stewards of this diversity we hold a responsibility to care for and exhibit these animals to the best of our knowledge and ability. Therefore, while they are in our care we also have a responsibility to learn as much as we possibly can from our animals so that they may be best cared for in captivity and conserved in their natural habitats. We learn about our animals from careful scientific study, which may include behavioral observations but also may include scientific study in other disciplines, including genetics, anatomy, physiology, veterinary medicine, and nutrition. Many of the scientists in these disciplines need biological samples in order to complete their studies. Biological samples, or biomaterials, may be defined as any organic piece or derivative of an animal (e.g., tissues, skeletal material, hair, feces). Biological samples are used in AZA collections every day by highly trained veterinary staff and are also provided to university scientists for research purposes. Properly collected and stored samples may be available for use by scientists for many decades into the future.

The responsibilities associated with the curation of an animal in our care also extend to the curation of the biological materials obtained from these animals. The curation of biological samples begins after their collection and includes the samples' care and management during storage and administration of their use and distribution. Protocols for the collection and storage of biological samples are developed by our conservation program research and veterinary advisors and by scientists interested in using the samples to conduct their studies. The use and distribution of biological samples may be recommended by conservation programs but ultimately are the responsibility of each AZA institution. Many resources are available to AZA members for the management of biological samples. This chapter reviews two of these resources: the Biomaterials Banking Advisory Group (BBAG) and the Zoological Information Management System (ZIMS). The management of samples may also extend to our partners (e.g.

researchers and repositories). The common elements of a material transfer agreement (MTA) and a perspective from the university environment are also discussed.

BBAG

Scientific advisory groups consist of experts in a particular field who help support, network, and coordinate the research activities of AZA members (http://www.aza.org). In 2002, the BBAG was formed. The mission of BBAG is to assist and advise taxonomy advisory groups, species survival programs (SSPs), and AZA institutions in the development of protocols and policies as they relate to the collection, storage, and distribution of biomaterials at AZA zoos and aquariums. Many AZA conservation programs have outlined a set of research goals and priorities for their associated taxa or programs. Many of these research priorities rely on the availability of biological samples for their completion. A new initiative for BBAG is to work with selected conservation programs in the development of Biomaterials Action Plans. The Biomaterials Action Plan is designed to take a program's research goals and priorities to the next level by specifically outlining the protocols and policies needed to complete them and identifying resources and partners. The steps needed to complete a Biomaterials Action Plan have been outlined in Wildt et al. (1997) and Wildt (1992) and include the following:

- Summary of conservation program
- Justification for Biomaterials Action Plan
- Current status of North American captive population
- Current status of biological resource management of conservation program
- Current advisors and affiliations
- Current state of knowledge
- Ex situ programs in range countries
- Technical guidelines for the collection, storage, use, and ownership of biological resources
- Potential resources and funding

BBAG is working with the Gorilla SSP and the Cheetah SSP to develop their action plans. One of the sections included in the plan is to introduce programs to the functions provided by ZIMS for their programs and for the management of their biological samples.

ZIMS

Since 1974, the International Species Information System (ISIS) has served animal management and conservation goals by supporting the most comprehensive and reliable source of information on animals and their environments. ISIS offers its members the tools to work together across the world by providing the following:

- Access to key global community information
- Access to knowledge about each others' inventories and plans
- World-standard recordkeeping system
- Global base for developing cooperation

Today there are more than 675 members in 73 countries using the current suite of ISIS software, and although the current software products have served our community well for the past eighteen years, the future offers new opportunities. One of the opportunities being facilitated by ISIS is the ZIMS Project. The mission of the ZIMS Project is to develop, install, and maintain a comprehensive information system for activities associated with the management and care of animals in zoological parks and aquariums and for the zoological community in general (http://www.isis.org).

What Is ZIMS?

- ZIMS is a new software package replacing the current suite of ISIS software (e.g., Animal Records Keeping System [ARKS], Medical Animal Records Keeping System [MedARKS], Single Population Animal Records Keeping System [SPARKS], Eggs).
- ZIMS provides its users with shared data on almost 2 million animals in seventy-three countries.
- ZIMS provides improvements for every facet of animal care activities.
- ZIMS translates into increased productivity by streamlining workflow, facilitating communication and access to information.
- ZIMS improves decision making by providing access to accurate, up-to-date information (e.g., best practices).
- ZIMS provides opportunities for collaboration by allowing unprecedented information exchange.
- ZIMS is a biosurveillance tool used to alert health officials as zoonotic disease outbreaks are detected around the world.

Building any new software system is a complex process. The ZIMS Project is no exception and is made additionally complex by the diversity of zoological systems and institutional stakeholders around the world. ZIMS is one of the most ambitious online projects ever attempted; this project will span the entire globe and could not be completed without the help of an extensive volunteer network. More than 500 subject matter experts have dedicated thousands of hours developing more than 250 data standards, reviewing design documents, creating structure for the project, monitoring electronic mailing lists, and attending workshops around the globe. This effort has resulted in the design of more than 500 data entry screens, customized to each area of recordkeeping. One group of subject matter experts formed for the development of ZIMS is the Sample Focus Group. This group consists of veterinarians, veterinary technicians, pathologists, researchers, and system analysts from around the globe working together to develop the functions necessary for the curation of samples. A preview of some of the screens developed by this group follows.

ZIMS Preview

An overview of a basic ZIMS screen is presented in Lehn et al. (2007). The left-hand side of the screen provides a dynamic menu system. Shown in the example provided in figure 1 in Lehn et al. is the Animal Menu Option, presenting the user with a list of husbandry and health menu options. The Banner is located at the top of the screen and provides specific animal information (e.g. global accession number, the scientific and common names, and the preferred ID). Breadcrumbs, a navigational tool that allows a user to access previous pages displayed, is located immediately below the Banner. The remainder of the page is dedicated to displaying the Animal Details, including expandable and collapsible wire frame boxes that provide additional information (e.g., pedigree, current enclosures).

The curation of a sample includes many different activities, all of which are facilitated through ZIMS (see figure 2 in Lehn et al., 2007). The Manage Samples page provides the user with a central location to record information related to the management of a sample (e.g., receiving, shipping, splitting, analyzing, handling, adding a sample to or removing a sample from a bank, disposing of a sample, and tests performed on a sample).

Recording information on where a sample is stored first requires the configuration of the storage unit (see figure 3 in Lehn et al., 2007). ZIMS provides a tree-like structure to identify the specific storage tracking levels (e.g., shelf, tray, rack) along with profile information on the storage unit such as the unit name, type, temperature, manufacturer, date placed into service, and responsible department. Additionally, specific characteristics such as whether the unit has an alarm and when and how it was maintained can be recorded.

Once the storage unit is configured, a collected sample can be banked from the Manage Samples page (see figure 4 in Lehn et al., 2007). Information recorded on the Bank Sample page includes the specific freezer (or other storage unit) where the sample will be stored, who banked the sample, and the reason for banking the sample.

Additionally, ZIMS supports the recording of information related to the ownership and rights of a sample, such as the physical location of a sample, the legal owner, the terms of the transaction (whether the sample has been loaned or donated), and the rights to update the record; ZIMS also allows the user to attach the specific MTA (see figure 5 in Lehn et al., 2007).

MTAs

AZA institutions own many of the animals in their collection and, by extension, the biological samples of those animals; however, animals may also be owned by another AZA institution, the U.S. government, or a foreign government. Ownership is not only a legal concept; with regard to the animals and biological samples in our care, it is also an ethical concept. The rights and responsibilities associated with the ownership of an animal also extend to the management of its biological samples. The rights associated with owning biological samples include the authority to determine how the samples are used and who will have access to them. Proper curation of samples is an important responsibility associated with ownership. Curation includes the appropriate storage of samples and their associated data, the retrieval of samples and data in an efficient manner, and the employment of qualified personnel for sample acquisition, maintenance, and disposition. When a sample is transferred to a researcher or to another partner, care should be taken to ensure that all institutional responsibilities associated with the sample are met. The distribution of biological samples should include an MTA, which outlines the rights and responsibilities for all parties participating in the transfer.

The MTA specifies the conditions under which a biological sample is transferred from a zoological park or aquarium to a researcher for a specific project. The MTA outlines the rights and responsibilities of each party related to the transfer of material.

- Use restrictions:
  No commercial use
  No third party
- Disposal
- Liability
- Fees and shipping
- Permit responsibilities
- Acknowledgment
- Ownership
- Proper handling of samples
- Signatures

The MTA: Managing Biological Materials in the University Environment

The management of biological materials is important to a research university for a variety of reasons. These materials are essential to life science research, and good management is key to maintaining the integrity of collections and ensuring legal compliance. In addition, universities have related intellectual property obligations that necessitate good management of these materials.

Universities manage a wide range of biomaterials including living and dead whole animals and plants, colonies, tissues and organs, cell lines, seeds, plasmids, and DNA, and these materials are contained in many ways, including test tubes, cages, and fields.

Underlying the university's management policies are its values: the primacy of the educational and research mission, ethical behavior, research integrity, academic freedom, and the public good. The cornerstones of these policies as they relate to biomaterials include clear, defined ownership, effective possession and control, respect for others' rights, and a reliance on bailment contracts of property. Bailments are a legal mechanism of long standing in which the owner of tangible property transfers possession but not ownership to another. The MTA is a bailment contract that is the heart of the university's policy on biomaterial management.

A basic premise of the university biomaterial policy is that all biomaterials possessed by the university are owned by the university or covered by an MTA of another party. All biomaterials that are physically transferred out of the university are covered by the university's standard MTA. The MTA asserts and clarifies ownership rights, allows good recordkeeping, controls dissemination, and limits uses to those in accord with the university's mission and values.

An MTA, like any bailment, has several typical elements: It defines the parties and the material and also defines acceptable handling and use. The MTA usually limits distribution to third parties and describes the disposition of progeny, derivatives, parts, and the protocol for termination. At a university, the MTA usually allows but imposes stipulations on intellectual property developed with the materials and for-profit uses.

To be an effective tool in an educational and research environment, MTAs should have certain properties: They should be simple (one or two pages) and standardized whenever possible. MTAs should not overreach the owner's rights into those of the recipient. The MTA process should be straightforward and streamlined as much as possible, and excessive red tape and delays should be eliminated. Overarching or umbrella MTAs and Web-based point-and-click MTAs should be used when feasible. Ideally, an institution will have at least one staff person dedicated to MTA management (if only on a part-time basis).

MTAs are not without problems. For example, ownership can be complex, making the MTA a complicated process, especially when the owners include foreign governments and different types of institutions. And there is a management burden of recording and monitoring each MTA. There has been a rise in the strings attached to incoming MTAs as overzealous owners seek too much control over the use of their materials. A further complication for AZA members is the overlay of wildlife laws such as the Convention on International Trade in Endangered Species of Wild Fauna and Flora and the Lacey Act.

Despite the challenges, the assertion of ownership in biomaterials, as embodied in MTAs, is here to stay and will continue to become an even more integral part of any institution that manages animals or plants, their progeny, derivatives, and parts. Depending on how MTAs and their underlying policies are implemented, this can have a significantly positive impact on the overall mission of institutions such as AZA members and on the welfare of the precious animals they seek to protect and conserve.

Summary

Biological samples collected from the animals housed in AZA institutional collections have the potential to provide a wealth of information to scientists today and tomorrow. If managed and cared for properly, these resources will be available for decades to come. It is the responsibility of each AZA institution to curate these samples to the best of their ability. Scientific advisory groups are available to advise AZA members, and the BBAG is composed of experts available to advise on the curation of biological samples. The ZIMS is also a tool to be used in the management of biological samples. This software package is overcoming barriers of language, culture, currency, and competition to revolutionize zoos' and aquariums' approach to animal and sample care. The management of samples also includes the tracking of ownership, and this should be an important consideration whenever material is transferred between parties. The MTA outlines the rights and responsibilities of all parties involved and is an important element in the management of biological samples.

Acknowledgments

We would like to thank the ISIS and its dedicated staff for contributions to this presentation. We extend special thanks to Nate Flesness, Hassan Syed, Michele Peters, and Nell Bekiares. We also thank the Sample Focus Group for their vision and for their incredible dedication and determination to bring the sampling functionality of ZIMS to our community.

References

Lehn, C., K. Bryning, R. DeSalle, and R. Cahoon, R. 2007. Who owns the Ark? and why does it matter? *Proceedings of the American Zoological Society.* http://www.aza.org/AZAPublications/Documents/2007Annualconf13.pdf.

Wildt, D. E. 1992. Genetic resource banking for conserving wildlife species: Justification, examples and becoming organized on a global basis. *Animal Reproduction Science* 28:247–257.

Wildt, D. E., W. F. Rall, J. K. Critser, S. L. Monfort, and U. S. Seal. 1997. Genome resource banks: Living collections for biodiversity conservation. *BioScience* 47(10):689–698.

Part IV

Genomic Technology Meets Conservation Biology

Genomic techniques have promised to expand and supplement the field of conservation genetics. The specific ways in which this expansion might be possible are the subjects of this section. The first chapter describes technical advances that can contribute to the expansion of conservation genetics to high-throughput science. In this chapter, George Amato and Rob DeSalle ask whether "conservomics" (the union of high-throughput technology with conservation genetics) is feasible and, if so, whether it will be useful. Judith Blake describes three specific ways in which modern genomics will contribute to conservation biology. Genomics changes our perception scale by making available whole genome sequences, the genome information gives us a more complete approach to genotype and phenotype questions, and comparative biology, of which conservation genetics is a part, can now be examined from a whole genome perspective. Norman Ellstrand examines the increasing trend of producing transgenic animals and plants. In chapter 16 he examines the problems and pitfalls of the infusion of this important technology in agriculture and its potential impact on the environment. Anne McLaren follows with a discussion of the role of assisted reproduction in conservation and suggests that although this approach has some limitations, the possibilities of gamete banking and genome storage can help make future conservation efforts of this nature more successful. Lastly, Ian Wilmut and Lesley Paterson describe the adult cell cloning of endangered animals and the potential importance of the approach to conservation biology.

George Amato and Rob DeSalle

14

Conservomics? The Role of Genomics in Conservation Biology

The suffixes "-omics" and "-genomics" have been added to a wide range of root words in the past decade to reflect the impact genomics has had on these areas of biological research. A good example of this suffix addition is sociogenomics (Robinson, 1999, 2002; Robinson et al., 2005), the study of social behavior as illuminated by genomic approaches. This chapter addresses whether we need to add the suffix to conservation biology. Specifically, we examine the genomic technologies that are most pertinent to conservation biology and then assess which technologies will have the biggest impacts on conservation of biological species.

Genomics is a difficult field to define precisely because it easily blends into other areas of biology, such as genetics, medical genetics, and molecular biology, to name a few. One working definition of genomics is "genetics writ large" (Harold, 2005), and that seems to be a pretty good starting point. Several major aspects of genomics are relevant to an understanding of its role in conservation biology. First, genomics is the culmination of three decades of technological development that has paved the way for the creation of high-throughput biology. Therefore for this review the high-throughput approaches of genomics must be discussed in the context of conservation biology. Second, genomics connects very disparate disciplines within biology, and therefore many aspects of genomics that could be useful to advancing conservation decision making must be explained and explored. Third, genomics has increased the need for and impact of quantitative approaches, an area called bioinformatics. Biologists are seeing an unprecedented amount of primary data. Methods for handling, analyzing, and interpreting the data are in demand in genomics, and this need can also be exploited by conservation biologists.

The scope of genomics is represented in figure 14.1. There are three major arms to modern genomics: data acquisition, data analysis (bioinformatics), and data storage (archiving). This figure also shows the breadth of approaches and technologies that are subsumed by genomics. These elegant approaches include sequencing of entire genomes, gene discovery via database searches, expressed

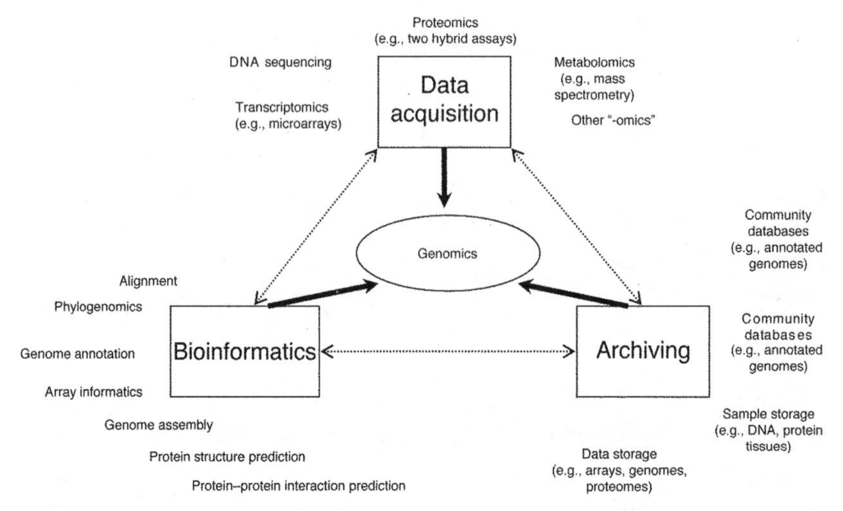

Figure 14.1 Diagrammatic representation of the roles of genomics in modern biology.

sequence tag approaches, physical and genetic mapping using markers, quantitative trait loci (QTL) mapping, characterization of the transcripts that come from the genome using microarrays and other methods, characterization of the interactions of proteins produced by the genome, candidate gene mapping of adaptive genes (such as disease resistance), and comparative mapping of adaptive genes. All these techniques and approaches have been used almost exclusively in medical genetics and model organism study, yet all of them could be relevant to the field of conservation genetics (Kohn et al., 2006; Ryder, 2005; Wayne and Morin, 2004).

Data Acquisition

High-Throughput Approaches

High throughput refers simply to the ability of modern techniques to generate large amounts of data by increasing the throughput of specimens analyzed. Most high-throughput approaches are clever adaptations of existing technologies. As an example, consider the evolution of the number of specimens analyzed in particular studies. In the early days of molecular population genetics and molecular applications to conservation studies, the typical number of specimens analyzed was around 10 because this was the number of lanes that could be run easily on an assay gel. With the advent of automated sequencing in the early 2000s this number jumped to 96 specimens because of the introduction of assay plates with twelve rows and eight columns of wells in the plates. The next jump was to 384 specimens (assay plates with twenty-four rows of sixteen wells in them), and as we will see, the newer sequencing technologies promise to allow jumps of several orders of magnitude in throughput. Throughput increases not only in DNA sequencing approaches but also in how transcripts from the genome, protein interactions, and other aspects of the genome are analyzed.

High-throughput sequencing has the potential to make the biggest impact on conservation genetics. Two aspects of high-throughput sequencing are useful in this context. The first is that whole genome sequences of flagship animals and plants can be produced that enhance the genetic analysis of endangered species (a picture of the complete and nearly complete genomes of eukaryotes can be accessed at the National Center for Biotechnology Information Web site at http://www.ncbi.nlm.nih.gov/mapview/MVgraph.html). Immediate examples of this approach are the dog genome project and its impact on wild canine biology (Kohn et al., 2006), the cow genome project and its impact on bison conservation genetics (Halbert et al., 2005), the *Populus* genome project and forestry management issues, and big cat genomics (O'Brien and Johnson, 2005). Some of the newer DNA sequencing techniques discussed in this chapter have the potential to allow whole genome sequences of any organism, so the "flagship approach" will probably become obsolete. However, these flagship species will continue to be important for purposes of genome annotation or comparative approaches.

The second aspect of high-throughput approaches involves the large-scale processing of many individuals of species relevant to conservation studies. Conservation studies often entail population-level analyses, and these high-throughput approaches can make vast contributions to increasing the throughput of conservation genetics. DNA barcoding is an approach relevant to conservation biology (DeSalle, 2006; Rubinoff, 2005) that relies on the high-throughput nature of DNA sequencing for large numbers of individuals. The barcoding initiative has benefited greatly from the infusion of high-throughput technology.

New DNA Sequencing Technologies

To date the most common sequencing approach has been the Sanger dideoxy approach. This approach relies on the construction of libraries from the genomic DNA of interest and subsequent sequencing of the individual clones. The sequence information is then assembled using informatics. This sequencing approach has been the mainstay of conservation genetics when population- or phylogenetic-level information is needed from one or a few genes to approach a conservation problem. These studies do not target whole genome sequences but rather require that only enough information from sequences be acquired to approach a conservation problem. In this context, the ability to handle hundreds of specimens is useful. The dideoxy approach is rapid, and about 2,000 sequences of 1-kb length can be handled by a single automated DNA sequencer in a day. However, the dideoxy approach is being supplanted by more rapid techniques. These newer techniques incorporate parallelization of the steps involved in sequencing and have increased the throughput of sequencing by two to three orders of magnitude. The following approaches have great promise for modern conservation genetics. They all start by the shearing up of genomic DNA to small enough size to manipulate in parallel fashion.

The most time- and space-consuming step in the Sanger approach is the cloning step. Therefore many new approaches take advantage of the ability to miniaturize the cloning process on beads by developing in vitro cloning and amplification approaches. A good example of parallelization of DNA sequencing is an approach developed by Solexa (currently owned by Illumina) called bridge polymerase chain reaction (PCR). This approach amplifies target DNA fragments on anchor primers that are attached to a surface. This process implements the physical isolation of many copies of a single fragment of DNA in miniature. A variation on this approach developed by Helicos eliminates the amplification step and fixes individual DNA fragments directly to the solid surface.

Another of these approaches is called emulsion PCR (Margulies et al., 2005). This approach takes individual DNA molecules and primer-coated beads and in aqueous bubbles encased by an oil phase. PCR is then performed so that the individual DNA fragment is amplified and each bead is coated with an individualized population of the same DNA fragment. The beads are then immobilized and sequenced using methods adapted to be massively parallel. The Margulies emulsion approach has been commercialized by 454 Life Sciences; the Shendure emulsion approach, which was originally called "polony sequencing," has been commercialized by Agencourt.

Once the DNA fragments have been affixed to a solid substance and the fragments are individualized to separate positions on the surface, the fragments are ready to be sequenced. Some of the aforementioned approaches use sequencing by synthesis. The Solexa and Quake approaches use this synthesis approach, which is similar but not identical to Sanger sequencing. The 454 Life Sciences sequencing method uses pyrosequencing approaches. This method of sequencing adds one nucleotide at a time and then detects and quantifies the number of nucleotides added to a given fragment at a given location on the solid surface through light emitted by the release of attached pyrophosphates (Ronaghi et al., 1996). Polony sequencing uses yet another approach to obtaining the sequenced from the affixed DNA fragments called sequencing by ligation.

All these methods have the limitation that the fragments immobilized on the solid surface usually are very short. Another approach to DNA sequencing is to take advantage of the microarray approaches that have been developed in the past decade to accomplish what is commonly called DNA resequencing. All microarray approaches are based on DNA hybridization. Instead of using RNA as a template for hybridization to the array, as is done in expression studies, in the DNA resequencing approach target DNA is used as the template for hybridization. Because the array can be spotted with tens of thousands of oligonucleotides that can act as tools to detect polymorphisms, array-based approaches can scan a genome for the state of thousands of single nucleotide polymorphisms (SNPs). The resequencing can be done in several ways. The specific position of the SNP where the variation occurs can also be called the interrogated site. That is, we are asking or interrogating what base is at the site in the variable SNP site. One approach is to use "gain of signal" characteristics of the hybridized array. In this approach, the relative hybridization of the target DNA labeled with a fluorescent dye is assessed as it hybridizes to each of the four possible nucleotides that could be at an interrogated site. So four oligonucleotides are synthesized, with a different base inserted in the interrogated site for each potential SNP. The oligonucleotide spot on the array that lights up best with the specific fluorescent dye is the base for the SNP polymorphic site. Another method is called the loss of hybridization signal approach. In this technique the target DNA is labeled with a particular color of fluorescent dye, and a reference or normal DNA is labeled with a second, differently colored dye. The two labeled DNAs are then hybridized to an array with potential SNP oligonucleotides. The decrease of target labeled fluorescence relative to normal labeled fluorescence indicates the presence of a sequence change at an interrogated site.

New High-Throughput DNA Polymorphism Detection Approaches

The earliest methods used to detect DNA polymorphisms were restriction fragment length polymorphism (RFLP) methods. Since the development of RFLP approaches, many ingenious approaches to detecting polymorphisms have been developed. Most rely on the prior detection of variation of single nucleotides in the genome, called SNPs (Aitken et al., 2004; Morin et al., 2004), or on short repeated sequences that cause length polymorphism, called microsatellites. One of the more novel approaches to detecting polymorphisms is denaturing high-performance liquid chromatography (DHPLC). This approach uses reversed-phase high-performance liquid chromatography to examine SNPs. In order to do DHPLC, one needs to synthesize oligonucleotides that can generate two fragments via PCR. One pair of oligonucleotides should generate a fragment of DNA that corresponds to the target DNA that includes an SNP site with a sequence for a known allele. The second fragment is called a "normal fragment," and it is an amplified product for the same SNP region but for a known allele at the SNP.

The two PCR products are denatured together and then allowed to reanneal. If the two PCR products have the same allele in them, then only a single kind of DNA duplex will be formed, called a homoduplex. On the other hand, if the target DNA and the normal DNA have the same allele, then three kinds of annealing products will be produced: homoduplexes of the target DNA product, homoduplexes of the normal DNA fragment, and a hybrid duplex or heteroduplex that contains the target and the normal DNA fragments, with a mismatch at the position of the SNP. The reannealed products are then run through a DHPLC column, where any homoduplexes are retained longer than the heteroduplexes. The stuff that comes out of the column is continually measured and quantified using ultraviolet (UV) absorption. UV is absorbed by DNA, so if there is a fraction from the column with no DNA in it, UV absorption will be low or nil. If DNA starts to come out of the column, then it will be detected as a spike of UV absorption. If the target DNA is identical to the normal DNA, then the UV absorption pattern will show only a single peak when the homoduplex comes through the column and passes through the UV absorption apparatus. On the other hand, if the target DNA has a different allele from the normal DNA, two peaks will be observed when the heteroduplex DNA passes through the UV absorption apparatus. Other methods that take advantage of the properties of DNA to detect polymorphisms are temperature gradient electrophoresis, high-resolution melting of entire amplicons, matrix-assisted laser desorption/ionization time-of-flight (MALDI-TOF) mass spectrometry, and single-strand conformation characterization.

Specific Adaptations of Genome Technology to Conservation Biology

DNA Barcoding

DNA barcoding is an initiative begun around 2003 that has targeted the sequencing of a DNA marker for the 1.7 million named species on the planet (Hajibabaei et al., 2006; Hebert et al., 2003a, 2003b; Kohler, 2007; Tautz et al., 2003). A consortium for DNA barcoding has been established, called the Consortium for the Barcode of Life (CBOL). The mission of CBOL is "to promote the exploration and development of DNA barcoding as a global standard for species identification. In pursuing this mission, CBOL promotes: the rapid compilation of high-quality DNA barcode records in a public library of DNA sequences, the development of new instruments and processes that will make barcoding cheaper, faster, and more portable, the participation of taxonomists and taxonomic research organizations in all regions and countries, and the use of DNA barcoding for the benefit of science and society" (http://www.barcoding.si.edu/; http://www.dnabarcoding.ca/). The initiative uses a short DNA sequence from a standardized position in the genome as a molecular diagnostic for species-level identification of animal, plant, and fungal life. Compared with the entire genome of most organisms, DNA barcodes are short and can be obtained rapidly and cheaply as a result of high-throughput approaches. The Folmer region (648 nucleotide base pairs long) at the 5′ end of the cytochrome c oxidase subunit 1 mitochondrial region (COI) has been tagged as the standard barcode for higher animals. Other genome regions are being explored as potential DNA markers for other groups of organisms. The database for DNA barcodes is rapidly growing and can be viewed at http://www.barcodinglife.org/). Several new bioinformatics tools have been developed to handle the large amounts of DNA barcode data that will be generated during the initiative (Blaxter and Floyd, 2003; Kelly et al., 2007; Rach et al., 2008; Ratnasingham and Hebert, 2007; Richardson et al., 2007).

Although there has been much discussion of the relevance of this approach to conservation (DeSalle, 2006; Rubinoff, 2005; Rubinoff et al., 2006), we can see two specific areas of conservation biology that will benefit from a comprehensive DNA barcoding database. First, an identification system for endangered organisms or parts of endangered organisms can be generated. This database will be incredibly useful in forensic applications to the trade and import of endangered species. Second, the approaches of DNA barcoding can be used to enhance taxonomy of endangered species and their close relatives. Note that we do not see DNA barcoding as the sole means to taxonomy (DeSalle, 2006). The former use is essential to executing laws crafted around endangered species lists, and the latter can assist in the formulation of these lists.

Assessing Biodiversity en Masse

Recent advances in environmental and medical microbiology that have been developed to assay microbial diversity in environmental samples afford an important and exciting inroad to conservation biology at the community level (Pearson, 2007). This approach is also known as the metagenomic approach (Allen and Banfield, 2005; Daniels, 2005; DeLong, 2005, 2007; Handelsman, 2005; Jones, 2007; Keller and Zengler, 2004; Ordovas and Mooser, 2006; Ward, 2006) because it allows the analysis of composition of metapopulations of organisms. There are two major ways in which environmental sampling of microbial communities have added to our knowledge of microbial communities. The first uses the high-throughput approaches to obtain genome sequences, even whole genomes of organisms from difficult to culture microbes. Several interesting novel microbes and, more importantly, metabolic pathways have been discovered this way (DeLong, 2007; Nealson and Venter, 2007; Pachter, 2007; Steward and Rapp, 2007). The second approach uses high-throughput approaches to determine the composition of microbial communities and is more pertinent to conservation biology goals. Initial use of the approach involved the cloning and rapid sequencing of large numbers of DNA fragments to obtain enough information to interpret species composition. A typical assay for a sample yielded on average a thousand data points for determining species composition. The approach has recently been adapted to the 454 Life Sciences sequencing technology discussed earlier (Harkins and Jarvie, 2007; Sogin et al., 2006). Adaptation of this approach to 454 Life Sciences technology can yield more than 200,000 data points and result in more precise estimates of species composition in microbial communities.

Although these approaches appear to be of high utility to microbial community biology and adaptable to communities of small-bodied eukaryotes, they seem to be at odds with the goals of conservation biology of larger multicellular organisms. However, new techniques that use synthetic markers at the ends of the primers used in the amplification step are a promising way to bring these approaches closer to large organism conservation analysis (Malmstrom et al., 2007). As an example, consider a two–base pair synthetic code that is added onto the 5′ end of the primers used for amplification. There are sixteen two–base pair combinations for the four bases in DNA: AA, AG, AT, AC, GA, GG, GT, GC, TA, TG, TT, TC, CA, CG, CT, and CC. Such a tagging system would allow the inclusion of sixteen different PCR reactions in a 454 Life Sciences run. Three base pair codes would allow 81 reactions, four base pair codes would allow 256 reactions, and so on, with n^4 reactions, where n is the number of base pairs in the code.

High-Throughput DNA Polymorphism Detection and Conservation Biology

Kohn et al. (2006:629) discussed this issue cogently. They concluded,

> The relevance of genetics to conservation rests on the premise that neutral marker variation in populations reflects levels of detrimental and adaptive genetic variation. . . . Genome sequence information and new technological and bioinformatics platforms now enable comprehensive surveys of neutral variation and more direct inferences of detrimental and adaptive variation in species with sequenced genomes and in "genome-enabled" endangered taxa. . . . A new conservation genetic agenda would utilize data from enhanced surveys of genomic variation in endangered species to better manage functional genetic variation.

Their review described the importance of finding, characterizing, and using neutral marker variation in SNPs or microsatellite variation. They discussed three important areas where genome-enabled biology and conservation needs intersect (Kohn et al., 2006:635). The basic approaches are all after the same thing in a genomic context: the detection of deviation from the genome-wide levels of genetic variation and differentiation that can be interpreted as the footprint of localized natural selection (Kohn et al., 2006). The first approach is the candidate gene approach, in which genes are selected from some a priori knowledge of the biology of an organism or group of organisms. These approaches usually are accomplished with QTL mapping techniques, and the classic example is the histocompatibility complex. The second approach is the selective sweep approach. This method looks for phenotypic traits or loci that leave a variation profile that indicates natural selection. This approach is complicated by the contribution of the strength of selection, population size, and genetic recombination. This approach has been used effectively to localize regions of the human genome that have experienced recent adaptive evolution (Charpentier et al., 2007). The final approach is the linkage disequilibrium mapping approach, which identifies QTLs or continuously varying traits. This approach entails breeding experiments, and this limits its applicability to organisms of conservation importance. Kohn et al. (2006) also pointed out that one of the more important applications of these approaches to conservation lies in the analysis of disease or pathogen susceptibility in endangered species. One aspect of modern genomics in the bioinformatics realm that has assisted in these approaches is profusion of computer- and Web-based tools. Several recent reviews have summarized the bioinformatics platforms available for these approaches (Balding, 2006; Excoffier and Heckel, 2006; Marjoram and Tavare, 2006).

Conclusions

Genomics has broadened and expanded many areas of modern biology. Conservation biology is in a position now to gain immensely from the fruits of the technological and bioinformatics explosions that have resulted from genomics (DeSalle and Amato, 2004; Kohn et al., 2006; Ryder, 2005). In this chapter we have described many genomics advances that have been applied in a conservation context. We are poised to enter a true conservomics era. However, genomics cannot and should not be applied indiscriminately. Other chapters in this book discuss at length the role of demographic factors in conservation biology, and the relative importance of genetics and demographics has long been a bone of contention. However, we strongly encourage the incorporation of genomic technology into conservation biology. It is a crisis discipline that can only benefit from the high-throughput nature of genomic analysis approaches.

Acknowledgments

We thank the Korein Foundation at the American Museum of Natural History (AMNH), the Lewis and Dorothy Cullman Program for Molecular Systematics Studies at the AMNH, and the Sackler Institute for Comparative Genomics at the AMNH.

References

Aitken, N., S. Smith, C. Schwarz, and P. A. Morin. 2004. Single nucleotide polymorphism (SNP) discovery in mammals: A targeted-gene approach. *Molecular Ecology* 13(6):1423–1431.

Allen, E. E. and J. F. Banfield. 2005. Community genomics in microbial ecology and evolution. *Nature Reviews. Microbiology* 3:66, 489–498.

Balding, D. J. 2006. A tutorial on statistical methods for population association studies. *Nature Reviews. Genetics* 7:781–791.

Blaxter, M. and R. Floyd. 2003. Molecular taxonomics for biodiversity surveys: Already a reality. *Trends in Ecology & Evolution* 18:268.

Charpentier, C., O. Tenaillon, C. Hoede, F. Clavel, and A. J. Hance. 2007. Localizing recent adaptive evolution in the human genome. *PLoS Genetics* 3:e90.

Daniels, R. 2005. The metagenomics of soil. *Nature Reviews. Microbiology* 3(6):470–478.

DeLong, E. F. 2005. Microbial community genomics in the ocean. *Nature Reviews. Microbiology* 3:459.

——. 2007. Sea change for metagenomics? *Nature Reviews. Microbiology* 5:326.

DeSalle, R. 2006. Species discovery versus species identification in DNA barcoding efforts: Response to Rubinoff. *Conservation Biology* 20(5):1545–1547.

DeSalle, R. and G. Amato. 2004. The expansion of conservation genetics. *Nature Reviews. Genetics* 5:702–712.

Excoffier, L. and G. Heckel. 2006. Computer programs for population genetics data analysis: A survival guide. *Nature Reviews. Genetics* 7:745–758.

Hajibabaei, M., D. H. Janzen, J. M. Burns, W. Hallwachs, and P. D. Hebert. 2006. DNA barcodes distinguish species of tropical Lepidoptera. *Proceedings of the National Academy of Sciences USA* 103:968–971.

Halbert, N. D., T. J. Ward, R. D. Schnabel, J. F. Taylor, and J. Derr. 2005. Conservation genomics: Disequilibrium mapping of domestic cattle chromosomal segments in North American bison populations. *Molecular Ecology* 14(8):2343–2362.

Handelsman, J. 2005. Metagenomics or megagenomics. *Nature Reviews. Microbiology* 3(6):457–458.

Harkins, T. and T. Jarvie. 2007, June. Metagenomics analysis using the Genome Sequencer FLX system (454 Life Sciences). *Nature Methods,* application notes iii–v.

Harold, F. M. 2005. Review of *A primer of genome science. ASM News* 71:492.

Hebert, P. D., A. Cywinska, S. L. Ball, and J. R. deWaard. 2003a. Biological identifications through DNA barcodes. *Proceedings. Biological Sciences* 270:313–321.

Hebert, P. D. N., S. Ratnasingham, and J. R. deWaard. 2003b. Barcoding animal life: Cytochrome c oxidase subunit 1 divergences among closely related species. *Philosophical Transactions of the Royal Society, London B.* 270(Suppl.):S96–S99.

Jones, S. 2007. Environmental microbiology: Detecting variation. *Nature Reviews. Microbiology* 5:327.

Keller, M. and K. Zengler. 2004. Tapping into microbial diversity. *Nature Reviews. Microbiology* 2:141–150.

Kelly, R. P., I. N. Sarkar, D. J. Eernisse, and R. DeSalle. 2007. DNA barcoding using chitons (genus *Mopalia*). *Molecular Ecology Notes* 7(2):177–183.

Kohler, F. 2007. From DNA taxonomy to barcoding: How a vague idea evolved into a biosystematic tool. *Mitteilungen aus dem Museum für Naturkunde Berlin, Zoologische Reihe* 83(Suppl.):44–51.

Kohn, M. H., W. J. Murphy, E. A. Ostrander, and R. K. Wayne. 2006. Genomics and conservation genetics. *Trends in Ecology & Evolution* 21:629–637.

Malmstrom, H., E. M. Svensson, M. T. Gilbert, E. Willerslev, A. Gotherstrom, and G. Holmlund. 2007. The use of coded PCR primers enables high-throughput sequencing of multiple homolog amplification products by 454 parallel sequencing. *PLoS ONE* 2:e197.

Margulies, M., M. Egholm, W. E. Altman, S. Attiya, J. S. Bader, L. A. Bemben, et al. 2005. Genome sequencing in microfabricated high-density picolitre reactors. *Nature* 437:376–380.

Marjoram, P. and S. Tavare. 2006. Modern computational approaches for analysing molecular genetic variation data. *Nature Reviews. Genetics* 7:759–770.

Morin, P. A., G. Luikart, R. K. Wayne, and SNP Workshop Group. 2004. SNPs in ecology, evolution and conservation. *Trends in Ecology & Evolution* 19(4):208–216.

Nealson, K. H. and J. C. Venter. 2007. Metagenomics and the global ocean survey: What's in it for us, and why should we care? *The ISME Journal* 1:185–187.

O'Brien, S. J. and W. E. Johnson. 2005. Big cat genomics. *Annual Review of Genomics and Human Genetics* 6:407–429.

Ordovas, J. M. and V. Mooser. 2006. Metagenomics: The role of the microbiome in cardiovascular diseases. *Nature Reviews. Genetics* 17:157–161.

Pachter, L. 2007. Interpreting the unculturable majority. *Nature Methods* 4:479–480.

Pearson, H. 2007. Microbe meeting promotes habitat conservation. *Nature* 447:127.

Rach, J., R. DeSalle, I. N. Sarkar, B. Schierwater, and H. Hadrys. 2008. Character-based DNA barcoding allows discrimination of genera, species and populations in Odonata. *Proceedings. Biological Sciences* 275(1632):237–247.

Ratnasingham, S. and P. D. N. Hebert. 2007. The Barcode of Life Data Systems (www.barcodinglife.org). *Molecular Ecology Notes* 48:284–290.

Richardson, D. E., J. D. Vanwye, A. M. Exum, R. K. Cowen, and D. L. Crawford. 2007. High-throughput species identification: From DNA isolation to bioinformatics. *Molecular Ecology Notes* 7:199–207.

Robinson, G. E. 1999. Integrative animal behaviour and sociogenomics. *Trends in Ecology & Evolution* 14:202–205.

———. 2002. Development. Sociogenomics takes flight. *Science* 297:204–205.

Robinson, G. E., C. M. Grozinger, and C. W. Whitfield. 2005. Sociogenomics: Social life in molecular terms. *Nature Reviews. Genetics* 6:257–270.

Ronaghi, M., S. Karamohamed, B. Pettersson, M. Uhlén, and P. Nyrén. 1996. Real-time DNA sequencing using detection of pyrophosphate release. *Analytical Biochemistry* 242:84–89.

Rubinoff, D. 2005. Utility of mitochondrial DNA barcodes in species conservation. *Conservation Biology* 20:1026–1033.

Rubinoff, D., S. Cameron, and K. Will. 2006. A genomic perspective on the shortcomings of mitochondrial DNA for "barcoding" identification. *Journal of Heredity* 97:581–594.

Ryder, O. A. 2005. Conservation genomics: Applying whole genome studies to species conservation efforts. *Cytogenetic and Genome Research* 108:6–15.

Sogin, M. L., H. G. Morrison, J. A. Huber, D. M. Da Welch, S. M. Huse, P. R. Neal, et al. 2006. Microbial diversity in the deep sea and the underexplored "rare biosphere." *Proceedings of the National Academy of Sciences* 103:12115–12120.

Steward, G. F. and M. S. Rapp. 2007. What's the "meta" with metagenomics? *The ISME Journal* 1:100–102.

Tautz, D., P. Arctander, A. Minelli, R. H. Thomas, and A. P. Vogler. 2003. A plea for DNA taxonomy. *Trends in Ecology & Evolution* 18:70–74.

Ward, N. 2006. New directions and interactions in metagenomics research. *FEMS Microbiology Ecology* 55:331–338.

Wayne, R. K. and P. A. Morin. 2004. Conservation genetics in the new molecular age. *Frontiers in Ecology and the Environment* 2(2):89–97.

Judith A. Blake

15

Genomics and Conservation Genetics

The Human Genome Project was launched with the ultimate goal of obtaining the complete sequence of the human genome. Along the way, this initiative has accelerated the scientific endeavor in novel and exciting ways. As would be anticipated from such a well-funded and compelling endeavor, the developments and results of this work have influenced many scientific fields, including conservation genetics. Of course, advances in technology were expected and essential. Scientific fields from biochemistry to biomedicine to computer science and engineering devoted resources to the development of informatics and analysis systems to maintain, manipulate, and interpret the flood of genome information. Fortunately, support for the sequencing and analysis of model organism genomes was included in the project. This expanded the scope of the human genome sequencing initiative to include representatives of a wide range of organisms and brought many biologists into the field of genomics who might not otherwise have incorporated genomic concepts into their research so readily.

Here I address three ways the field of genomics has furthered conservation genetics. The first is the new genomics paradigm that is based on our ability to think in terms of having a whole genome sequence, the set of instructions for development and maintenance of the organism. This new way of thinking changes the framework and scale of the potential knowledge we might obtain for an organism. The second is the emergence of a molecular and cellular system biology approach to the investigation of the information space from the genome to the phenotype. Such an approach adds to the understanding of the functional physiology of an organism in the context of the genomic contribution. Finally, the field of genomics has contributed to the development of a whole-genome approach to comparative biology. This extension of comparative biology techniques for genomes is particularly relevant to our understanding of the inheritance of genetic variability, the evolution of genomes, and the impact of genome evolution on cellular and physiological processes. And this knowledge helps place conservation genetics analyses in the genome evolution context.

Age of Genomes: Generating the Part Lists

The sequencing of complete genomes is such a new phenomenon that the impact of this accomplishment is still being integrated into our work. Before the availability of genome analyses, experimental studies focused on gene function, protein biochemistry, and the genetic basis of disease, all in regard to one or a small number of genes. Now, with the completed genomes available for increasing numbers of species, it is possible to seek correlation and causation of genetic change with function. The generation of complete genomes initiated an era of comparative genomics and made available to researchers complete part lists for a wide array of organisms. With the third genome completed, the autotrophic archaean *Methanococcus jannaschii,* researchers were able to compare the structural and cellular organization of Eukaryotes, Eubacteria, and Archaea and to demonstrate some aspects of common ancestry between the microbes that are independent of their shared features (Bult et al., 1996). Since this initial work in whole genome sequencing and analysis, the genomes of hundreds of viruses and microbes have been published (http://img.jgi.doe.gov/cgi-bin/pub/main.cgi, http://www.tigr.org/tigr-scripts/CMR2/CMR HomePage.spl), and research efforts have focused on the comparative genome analyses of these organisms in the investigation of the evolution of cellular structure and function (Fraser et al., 2000; Martin et al., 2003). The completion of several multicellular organisms, including the nematode *Caenorhabditis elegans* (The *C. elegans* Sequencing Consortium, 1998), the plant *Arabidopsis thaliana* (The *Arabidopsis* Genome Initiative, 2000), and the fly *Drosophila melanogaster* (Adams et al., 2000), has stimulated comparative genomic research on eukaryotes (Hedges, 2002; Rubin et al., 2000).

Developing New Tools

Genome sequencing projects have necessitated the ramping up of technologies for high-throughput sequencing and the generation of new algorithms for sequence assembly, verification, and analysis. With the first genomes sequenced (the microbes *Haemophilus influenza,* the minimal genome *Mycoplasma genitalium,* and the archaean microbe *Methanococcus jannaschii*), database systems were designed to hold multiple overlapping sequences and to provide annotation systems for defining features and substrings of the sequence. Query and visualization systems were developed to allow the analysis of multiple genome sequences that were approximately 2 million base pairs long.

As the ability to sequence genomes advanced and more, mostly microbial genomes were sequenced and assembled, the task of comparative genome analysis progressed accordingly (Eisen, 1998; Fraser et al., 2000; Glusman et al., 2001; Ivanova et al., 2003). Much work in the area of phylogenetic analysis was applicable to this task. The difference clearly was the scope of the project (Koonin et al., 2002; Nash et al., 1998; Navarro and Barton, 2003). The first articles were descriptive, comparing, for example, classes of genes within and between organisms (Ledent and Vervoort, 2001; Lee et al., 2002), but more complex evaluations of gene order and genome organization have followed (Friedman and Hughes, 2001; Huynen et al., 2001).

Having completed genomes and developed first-generation tools for genome comparative analysis, the next step is to map the genome and genome features to other biological information, that is, to relate the genome and the information it encodes to the extensive collection of biological knowledge gathered over many years of experimental research. In particular, our understanding of the genetic basis of phenotypes now can be investigated in the context of understanding the phenotype in the context of the whole genome. These investigations are very preliminary even for the species whose genome has been sequenced. For the multicellular species, very few genomes have been completed even in draft form, and even fewer are considered finished. However, the important point is that we can now think of genetic and genomic variability fully in terms of the entire genome.

Sequence as a Unifying Dimension

Fifty years ago, with the unraveling of the structure of DNA and the recognition of DNA as the molecular basis of inheritance, a new era of scientific investigation of nucleic acids and proteins was initiated. Molecular biologists, biochemists, physicists, and others sought to understand the molecular basis of physiological processes in relation to the inheritance of heritable traits. Geneticists investigated the genetics of mutant phenotypes and constructed linkage groups to elucidate the organization and structure of chromosomes as they sought to relate genes to the underlying DNA and protein sequence information. With the ability to sequence DNA in an efficient manner available in many laboratories as of the 1980s, biologists from microbiology to botany to mouse genetics were sequencing the genes and genomic fragments of many organisms using the same technologies, developing similar information systems, and sharing annotation and visualization tools.

An early and far-reaching decision by the genomics community was to require the deposition of sequence data into public databases as part of the International Nucleotide Sequence Database Collaboration. This collaboration consists of the DNA DataBank of Japan, the European Molecular Biology Laboratory, and GenBank at the National Center for Biotechnology Information. These three organizations exchange data on a daily basis. The result is the availability of sequence data from a multitude of organisms[1] and the development of tools such as BLAST (http://blast.wustl.edu/, http://www.ncbi.nlm.nih.gov/BLAST/; Altschul et al., 1990) and BLAT (see http://genome.ucsc.edu/; Kent, 2002) to compare and visualize sequence information. Scientists who might previously have communicated their endeavors and results primarily with others working on the same system as themselves have found that they can access data from multiple organisms. The currency of information exchange is the sequence data, and efforts have focused on the development of common exchange formats and comparative analysis systems.

Recently, applications such as the Ensembl Genome Browser (http://www.ensembl.org/), the Santa Cruz Genome Browser (http://genome.ucsc.edu/), and the National Center for Biotechnology Information Map Viewer (http://www.ncbi.nlm.nih.gov/mapview/) are enabling access, visualization, and comparative analysis beyond the generation of pairwise sequence comparisons. Regions of multiple genomes can be examined, and the results of various gene prediction and comparative genomic applications can be viewed simultaneously.

Refining the Language of Genomics and Molecular Biology

Science is a community endeavor and builds on the knowledge and experimental analysis that came before. To accomplish and make freely available our work, we must write up and publish our results. The emergence of sequence and genomic data and the development of shared technologies to generate and analyze these data has meant that biologists who had previously published only in journals devoted to their subdiscipline can now share their results with a wider audience. Although the language of sequence description, at least on a primary level, can be understood out of context, the language of molecular function and physiological processes is much more context dependent. In particular, although a person might comprehend and integrate different presentations of molecular descriptions in the literature, computers need rigorous and defined vocabularies of molecular concepts in order to successfully exchange and compare molecular data from multiple systems. Recent developments of ontologies for molecular biology, structured controlled languages for aspects of biological systems, are helping to bridge the language barrier for both people and computers so that the commonality of biological information first spotlighted by the sequence data warehouses can be compared and explored across multiple disciplines.

Ontologies provide the precise language (defined terms) for a domain of knowledge and also provide the classifications, or taxonomies, of terms. Ontologies not only enable more precise communication between scientists but also facilitate precise data exchange between computer systems and facilitate data integration and data inference systems. In the late 1990s the need for bio-ontologies as structured, controlled vocabularies became increasingly evident as the model organism genomes of yeast (*Saccharomyces*), worm (*Caenorhabditis*), and fly (*Drosophila melanogaster*) were being completed. Early analysis confirmed that many of the known human disease-causing genes were conserved in the fly genome (Fortini et al., 2000; Kornberg and Krasnow, 2000; Reiter et al., 2001), and therefore the investigations of fly geneticists and genomists were of greater relevance to biomedical researchers (Feany and Bender, 2000; Jackson et al., 1998). The need to exchange and compare results of research investigations was essential; the solution was not so much a technological one as a sociological one that depended on the use of standard defined terms to described experimental results across many experimental systems.

The Gene Ontology (GO) Consortium provides one of the solutions to the problem of refining language for genomics and molecular biology to facilitate data integration and exchange. The GO project was initiated by the model organism databases to provide a common, structured vocabulary for the biological domains of "molecular function," "biological process," and "cellular component" (The Gene Ontology Consortium, 2000, 2001; http://www.geneontology.org). These standards have been widely adopted by bioinformatics providers. Other bioinformatics standards are also being developed to describe features of expression information and classifications of species-specific anatomies (e.g., Microarray Gene Expression Data, http://www.mged.org/; http://www.ana.ed.ac.uk/anatomy/). These efforts to refine and standardize languages for genomics and molecular biology reflect the impact of the genomics revolution in bringing together formerly disparate experimental research communities and engendering an immediacy in language investigations in order to advance the understanding and use of the data produced in the genomics age.

System Approaches in Molecular Biology

As our understanding of the organization and variability of genomes advances, the next frontier is to understand the cellular systems, the tissue systems, the organ systems, and their interactions with one another and with the internal and external environment to form the functioning (and dysfunctioning) organism. The vision of system biology in this context is the search for knowledge of the levels of biological organization from the individual gene to the complete organism (Kitano, 2002; see also http://www.systemsbiology.org/). One example of this approach is the Physiome Project (Hunter and Borg, 2003), which seeks to provide a framework for modeling the human body, integrating information from the genome, cells, tissues, and organs in order to provide a system approach to investigate and interpret the human system. This integrative approach can inform the interpretation of morphological and behavioral studies and influence the evaluation of results from conservation genetics investigations.

The extension of the genome age to encompass functional genomics and system biology will continue to influence research in the area of conservation genetics. For example, knowledge about cellular and physiological systems, along with information about genetic variants, allows the interpretation of results such as those concerning social and reproductive behavioral differences in prairie and mountain voles that can be ascribed to the variation in hormone levels and to cellular mechanisms of hormone action (Lightman et al., 2002; Young et al., 2001). In another example, genome mapping technologies have been used to investigate the genetic components of morphological divergences in the three-spine stickleback species (Peichel et al., 2001). Understanding the genetic components of characters that differentiate phylogenetic groupings will clarify the genomic context of phenotypic variability and may influence the evaluation of characters for phylogenetic analysis.

Genomics and the Comparative Method

Comparative biology is a cornerstone for understanding the structure and evolution of biological entities. With the complete sequencing of genomes come the complete "part lists" for organisms, and a truly comparative analysis of systems can be undertaken. For instance, a comparative study of development in plants and animals revealed, among other results, that comparable developmental processes in plants and animals are nonhomologous (Meyerowitz, 2002).

The recent completion of the genome of the laboratory mouse has opened up a flurry of comparative genomic investigations between mouse and human because mouse is the premier model organism for the study of mammalian genetics and genomics (Eames et al., 2003; Mouse Genome Sequencing Consortium, 2002; Waterston et al., 2002). Studies of genome rearrangements between mouse and human (Pevzner and Tesler, 2003) and the investigation of conservation of splicing events between mouse and human (Thanaraj et al., 2003) are examples of ongoing investigations. With the whole list of genes and genomic features available, phylogenetic analysis on the basis of comparative genomics is increasingly common (e.g., Cotton and Page, 2002; Ledent and Vervoort, 2001; Thomas and Touchman, 2002).

Comparative genome analysis will continue to provide important insights into the understanding of the evolution of cellular, physiological, and organismal systems. The computational challenges for whole-genome analysis are intense, but attempts are under way to provide new tools and bioinformatics resources to this work (Dubchak and Pachter, 2002; Lee et al., 2002). Recently, investigations have focused on comparative genome analysis of vertebrate systems to provide new insights into vertebrate origins and evolution (Dehal et al., 2002; Hedges and Kumar, 2002) and have extended investigations of chromosomal evolution to the whole genome level (Nash et al., 1998; Navarro and Barton, 2003). The impact of comparative genomics on conservation genetics is in the continuing evaluation and understanding of the evolution of organisms.

Summary: Contributions of Genomics to the Field of Conservation Genetics

The nascent field of genomics burst onto the scientific scene in the 1980s, bringing new energies and insights to most fields of biological research, including the field of conservation genetics. Although most conservation genetics research continues to focus on the investigation of genetics of species and populations, the conceptualization of a complete genome, whether or not the genome is completed for a given organism, changes the framework of our analyses. The development of genomic and bioinformatics tools that support comparative genomics research have also enhanced the application of a genomic perspective to conservation genetics endeavors.

Notes

1. Currently about 100,000 species have some sequence data deposited in the sequence resources. See http://www.ncbi.nlm.nih.gov/Taxonomy/taxonomyhome.html/index.cgi for more information on this topic.

References

Adams, M. D., S. E. Celniker, R. A. Holt, C. A. Evans, J. D. Gocayne, P. G. Amanatides, et al. 2000. The genome sequence of *Drosophila melanogaster*. *Science* 287:2185–2195.

Altschul, S., W. Gish, W. Miller, E. Myers, and D. J. Lipman. 1990. Basic local alignment search tool. *Journal of Molecular Biology* 215:403–410.

The *Arabidopsis* Genome Initiative. 2000. Analysis of the genome sequence of the flowering plant *Arabidopsis thaliana*. *Nature* 408:796–815.

Bult, C. J., O. White, G. J. Olsen, L. Zhou, R. D. Fleischmann, G. G. Sutton, et al. 1996. Complete genome sequence of the methanogenic archaeon, *Methanococcus jannaschii*. *Science* 273:1058–1073.

The *C. elegans* Sequencing Consortium. 1998. Genome sequence of the nematode *C. elegans*: A platform for investigating biology. *Science* 282:2012–2018.

Cotton, J. A. and R. D. M. Page. 2002. Going nuclear: Gene family evolution and vertebrate phylogeny reconciled. *Proceedings of the Royal Society of London B* 269:1555–1561.

Dehal, P. et al. 2002. The draft genome of *Ciona intestinalis:* Insights into chordate and vertebrate origins. *Science* 298:2157–2167.

Dubchak, I. and L. Pachter. 2002. The computational challenges of applying comparative-based computational methods to whole genomes. *Briefings in Bioinformatics* 3:18–22.

Eames, R. D., L. Goodstadt, E. E. Winter, and C. P. Ponting. 2003. Comparison of the genomes of human and mouse lays the foundation of genome zoology. *Human Molecular Genetics* 12:701–709.

Eisen, J. A. 1998. Phylogenomics: Improving functional predictions for uncharacterized genes by evolutionary analysis. *Genome Research* 8:163–167.

Feany, M. B. and W. W. Bender. 2000. A *Drosophila* model of Parkinson's disease. *Nature* 404:394–398.

Fortini, M. E., M. P. Skupski, M. S. Boguski, and I. K. Hariharan. 2000. A survey of human disease gene counterparts in the *Drosophila* genome. *Journal of Cell Biology* 150:F23–F29.

Fraser, C. M., J. Eisen, R. D. Fleischmann, K. A. Ketchum, and S. Peterson. 2000. Comparative genomics and understanding of microbial biology. *Emerging Infectious Diseases* 2000:505–512.

Friedman, R. and A. L. Hughes. 2001. Gene duplication and the structure of eukaryotic genomes. *Genome Research* 11:373–381.

The Gene Ontology Consortium. 2000. Gene ontology: Tool for the unification of biology. *Nature Genetics* 25:25–29.

——. 2001. Creating the gene ontology resource: Design and implementation. *Genome Research* 11:1425–1433.

Glusman, G., I. Yanai, I. Rubin, and L. Lancet. 2001. The complete human olfactory subgenome. *Genome Research* 11:685–702.

Hedges, S. B. 2002. The origin and evolution of model organisms. *Nature Reviews. Genetics* 3:838–849.

Hedges, S. B. and S. Kumar. 2002. Vertebrate genomes compared. *Science* 297:1283–1285.

Hunter, P. J. and T. K. Borg. 2003. Integration from proteins to organs: The Physiome Project. *Nature Reviews. Molecular Cell Biology* 4:237–243.

Huynen, M. A., B. Snel, and P. Bork. 2001. Inversions and the dynamics of eukaryotic gene order. *Trends in Genetics* 17:304–306.

Ivanova, N., A. Sorokin, I. Anderson, N. Galleron, B. Candelon, V. Kapatral, et al. 2003. Genome sequence of *Bacillus cereus* and comparative analysis with *Bacillus anthracis*. *Nature* 423:87–91.

Jackson, G. R., I. Salecker, X. Dong, X. Yao, N. Arnheim, P. W. Faber, et al. 1998. Polyglutamine-expanded human huntingtin transgenes induce degeneration of *Drosophila* photoreceptor neurons. *Neuron* 21:633–642.

Kent, W. J. 2002. BLAT: The BLAST like alignment tool. *Genome Research* 12:656–664.

Kitano, H. 2002. Systems biology: A brief overview. *Science* 295:1662.

Koonin, E. V., Y. I. Wolf, and G. P. Karev. 2002. The structure of the protein universe and genome evolution. *Nature* 420:218.

Kornberg, T. B. and M. A. Krasnow. 2000. The *Drosophila* genome sequence: Implications for biology and medicine. *Science* 287:2218–2220.

Ledent, V. and M. Vervoort. 2001. The basic helix–loop–helix protein family: Comparative genomics and phylogenetic analysis. *Genome Research* 11:754–770.

Lee, Y., R. Sultana, G. Pertea, J. Cho, S. Karamycheva, J. Tsai, et al. 2002. Cross-referencing eukaryotic genomes: TIGR orthologous gene alignments (TOGA). *Genome Research* 12:493–503.

Lightman, S. L., T. R. Insel, and C. D. Ingram. 2002. New genomic avenues in behavioural neuroendocrinology. *European Journal of Neuroscience* 16:369–372.

Martin, M. J., J. Herrero, A. Mateos, and J. Dopazo. 2003. Comparing bacterial genomes through conservation profiles. *Genome Research* 13:991–998.

Meyerowitz, E. M. 2002. Plants compared to animals: The broadest comparative study of development. *Science* 295:1482–1485.

Mouse Genome Sequencing Consortium. 2002. Initial sequencing and comparative analysis of the mouse genome. *Nature* 420:520–562.

Nash, W. G., J. Wienberg, M. A. Ferguson-Smith, J. C. Menninger, and S. J. O'Brien. 1998. Comparative genomics: Tracking chromosome evolution in the family Ursidae using reciprocal chromosome painting. *Cytogenetics and Cell Genetics* 83:182–192.

Navarro, A. and N. H. Barton. 2003. Chromosomal speciation and molecular divergence–accelerated evolution in rearranged chromosomes. *Science* 300:321.

Peichel, C. L., K. S. Nereng, K. A. Ohgi, B. L. Cole, P. F. Colosimo, C. A. Buerkle, et al. 2001. The genetic architecture of divergence between threespine stickleback species. *Nature* 414:901–905.

Pevzner, P. and G. Tesler. 2003. Genome rearrangements in mammalian evolution: Lessons from human and mouse genomes. *Genome Research* 13:37–45.

Reiter, L. T., L. Potocki, S. Chien, M. Gribskov, and E. Bier. 2001. A systematic analysis of human disease–associated gene sequences in *Drosophila melanogaster*. *Genome Research* 11:1114–1125.

Rubin, G. M. et al. 2000. Comparative genomics of the eukaryotes. *Science* 287:2204–2215.

Thanaraj, T. A., F. Clark, and J. Muilu. 2003. Conservation of human alternative splice events in mouse. *Nucleic Acids Research* 31:2544–2552.

Thomas, J. W. and J. W. Touchman. 2002. Vertebrate genome sequencing: Building a backbone for comparative genomics. *Trends in Genetics* 18:104–108.

Waterston, R. H. et al. 2002. Initial sequencing and comparative analysis of the mouse genome. *Nature* 420:520–562.

Young, L. J., M. M. Lim, B. Gingrich, and T. R. Insel. 2001. Cellular mechanisms of social attachment. *Hormones and Behavior* 40:133–138.

Norman C. Ellstrand

16

Crop Transgenes in Natural Populations

Public discussion of concerns surrounding the products of crop biotechnology has increased steadily over the past few years. One of the most frequently cited scientifically based concerns is that hybridization between engineered crops and wild plants may permit transgenes to escape into the environment (Ellstrand, 2003; Ellstrand and Hoffman, 1990; Hancock et al., 1996; Snow and Moran-Palma, 1997). The basis for this concern is that once transgenes move into unmanaged populations, they have the opportunity to multiply and spread via sexual reproduction. If the transgenes have unintended, unwanted effects in the wild, they may be difficult or impossible to contain. Is it possible to identify potential hazards created by the escape of crop transgenes and the likelihood that these impacts will occur? To a large extent, experience from traditionally improved crops provides a point of departure for answering this question.

In this chapter, I will review and evaluate what has come to be known as the gene flow issue: the potential for engineered crops to mate with wild relatives and the potential for the transfer of transgenes into those populations. I first focus on what is known from traditional agriculture concerning crop-to-wild gene flow in current agroecosystems and its consequences. Then, I discuss whether and how transgenes are different from the crop genes with which we have experience. Finally, I examine whether there is cause to worry about crop transgenes in natural populations.

Crop-to-Wild Gene Flow

Gene flow is defined by population geneticists as the successful migration of alleles from one population to another (Slatkin, 1987). In the case of plants, that migration may be by pollen or seed. Gene flow is a powerful evolutionary force that can sometimes counterbalance natural selection. That is, moderate rates of incoming gene flow (about 1–5 percent per generation) are expected to be sufficient to introduce and maintain alleles when the local selective coefficient against them is of the same magnitude (i.e., 1–5 percent) (Slatkin, 1987). Alternatively, when gene flow and selection work in concert, that is, the immigrant alleles are advantageous, gene flow will help accelerate the establishment and spread of those beneficial alleles (Wright, 1969).

In the case of crops, we are particularly interested in the migration of pollen that is moved by wind or animals from one crop population to the next. Plant breeders worry about what they call pollen contamination, which is gene flow from sources outside their breeding fields, and go to great lengths to ensure that their breeding sites are isolated from other plantations of the same species (Kernick, 1961).

Pollen may also move alleles from a crop to a population of a wild relative. Sunflower (*Helianthus annuus*) is one of the world's most important oilseed crops, and in the United States it is one of the top ten crops in terms of area planted (FAOSTAT Statistics Database, http://faostat.fao.org/). The wild form of the same species is a noxious weed in parts of the United States (Whitson et al., 1996) and can act as a weed of the crop form. If the crop grows close to the weed, experiments have shown that pollinating insects may move some of the crop's pollen into the weed populations, resulting in hybridization (Arias and Rieseberg, 1994). In fact, the gene flow rates occurred at levels that are biologically significant (much higher than the mutation rates). Hybrids are fertile under field conditions, although they are less fit than "pure" wild individuals (Cummings et al., 1999; Snow et al., 1998). The establishment of such hybrids is the first step in the establishment of crop alleles in natural populations. Population genetic analysis of wild sunflower populations that have been in contact with cultivated sunflower populations has demonstrated that crop alleles have established and spread through those wild populations (Linder et al., 1998; Whitton et al., 1997).

Therefore, if a field of transgenic sunflowers were grown near wild sunflowers, it is likely that the crop alleles would soon find themselves in the weed. If the transgenic trait were for herbicide resistance, that trait could provide a benefit to the weed, with the unintended result of rendering that herbicide ineffective for any crop in which wild sunflower is a problem. In fact, hybridization between domesticated sunflower and wild sunflower has already been implicated in the evolution of increased weediness in the wild plants (Heiser, 1978).

Whether crop-to-wild gene flow should be a topic of concern for transgenic crops is part of the broader question of whether this is a topic of concern for all crops, transgenic or not. To get an answer to that question, three factors must be addressed: Is spontaneous crop-to-wild gene flow a general feature of many important crops? Does that gene flow occur to a biologically significant extent between those forms? and Has such gene flow been known to cause problems? In the case of sunflower, we have seen that the answer to all three questions is affirmative. Is sunflower an exception or an example of a common characteristic of crop plants? Table 16.1 lists the world's thirteen most important food crops in terms of area planted. Whether they are known to spontaneously hybridize with wild relatives was examined in a recent review article (Ellstrand et al., 1999). The authors of that article chose those crops because their impact on human well-being made them likely to be among the best-studied domesticated plants. Some of the crops on that list comprise more than one species; as many species as were necessary were reviewed to account for a substantial majority of the area harvested (table 16.1). The number of crops reviewed was limited by considerations of time and space. The case studies represent a heterogeneous group of eighteen tropical, subtropical, and temperate species belonging to five different families.

The AGRICOLA (contents of the National Agricultural Library) and BioSciences Information Service of Biological Abstracts (BIOSIS) Previews bibliographic databases were points of departure for the literature search for information on whether the reviewed crops naturally hybridize with wild species. Additionally, one or more experts were consulted (see "Acknowledgments" in Ellstrand et al., 1999) to review the conclusions, to suggest other experts, and to identify gaps in the treatment. I have updated that information for the purposes of this chapter.

Table 16.2 summarizes what is known about natural hybridization between the thirteen most important food crops and their wild relatives. The first question we can ask is whether there is evidence of crop-to-wild gene flow for a given crop. All but one (groundnut) of the reviewed crops apparently hybridize naturally with wild relatives in some part of their agricultural distribution. For two species (barley and finger millet), the evidence is based solely on the identification of plants morphologically intermediate to the putative

Table 16.1 The world's 13 most important food crops in terms of area planted.

Rank	Crop	Scientific name	Area planted (kilohectares)[a]
1	Wheat	*Triticum aestivum*[b]	228,131
		T. turgidum[b]	
2	Rice	*Oryza sativa*	149,555
		O. glaberrima	
3	Maize	*Zea mays mays*	143,633
4	Soybean	*Glycine max*	67,450
5	Barley	*Hordeum vulgare*	65,310
6	Cotton	*Gossypium hirsutum*	51,290
		G. barbadense	
7	Sorghum	*Sorghum bicolor*	45,249
8	Millet	*Eleusine coracanac*	38,077
		Pennisetum glaucum[b]	
9	Beans	*Phaseolus vulgaris*[b]	28,671
10	Rapeseed	*Brassica napus*	24,044
		B. rapa	
11	Groundnut	*Arachis hypogaea*	23,647
12	Sunflower	*Helianthus annuus*	19,628
13	Sugar cane	*Saccharum officinarum*[b]	19,619

[a]Estimated area of production for 1997 from the 1998 FAOSTAT Web site, June 15, 1998.

[b]Other taxa account for only a small portion of world production of this crop.

Table 16.2 Spontaneous hybridization, weed evolution, and extinction risk for the world's most important food crops and their wild relatives.

Rank	Crop	Evidence of hybridization[a]	Implicated in weed evolution	Implicated in extinction risk
1	Wheat	C	No	No
		C	Yes	No
2	Rice	C	Yes	Yes
		C	No	No
3	Maize	C	No	No
4	Soybean	C	Yes	No
5	Barley	m	No	No
6	Cotton	C	No	Yes
		C	No	No
7	Sorghum	C	Yes	No
8	Millet	m	No	No
		C	Yes	No
9	Beans	C	Yes	No
10	Rapeseed	C	No	No
		C	No	No
11	Groundnut	None	n/a	n/a
12	Sunflower	C	Yes	No
13	Sugar cane	C	No	No

C = more substantial evidence of hybridization; m = morphological intermediacy only.

parents in regions where domesticated and wild forms are sympatric. This evidence is generally considered adequate for assigning hybridity but is weak compared with genetic evidence (Stace, 1975). Nonetheless, more often than not, the identification of hybridity by morphological intermediacy has been supported by additional genetic evidence (Rieseberg and Wendel, 1993). For most of the cases enumerated in table 16.2 (fifteen of the eighteen species examined), there is strong evidence of hybridity based on genetic information, confirming that natural hybridization occurs between those crops and their wild relatives.

The second question to ask is whether gene flow rates are biologically significant, because the occurrence of hybridization does not guarantee that the gene flow rate is high. Biologically significant gene flow occurs when the immigration rate is an order of magnitude greater than the mutation rate (Ellstrand et al., 1999). There are a growing number of studies measuring hybridization rates between crops and their wild relatives. Typically, these experiments involve a stand of crop plants surrounded by synthetic populations of a wild relative. The plants are chosen so that the crop bears an allele that is absent in the wild plants. Progeny testing of seeds harvested from the wild plants reveals the fraction that bear the allele specific to the domesticate, giving an estimate of the hybridization rate.

A few experiments have shown certain wild–domesticate pairs to be fully incompatible. Field experiments were conducted to test whether potato (*Solanum tuberosum*) spontaneously mates with the related species *Solanum nigrum* and *Solanum dulcamara*. McPartlan and Dale (1994) planted both species around a plot of potatoes transgenic for kanamycin resistance. Seeds were harvested from the weeds and germinated. Thousands of seedlings were screened; not one showed resistance to kanamycin. The measured hybridization rate was zero.

However, reports of experimentally measured crop-to-wild hybridization rates of zero are exceptional. In many cases, hybridization rates exceed 1 percent for cultivated–wild pairs as diverse as domesticated sunflower and wild sunflower (Arias and Rieseberg, 1994), sorghum and johnsongrass (Arriola and Ellstrand, 1996), squash and wild gourd (Kirkpatrick and Wilson, 1988), domesticated radish and wild radish (Klinger et al., 1991), and various cultivars of rice and red rice (Langevin et al., 1990). Hybridization rates greater than 1 percent have sometimes been detected over distances of hundreds of meters (Arias and Rieseberg, 1994; Kirkpatrick and Wilson, 1988; Klinger et al., 1991). At shorter distances, crop-to-wild hybridization rates may exceed 50 percent (Arias and Rieseberg, 1994; Klinger et al., 1991; Langevin et al., 1990; Rabinowitz et al., 1990). But in other cases, hybridization rates have been found to be quite low, even at short distances. Till-Bottrand et al. (1992) measured bilateral hybridization rates between foxtail millet (*Setaria italica*) and selected wild *Setaria* species growing in an experimental garden. At distances that were mostly less than 100 cm, they detected hybridization, but at rates lower than 1 percent (from 0 to 0.58 percent). Likewise, crop-to-wild gene flow rates may vary tremendously between plants and plots. For example, in an experiment involving a range of distances and target population sizes, gene flow from cultivated radish to wild radish varied from 0 to 79 percent, depending on treatment (Klinger et al., 1992). Clearly, crop-to-wild gene flow rates often are biologically significant, greatly exceeding mutation rates.

The third question to ask is whether crop-to-wild gene flow has had adverse consequences. The two most commonly discussed potential problems created by crop-to-wild gene flow are the evolution of more aggressive weeds, as discussed earlier in the case of sunflower, and the increased risk of extinction of an endangered species if pollen from a large crop plantation swamps out the tiny amount of pollen produced in nearby small populations of a rare species.

Are these hazards hypothetical? Let's return to the review of the world's thirteen most important crops. Increased weediness has evolved after hybridization for six of the thirteen crops examined. An additional example is the recent evolution of a weedy rye derived from natural hybridization between cultivated rye (*Secale cereale*) and wild *Secale montanum* in California. The weed's hybrid origin has been confirmed by genetic analysis (Sun and Corke, 1992). The weed has become such a serious problem that "farmers have abandoned efforts to grow cultivated rye for human consumption" (National Research Council, 1989:12).

Likewise, hybridization has been implicated as an agent of increased extinction risk in four of the thirteen crops examined. For example, natural hybridization with cultivated rice has been implicated in the near extinction of the endemic Taiwanese taxon *Oryza rufipogon* ssp. *formosana*. Collections of this wild rice over the last century show a progressive shift toward characters of the cultivated species and a coincidental decrease in fertility of seed and pollen (Kiang et al., 1979). Indeed, throughout Asia, typical specimens of other subspecies of *O. rufipogon* and the wild *Oryza nivara* are now rarely found because of extensive hybridization with the crop (Chang, 1995).

Crop-to-wild gene flow appears to be a common feature of domesticated plants. It can occur at evolutionarily significant levels, and it has been implicated in creating problems of human concern. Although those problems appear to be quite rare relative to the great amount of gene flow that occurs from crops to weeds, their impacts have been significant.

Enter Transgenic Crops

Should we expect transgenic crops to be different from traditionally improved crops? Regrettably, the answer is no. There is every reason to believe that transgenic plants will hybridize with wild relatives when they encounter them. We can predict that they will sometimes hybridize at rates that are biologically significant, and some small fraction of those hybrids will cause problems.

Can we expect transgenes in natural populations to behave in some way that is different from nonengineered genes? This is a more difficult question. The first generation of engineered crops has primary phenotypes that are roughly equivalent to those of conventionally improved varieties. That is, whether engineered or not, herbicide tolerance is herbicide tolerance, pest resistance is pest resistance, and so on. So it should be possible to predict the ecological consequences of, say, increased pest resistance in a wild plant, at least with some experimental work (Louda, 1999). But the type of genes used to engineer crops for pest resistance might be novel relative to those used by conventional breeders. For example, unengineered virus resistance has been accomplished using alleles preexisting in crops and their close relatives. But engineered virus resistance in plants usually is accomplished by inserting genes that produce the coat protein of the disease organism (Fitchen and Beachy, 1993). It is not clear whether alleles that accomplish the same primary phenotype by a different method will have different secondary phenotypes. Therefore, engineered genes may not be precisely equivalent to sexually introduced alleles in their effects on plant phenotype.

When to Worry?

Clearly, transgenes are going to enter natural populations. Like conventional crop alleles, those transgenes, on rare occasion, will create problems. Are there any ways to anticipate some of these problems? Let's first consider the standard problems associated with crop-to-wild hybridization: evolution of increased weediness and increased risk of extinction by hybridization.

Evolution of increased weediness depends, in part, on whether the transgene increases the recipient's fitness. This should be largely a function of the transgene's primary phenotype. Most of the first generation of transgenic crops have phenotypes that would serve as a benefit in a weed: resistance to pests and herbicides. If necessary, field experiments can demonstrate whether the transgenes confer a fitness advantage relative to genetically "pure" wild plants.

However, increased fitness is a necessary but not sufficient condition for increased weediness. In fact, certain alleles may confer a fitness benefit at one stage of a life cycle and a cost at another, possibly even resulting in the extinction of a population that is receiving them by gene flow (Muir and Howard, 1999). Indeed, predicting whether an introduced exotic species will become a noxious invasive or go extinct remains a largely unsolved challenge to those who study such species (Williamson, 1999). Nonetheless, information about the magnitude of a fitness change, the phenotype of the transgene, and the ecology of the transgenic crop and its relatives can be used for decision making. In the case of products that appear to have some risk, two of the many options could include placing a moratorium on the crop or monitoring it after commercialization. For example, genetically engineering (or otherwise creating) sunflowers for tolerance to a certain herbicide may not be worth the effort if that transgene is likely to move into the weed populations and render the herbicide obsolete.

Much more consideration has been given to the problem of extinction by hybridization. This is a general problem for both plants and animals. Extinction by hybridization can proceed very rapidly under the appropriate circumstances, whether crop alleles are neutral, beneficial, or detrimental. The topic has received a tremendous amount of theoretical attention in the past fifteen years (e.g., Ellstrand and Elam, 1993; Hopper, 1995; Huxel, 1999; Levin, 2001; Levin et al., 1996; Rhymer and Simberloff, 1996; Ribeiro and Spielman, 1986; Wolf et al., 2001).

The problem of extinction of wild relatives depends on a number of factors, especially the hybridization rate between a crop and its wild relative. Some obvious examples of transgenes increasing the hybridization rate include bringing a crop into closer proximity to a wild relative or somehow increasing a crop's outcrossing ability (cf. Bergelson et al., 1998). Therefore the theoretical models can serve as a point of departure for making a decision about whether a transgenic crop (or any other crop, for that matter) might increase a wild species' risk of extinction through increased hybridization. Our experience with nonengineered crops provides information for making informed choices that should reduce certain risks. But are there risks that cannot be easily anticipated? Consider pleiotropy, the control of a single gene of several distinct and seemingly unrelated effects. A good example of pleiotropy is associated with the allele that causes albinism in humans. That allele prevents the creation of a well-characterized biochemical end product, melanin. Because melanin is the compound that accounts for much of our coloration, homozygous individuals have white skin and white hair. However, homozygous individuals have other characteristics that might not be readily anticipated: skin roughness, high frequency of skin cancer, nearsightedness, structural abnormalities of the eye, and hearing problems (Mange and Mange, 1990). Pleiotropy is also well known for plant genes. More than 100 documented examples are in the databases of the "Demeter's Genomes" Web site (http://www.gramene.org/).

Furthermore, unanticipated, nonadditive phenotypes are extremely common in plant hybrids. Studies of artificially created plant hybrids typically show that about 45 percent of a hybrid's characters are intermediate between the parents, another 45 percent are the same as one parent or the other, and the remaining 10 percent are actually more extreme than either parent (Rieseberg and Ellstrand, 1993). Although a transgene's pleiotropic effects may be well characterized in a transgenic crop, it is hard to predict how they will play out in a wild genome. Even harder to predict is how an absolutely novel phenotype, such as the production of a pharmaceutical compound, would be expressed in the context of a wild genome.

Nonetheless, unanticipated effects are unanticipated effects. Lessons from traditional breeding serve as warnings for new methods of improving domesticated organisms. When honeybees from Africa were brought to a honeybee breeding program in Brazil, no one anticipated that their alleles would escape and enter the natural populations of honeybees in the New World, creating human hardship (Camazine and Morse, 1988). Likewise, when a method of creating hybrid varieties of sugar beets started in Europe, no one anticipated that hybridization with wild beets would lead to the evolution of a new weed, resulting in multi–billion-dollar losses and altering the dynamics of the world sugar trade (Boudry et al., 1994).

The problem with unpredictable effects is that they are, well, unpredictable.

A Need for Thought Experiments

I advocate a mindful approach to risk assessment. Those who are developing new products may be caught up in the enthusiasm over their potential benefits, but the same people work with the plants and may be best qualified to anticipate what it takes to create the best possible product. Being well acquainted with the plants is a necessary first step. Another necessity is to consider those plants in a variety of contexts, particularly the environmental and social contexts (Rissler and Mellon, 1996; Scientists' Working Group on Biosafety, 1998).

Let's take "Terminator" as an example. "Terminator" is a theoretical genetic construct designed to protect intellectual property (Crouch, 1998). Plants with the gene activated produce seed that is fully mature but dead. The construct has dominant expression.

Let's first examine what happens if the transgene spreads between different crops. If I'm growing organic corn downwind from your field of "Terminator" plants, viable "Terminator" pollen will fertilize some of my seed. If I save my seed to replant, then some fraction of my seeds is going to be dead, and that's going to make me unhappy.

Now, let's examine how "Terminator" might affect natural populations. If the wild relative receiving pollen from activated "Terminator" plants is considered a noxious weed by farmers, then that weed will bear sterile hybrid progeny. In this case, "Terminator" would be considered beneficial. Of course, some people may use that weed as a potherb, a traditional remedy, or an ornamental. In this case, "Terminator" would be considered detrimental. If the wild relative receiving pollen from activated "Terminator" plants is an endangered species, that endangered relative will bear sterile hybrid progeny. This is another case in which "Terminator" would be considered detrimental. This thought experiment demonstrates that identifying some of the unintended consequences of genetic engineering are straightforward. It also suggests that at least some risks are context dependent and that those conducting risk analysis need to think widely about the consequences of a new product (National Research Council, 1996).

Is it possible to say anything about pleiotropic effects or other unanticipated effects? Probably not. As proposed, the "Terminator" construct involves a minimum of five separate genetic elements. It is hard to imagine that there won't be some unanticipated effects, especially in hybrid progeny.

Conclusions

The lessons from traditional agriculture show that gene flow from crops to their wild relatives occurs, is sometimes biologically significant, and can create the headaches of weed evolution and extinction of wild relatives. We know enough about transgenic plants to predict that they will not be much different from traditionally improved plants, but that's not necessarily good news. Although it will certainly be possible to anticipate some potential problems, it is impossible to anticipate them all. This uncertainty might make some scientists nervous, but it is a good cure for our hubris. Perhaps H. L. Mencken said it best: "Penetrating so many secrets, we cease to believe in the unknowable. But there it sits nevertheless, calmly licking its chops."

The good news is that the controversies over transgenic products require scientists to come out of the lab or the field to educate both the public and each other, raising the scientific awareness of all of us.

Acknowledgments

I thank Oliver Ryder for his encouragement to write this chapter. I wrote this chapter while receiving support from U.S. Department of Agriculture Grant #00-33120-9801. Some of the work done to accumulate the data in my tables was supported by a Fulbright Award to Sweden and grants from the U.S. Department of Agriculture and Skogs-och Jordbrukets Forskningsråd of Sweden.

References

Arias, D. M. and L. H. Rieseberg. 1994. Gene flow between cultivated and wild sunflowers. *Theoretical and Applied Genetics* 89:655–660.

Arriola, P. E. and N. C. Ellstrand. 1996. Crop-to-weed gene flow in the genus *Sorghum* (Poaceae): Spontaneous interspecific hybridization between johnsongrass, *Sorghum halepense,* and crop sorghum, *S. bicolor. American Journal of Botany* 83:1153–1160.

Bergelson, J., C. B. Purrington, and G. Wichmann. 1998. Promiscuity in transgenic plants. *Nature* 395:25.

Boudry, P., M. Mrchen, P. Saumitou-Laprade, P. Vernet, and H. T. van Dijk. 1994. Gene escape in transgenic sugar beet: What can be learned from molecular studies of weed beet populations? In *Proceedings of the 3rd International Symposium on the Biosafety Results of Field Tests of Genetically Modified Plants and Microorganisms,* ed. D. D. Jones, 75–87. Oakland: University of California Division of Agriculture and Natural Resources.

Camazine, S. and R. A. Morse. 1988. The Africanized honeybee. *American Scientist* 76: 465–471.

Chang, T. T. 1995. Rice. In *Evolution of crop plants,* ed. J. Smartt and N. W. Simmonds, 147–155. 2nd ed. Harlow, U.K.: Longman.

Crouch, M. L. 1998. *How the "Terminator" terminates: An explanation for the non-scientist of a remarkable patent for killing second generation seeds of crop plants.* Edmonds, Wash.: The Edmonds Institute.

Cummings, C. L., H. M. Alexander, and A. A. Snow. 1999. Increased pre-dispersal seed predation in sunflower crop–wild hybrids. *Oecologia* 121:330–338.

Ellstrand, N. C. 2003. *Dangerous liaisons? When crops mate with their wild relatives.* Baltimore: Johns Hopkins University Press.

Ellstrand, N. C. and D. R. Elam. 1993. Population genetic consequences of small population size: Implications for plant conservation. *Annual Review of Ecology and Systematics* 24:217–242.

Ellstrand, N. C. and C. A. Hoffman. 1990. Hybridization as an avenue for escape of engineered genes. *BioScience* 40:438–442.

Ellstrand, N. C., H. C. Prentice, and J. F. Hancock. 1999. Gene flow and introgression from domesticated plants into their wild relatives. *Annual Review of Ecology and Systematics* 30:539–563.

Fitchen, J. H. and R. N. Beachy. 1993. Genetically engineered protection against viruses in transgenic plants. *Annual Review of Microbiology* 47:739–763.

Hancock, J. K., R. Grumet, and S. C. Hokanson. 1996. The opportunity for escape of engineered genes from transgenic crops. *HortScience* 31:1080–1085.

Heiser, C. B. 1978. Taxonomy of *Helianthus* and the origin of domesticated sunflower. In *Sunflower science and technology,* ed. J. F. Carter, 31–54. Madison, Wis.: American Society of Agronomy, Crop Science Society and Soil Science Society of America.

Hopper, S. D. 1995. Evolutionary networks: Natural hybridization and its conservation significance. In *Nature conservation 4: The role of networks,* ed. D. A. Saunders, J. L. Craig, and E. M. Mattiske, 51–66. Chipping Norton, N.S.W.: Surrey Beatty & Sons.

Huxel, G. R. 1999. Rapid displacement of native species by invasive species: Effect of hybridization. *Biological Conservation* 89:143–152.

Kernick, M. D. 1961. Seed production of specific crops. In *Agricultural and horticultural seeds,* 181–547. Rome: FAO.

Kiang, Y. T., J. Antonovics, and L. Wu. 1979. The extinction of wild rice (*Oryza perennis formosana*) in Taiwan. *Journal of Asian Ecology* 1:1–9.

Kirkpatrick, K. J. and H. D. Wilson. 1988. Interspecific gene flow in Cucurbita: *C. texana* vs. *C. pepo. American Journal of Botany* 75:519–527.

Klinger, T., P. E. Arriola, and N. C. Ellstrand. 1992. Crop–weed hybridization in radish (*Raphanus sativus* L.): Effects of distance and population size. *American Journal of Botany* 79:1431–1435.

Klinger, T., D. R. Elam, and N. C. Ellstrand. 1991. Radish as a model system for the study of engineered gene escape rates via crop–weed mating. *Conservation Biology* 5:531–535.

Langevin, S., K. Clay, and J. Grace. 1990. The incidence and effects of hybridization between cultivated rice and its related weed red rice (*Oryza sativa* L.). *Evolution* 44:1000–1008.

Levin, D. A. 2001. The congener as an agent of extermination and rescue of rare species. In *Evolutionary conservation biology,* ed. R. Ferriere, U. Dieckmann, and D. Couvet, 344–355. Laxenburg, Austria: International Institute for Applied Systems Analysis.

Levin, D. A., J. Francisco-Ortega, and R. K. Jansen. 1996. Hybridization and the extinction of rare plant species. *Conservation Biology* 10:10–16.

Linder, C. R., I. Taha, G. J. Seiler, A. A. Snow, and L. H. Rieseberg. 1998. Long-term introgression of crop genes into wild sunflower populations. *Theoretical and Applied Genetics* 96:339–347.

Louda, S. 1999. Insect limitation of weedy plants and its ecological implications. In *Ecological effects of pest resistance genes in managed ecosystems,* ed. P. L. Traynor and J. J. Westwood, 43–48. Blacksburg, Va.: Information Systems for Biotechnology.

Mange, A. P. and E. J. Mange. 1990. *Genetics: Human aspects.* 2nd ed. Sunderland, Mass.: Sinauer.

McPartlan, H. C. and P. J. Dale. 1994. An assessment of gene transfer by pollen from field-grown transgenic potatoes to non-transgenic potatoes and related species. *Transgenic Research* 3:216–225.

Muir, W. M. and R. D. Howard. 1999. Possible ecological risks of transgenic organism release when transgenes affect mating success: Sexual selection and the Trojan gene hypothesis. *Proceedings of the National Academy of Sciences, USA* 92:13853–13856.

National Research Council. 1989. *Field testing genetically modified organisms: Framework for decisions.* Washington, D.C.: National Academy Press.

——. 1996. *Understanding risk.* Washington, D.C.: National Academy Press.

Rabinowitz, D., C. R. Linder, R. Ortega, D. Begazo, H. Murguia, D. S. Douches, et al. 1990. High levels of interspecific hybridization between *Solanum sparsipilum* and *S. stenotomum* in experimental plots in the Andes. *American Potato Journal* 67:73–81.

Rhymer, J. M. and D. Simberloff. 1996. Extinction by hybridization and introgression. *Annual Review of Ecology and. Systematics* 27:83–109.

Ribeiro, J. M. C. and A. Spielman. 1986. The satyr effect: A model predicting parapatry and species extinction. *American Naturalist* 128:513–528.

Rieseberg, L. H. and N. C. Ellstrand. 1993. What can molecular and morphological markers tell us about plant hybridization? *Critical Reviews in Plant Sciences* 12:213–241.

Rieseberg, L. H. and J. F. Wendel. 1993. Introgression and its consequences in plants. In *Hybrid zones and the evolutionary process,* ed. R. Harrison, 70–109. New York: Oxford University Press.

Rissler, J. and M. Mellon. 1996. *The ecological risks of engineered crops.* Cambridge, Mass.: MIT Press.

Scientists' Working Group on Biosafety. 1998. *Manual for assessing ecological and human health effects of genetically engineered organisms.* Edmonds, Wash.: The Edmonds Institute.

Slatkin, M. 1987. Gene flow and the geographic structure of natural populations. *Science* 236:787–792.

Snow, A. A. and P. Moran-Palma. 1997. Commercialization of transgenic plants: Potential ecological risks. *BioScience* 47:86–96.

Snow, A. A., P. Moran-Palma, L. H. Rieseberg, A. Wszelaki, and G. J. Seiler. 1998. Fecundity, phenology, and seed dormancy of F_1 wild–crop hybrids in sunflower (*Helianthus annuus,* Asteraceae). *American Journal of Botany* 85:794–801.

Stace, C. A. 1975. *Hybridization and the flora of the British Isles.* London: Academic Press.

Sun, M. and H. Corke. 1992. Population genetics of colonizing success of weedy rye in northern California. *Theoretical and Applied Genetics* 83:321–329.

Till-Bottrand, I., X. Reboud, P. Braband, M. Lefranc, B. Kherissi, F. Vedel, et al. 1992. Outcrossing and hybridization in wild and cultivated foxtail millets: Consequences for the release of transgenic crops. *Theoretical and Applied Genetics* 83:940–946.

Whitson, T. D., L. C. Burrill, S. A. Dewey, D. W. Cudney, B. E. Nelson, R. D. Lee, et al. 1996. *Weeds of the West.* 5th ed. Jackson, Wyo.: Western Society of Weed Science.

Whitton, J., D. E. Wolf, D. M. Arias, A. A. Snow, and L. H. Rieseberg. 1997. The persistence of cultivar alleles in wild populations of sunflowers five generations after hybridization. *Theoretical and Applied Genetics* 95:33–40.

Williamson, M. 1999. Invasions. *Ecography* 22:5–12.

Wolf, D. E., N. Takebayashi, and L. H. Rieseberg. 2001. Predicting the risk of extinction through hybridization. *Conservation Biology* 15:1039–1053.

Wright, S. 1969. *Evolution and the genetics of populations. Vol. 2. The theory of gene frequencies.* Chicago: University of Chicago Press.

Anne McLaren

17

The Role of Assisted Reproduction in Animal Conservation

In the 1960s and 1970s, the exponential increase in world population at last penetrated social consciousness. The regulation of human fertility, the prevention of unwanted pregnancies, became a major target of scientific funding bodies. In the 1980s and 1990s, however, social consciousness tended to shift away from the world population problem and toward problems of the environment, sustainable development, and biodiversity. At the same time, the motivation for research of human reproduction shifted away from the regulation of fertility toward the alleviation of infertility. Louise Brown, born in 1978, was the first baby conceived by in vitro fertilization (IVF) and embryo transfer. IVF was initially used only as a treatment for women with blocked fallopian tubes, but its use rapidly expanded. For women and couples in rich countries, the reproductive options became ever greater. People could exercise greater control over whether to have babies, how many, and when. If they had an infertility problem, they could try IVF, intracytoplasmic sperm injection (ICSI), hormone-induced ovulation, or donor insemination. Ultrasound scanning, amniocentesis, chorionic villus sampling, and preimplantation genetic diagnosis (PGD) could be used to predict birth defects.

The rapid spread of IVF, in particular, led to a need for some regulation, and in the United Kingdom the government set up the Warnock Committee on Human Fertilization and Embryology, which published its report in 1984. This was followed six years later by an act in the British Parliament. The Warnock Committee used the term "assisted reproduction" rather than "artificial reproduction," which had been fashionable previously. The different aspects of assisted reproduction, and how it might assist in conservation, are the topic of this chapter.

Artificial Insemination

Artificial insemination (AI) was pioneered in farm animals and is used widely in the cattle industry. It is also used extensively in zoos and is particularly successful in bovids and felines, and even, happily, in pandas.

The usefulness of AI increases enormously if sperm can be stored. Cryopreservation works very well for cattle and human sperm and is used for many other species also. For more difficult species, such as the pig, research is in progress to develop more effective techniques of cryopreservation, such as vitrification. Today there are efficient methods for freezing mouse sperm, although for many years it was regarded as difficult or impossible to cryopreserve. There seems no good reason why protocols for freezing sperm of any mammalian species could not be devised. For endangered species, this is tremendously important, not just to facilitate captive breeding but, more particularly, to allow the genetic management of rare species, since frozen sperm can readily be transported from one part of the world to another.

Eggs

Eggs are a much tougher proposition. In cows, humans, and many other mammals, the endocrinology of the reproductive cycle and oogenesis are understood sufficiently well to allow timed ovulation and mating. Through superovulation, the number of eggs maturing can be increased dramatically, but litter size does not increase proportionately, so the eggs that are produced must be recovered and fertilized in vitro. If eggs could be cryopreserved, it would greatly assist the logistics of IVF, but so far the large volume of cytoplasm in the egg has proved to be a problem for freezing. Even in the human, little more than a dozen babies have been born from frozen eggs, and there is a real risk of inducing chromosome abnormalities. The chromosomes are not contained within a nucleus, and because the second metaphase spindle tends to disintegrate on freezing and thawing, the chromosomes may become scattered. In immature oocytes, the chromosomes are contained within a nucleus, but there are problems in maturing immature oocytes in vitro.

Embryos

Fertilized eggs or preimplantation embryos can be cryopreserved much more successfully than unfertilized eggs. IVF followed by embryo freezing is a technology that should certainly be exploited for endangered species wherever eggs and sperm are available, even if no surrogate mother is in prospect.

IVF requires many fewer sperm than AI, in which sperm must be reasonably motile for fertilization to occur. Fertilization can also be achieved by injecting a single sperm under the zona pellucida (subzonal insemination). ICSI, in which a single sperm is injected into the cytoplasm of the egg, has proved remarkably successful in both human and mouse, even with nonmotile or immature sperm. Using ICSI, it has been possible to achieve successful fertilization with sperm that has been preserved for long periods of time in alcohol (Tateno et al., 1998) or even freeze-dried (Wakayama et al., 1998) and kept at room temperature. This suggests that there may be cheaper and easier ways than cryopreservation to preserve the genetic material of highly endangered species for posterity.

A combination of in vitro maturation and ICSI has been used in the lowland gorilla (Kurz et al., 1999). Oocytes were removed from the antral follicles in the ovaries of a dead female, and the oocytes were matured in vitro and used for ICSI. Five cleavage-stage embryos were cryopreserved, awaiting a suitable surrogate for embryo transfer.

Embryo Transfer

From the point of view of captive breeding using assisted reproduction techniques, the biggest problem of all is the transfer of embryos to the uterus. The first calf from transferred embryos was born in 1950, but it is still not an easy technique. The surrogate mother has to be at the right stage of the cycle, and her own uterus must be receptive to supporting a pregnancy. For an endangered species, if females of the same species are available, AI may well be a better option.

If a species is endangered, it is unlikely that suitable surrogate females of the same species will be available. An alternative would be to use a surrogate female of a closely related species, but not many successes have been achieved. Reciprocal embryo transfers even between sheep and goat fail, possibly for immunological reasons. Additional progesterone is thought to minimize immune responses in the uterus; alternatively, it might be possible by genetic manipulation of the immune system to create for each group of species a sort of universal recipient whose uterus would be hospitable to any embryo. This approach would be analogous to growing transgenic pigs as xenograft donors for transplantation of organs to humans.

Embryo Splitting

Splitting of the preimplantation embryo into two or more cells or groups of cells increases the yield in cows and sheep and probably would also do so in their wild relatives. The technique has not been widely used in farm animal breeding because the young tend to be oversized at birth. For other mammalian species, the chances of getting one live-born offspring would be fewer if the early embryo was divided into two or more parts than if the single intact embryo was transferred, so this form of cloning is unlikely to have a wide application in conservation.

Nuclear Transfer Cloning

Details of nuclear transfer techniques and its potential value to conservation are described in chapter 18, so here I wish only to mention some possibilities and problems, with one or two examples.

With the technology at its present level, one would need several, perhaps many enucleated oocytes for each live-born clone. This means that for purposes of reproduction, unless there is absolutely no sperm available, carrying out ICSI on the available eggs would be a better approach than enucleating them.

Within a species, nuclear transfer cloning can be used to rescue a rare breed. For example, in the Enderby Island breed of cattle, adapted to sub-Antarctic conditions, there was only one elderly cow left alive and some frozen sperm. Somatic cells from the cow were cultured, and their nuclei were transferred to hundreds of enucleated eggs taken from females of standard cattle breeds (Wells et al., 1999). To date, one genuine Enderby female calf has been born and will be inseminated with frozen sperm in due course.

Between species, nuclear transfer is more of a problem. Lawrance Smith's group in Canada has transferred nuclei from *Bos indicus* embryos to enucleated *Bos taurus* oocytes, and calves have been born, but *Bos indicus* and *Bos taurus* are sometimes classified as subspecies rather than as separate species. What was particularly interesting in this experiment was that the mitochondria proved to be predominantly of the *taurus* type (Meirelles et al., 1999). Although some indicus mitochondrial DNA persisted, it was not preferentially replicated in the presence of the *indicus* nuclear genotype.

Neal First's group has shown that somatic nuclei from a number of different species injected into enucleated cow eggs will support development up to the morula or blastocyst stage (Dominkó et al., 1999). It seems unlikely that the embryos would develop further, but it is noteworthy that cow nuclei into cow oocytes developed no better than sheep or pig nuclei to cow, and monkey to cow proved the most successful of all. In a 1998 Internet announcement, Advanced Cell Technology claimed to have derived human embryonic stem cells from human somatic nuclei injected into enucleated cow eggs, but no further information has been published.

Finally, a heroic attempt is being made by Alan Trounson's group to save the northern hairy-nosed wombat, of which only about eighty-five individuals remain. They are culturing and cryopreserving fibroblast cells from as many of the hairy-nosed creatures as possible to preserve what genetic diversity remains, with the eventual aim of transferring the cultured cell nuclei into enucleated oocytes of the closely related common wombats, who, it is hoped, would also function as adequate surrogate mothers (Wolvekamp et al., 2000).

Conclusions

The approaches outlined in this chapter depend crucially on adequate knowledge of the reproductive biology of the species concerned. For most endangered mammalian species, this knowledge is minimal, sometimes nonexistent. Extrapolations may be made from closely related domestic or laboratory species, but these are often misleading. For nonmammalian vertebrates, the situation is even worse, and for most invertebrates our lack of understanding of their reproductive biology is matched only by our ignorance of how many species exist and which are endangered. There is an urgent need for more funding of research into reproductive biology, on as wide a comparative basis as possible.

Despite all our efforts, many extinctions are going to occur over the next few decades. DNA sequencing is rapidly becoming routine, and a record of the genome sequence of an extinct species could at least allow its genetic relationships to be explored and an evolutionary pathway established. It is imperative that DNA samples or appropriately preserved cells and tissues be stored for all species know to be endangered. To this end, we are establishing a Web site in which relevant information is collected and significant gaps identified (Ryder et al., 2000a, 2000b). This project forms parts of the 2001–2002 International Biodiversity Observation Year, initiated by DIVERSITAS. In the United Kingdom a parallel initiative, the Frozen Ark, is to be established in association with the

Natural History Museum of London and bodies elsewhere in the world that are interested in tissue and DNA preservation.

Acknowledgment

I am grateful to the Wellcome Trust for financial support.

References

Dominko, T., M. Mitalipova, B. Haley, Z. Beyhan, F. Memili, B. McKusick, et al. 1999. Bovine oocytes cytoplasm supports development of embryos produced by nuclear transfer of somatic cell nuclei from various mammalian species. *Biology of Reproduction* 60:1496–1502.

Kurz, S. G., M. R. Healy, N. M. Loskutoff, C. S. Brown, E. G. Crichton, A. M. Barnes, et al. 1999. In-vitro maturation and intracytoplasmic sperm injection of western lowland gorilla oocytes. *Theriogenology* 51:361 (abstr.).

Meirelles, F. V., V. Bordignon, Y. F. Watanabe, M. R. Watanabe, A. Dayan, R. B. Lobo, et al. 1999. Zygote reconstructions among *Bos indicus* and *Bos taurus* cattle and consequences on mitochondrial inheritance. *Theriogenology* 51:2989 (abstr.).

Ryder, O. A., A. McLaren, S. Brenner, Y.-A. Zhang, and K. Benirschke. 2000a. DNA banks for endangered animal species. *Science* 288:275–277.

Ryder, O. A., A. McLaren, S. Brenner, Y.-A. Zhang, and K. Benirschke. 2000b. Preservation of DNA from endangered species. *Science* 289:725–727.

Tateno, H., T. Wakayama, W. S. Ward, and R. Yanagimachi. 1998. Can alcohol retain the reproductive and genetic potential of sperm nuclei? Chromosome analysis of mouse spermatozoa stored in alcohol. *Zygote* 6:233–238.

Wakayama, T., A. F. C. Perry, M. Zuccotti, K. Johnson, and R. Yanagimachi. 1998. Full term development of mice from enucleated oocytes injected with cumulus cell nuclei. *Nature* 394:369–374.

Wells, D. N., P. M. Misica, H. R. Tervit, and W. H. Vivanco. 1999. The use of adult somatic cell nuclear transfer to preserve the last surviving cow of the Enderby Island cattle breed. *Theriogenology* 51:217 (abstr.).

Wolvekamp, M. C. J., M. Cleary, J. Shaw, S.-L. Cox, G. Jenkin, and A. Trounson. 2000. Novel approach to save the critically endangered northern hairy-nosed wombat (*Lasiorhinus krefftii*). *Theriogenology* 53:345 (abstr.).

Ian Wilmut and Lesley Paterson

18

Conservation and Cloning: The Challenges

The birth of Dolly the cloned sheep created many new opportunities in medicine, biology, and research, many of which are discussed elsewhere. There has also been a great deal of discussion about the use of cloning techniques to create copies of endangered or even extinct species. In order to consider this suggestion, we will first describe briefly the method of nuclear transfer and summarize the achievements and limitations of the present procedures. The resources and techniques needed if cloning is to be used in conservation will become apparent from these descriptions.

Cloning Procedures

Cloning is the technique of transferring a nucleus from a body cell to another cell, usually an unfertilized egg, previously stripped of its own genetic material. The resulting embryo is cultured in vitro for several days, and if development is normal, it is implanted into the uterus of a surrogate mother to develop to term. The offspring produced from cloning have the same genetic makeup as the implanted nucleus, not of the surrogate mother.

Cloning is evidently possible but has yet to become a practical procedure. Although the cloning successes are often remembered—Megan, Morag, and Dolly the cloned sheep (Campbell et al., 1996; Wilmut et al., 1997), cloned mice (Wakayama et al., 1998), cloned cows (Kato et al., 1998), cloned goats (Baguisi et al., 1999), and cloned pigs (Onishi et al., 2000)—for every cloned animal that is born alive, a far greater number of embryos or fetuses have died at some stage during their development.

In animals in which cloning has been successful, the method used is far from optimal. Following nuclear transfer, less than 3 percent of the newly created embryos become live offspring (Westhusin et al., 2001). After implantation of the embryos into a surrogate mother, problems continue to arise as many embryos and fetuses are lost (Wilmut et al., 1997). For the females that are able to carry their offspring to term, a high proportion of the young die shortly after birth (Campbell et al., 1996; Wilmut et al., 1997). There is also the problem of large offspring syndrome. When embryos are created in vitro, the offspring often have a significantly higher birth weight. This can lead to complications during delivery and can also be linked to developmental abnormalities of the organs (see Young et al., 1998).

The reasons for the developmental failure of most reconstructed embryos are being researched. More information is needed on the science behind cloning before the method can be improved: How does the oocyte's cytoplasm interact with the inserted nucleus? How are the inserted genes reprogrammed? What is the most efficient method to re-activate the reconstructed embryo? What are the causes of large offspring syndrome, and how can fetal development problems be resolved? These and many other questions remain to be answered before cloning productivity can be increased.

Resources Needed

Many resources are needed before cloning can be attempted: a source of mature oocytes, a source of donor cells, the ability to manipulate and coordinate the cell cycles of the donor cells and oocytes, methods for nuclear transfer and activation of the embryo, a means for embryo culture, and a supply of suitable surrogate mothers.

To obtain sufficient quantity of mature oocytes for cloning procedures, a method must be developed to induce the females to produce a very large number of these oocytes (superovulation). Once ovulation occurs, a means of harvesting the oocytes is needed. Furthermore, the mature oocytes must be at a very specific stage of their cell cycle at the time of nuclear transfer. One essential part of successful cloning is that the oocyte cytoplasm is able to reprogram the genes of the donor cell nucleus. Evidence shows that there is a narrow window of opportunity during which the oocyte cytoplasm is able to do this (Campbell et al., 1996). This window occurs during metaphase of the second meiotic division, when half of the chromosomes have been packaged into the first polar body and the remaining half of the chromosomes are lined up in the oocyte on a spindle. Finally, micromanipulation equipment is needed in order to remove the genetic material from the oocyte (enucleation).

The chosen donor cells must be isolated without injury and then cultured in vitro. There are likely to be differences between species in the needs for tissue culture. Despite extensive research, there are no fully effective methods for culturing pig fibroblasts. One key to cloning is to force the donor cells to leave their growth phase and enter a resting state, becoming quiescent (Campbell et al., 1996; Wilmut et al., 1997). A method is therefore needed to culture and manipulate the cells in vitro to reach this appropriate stage in the cycle. Evidence shows that inducing quiescence allows the cells to become reprogrammed.

The actual process of nuclear transfer can be achieved by one of two procedures. One technique fuses the cells together via an electrical pulse or addition of chemicals, the other removes the donor nuclei and microinjects them into the oocyte. At this stage, a method is needed to activate the newly formed embryo and initiate development, as would occur during normal fertilization. The means to sustain the embryos in vitro is essential, and after a period of cell division (assuming development proceeded normally), they can be implanted into surrogate females. Reconstructed embryos that are not implanted can be stored by cryopreservation (freezing); otherwise, a very large number of recipient females are necessary. Methods for the cryopreservation of cattle and sheep embryos have been developed (Rall, 1992), but other species, such as pigs, have proven to be very difficult. Again, there are important differences between species, suggesting that much more research may be needed for each one.

Even with all the necessary resources, successful cloning entails more than merely following a simple set of instructions. The cloning of every new species presents a unique set of problems. It will take a significant amount of time, effort, and money to develop the materials and methods needed for each step of the process. With respect to endangered and extinct species, the main obstacle is acquiring the requisite number of oocytes. At this time, the success rate for cloning is low, and as a result it generally takes many attempts to produce a single live offspring. In the recent experiment to clone an endangered wild ox (a gaur), 692 embryos were reconstructed. In this case, there was no practical difficulty because the recipient oocytes were from domestic cattle. Nevertheless, it is difficult to imagine such an effort being possible if the oocytes had to be recovered from a wild animal.

The large animals that have been cloned to date are all livestock. Because of their commercial interest, much work had already been completed on their reproductive and developmental biology before cloning was attempted. A very large number of donor cells, oocytes, and surrogate mothers are essential just to develop the methods, and these are not available for endangered and extinct species. At present, the only alternative therefore is to use females from other species to provide oocytes for interspecies nuclear transfer and to act as surrogate mothers.

Transfer Experiments Between Species

Research conducted on nuclear transfer using cells and oocytes from different species has achieved only limited success. Using the same basic techniques that made Dolly, the nuclei from the skin cells of an endangered argali wild sheep were transferred to cattle and domestic sheep enucleated oocytes (White et al., 1999). The newly formed wild sheep–cattle embryos showed some development in vitro but survived for only a short period of time. Of the reconstructed sheep–sheep embryos that were implanted, two ewes became pregnant, one with a domestic sheep–domestic sheep embryo and one with a wild sheep–domestic sheep hybrid embryo. Unfortunately, both fetuses were lost two months into the pregnancy. From these results alone, it appears that closely related species are needed to increase the chances of success.

Researchers in Wisconsin have reported that cows could provide the oocytes needed for nuclear transfer in all mammalian species. Reconstructed embryos were made using skin cell nuclei from a selection of mammalian species (cow, sheep, pig, monkey, rat) and enucleated cow oocytes (Dominiko et al., 1998; Mitalipova et al., 1998). The embryos underwent some very early developmental cell division in vitro, and 10 to 50 percent of the embryos reached the blastocyst stage (hollow balls of 100 or so cells that are sufficiently developed to implant). The researchers concluded that cow oocytes could act as a universal recipient for mammalian nuclear transfer (Mitalipova et al., 1998). However, additional research is necessary to determine whether the reconstructed hybrid embryos can develop to term. The species difference between the oocyte cytoplasm and nuclear genes may well prevent the embryo from progressing further if implanted into recipient females.

The only live birth produced from an interspecies reconstructed embryo is Noah. Noah was created from the skin cell nucleus of a gaur and an enucleated oocyte from its relative, the domestic cow (Lanza et al., 2000). To do so, 692 skin cell nuclei were fused with cow eggs, stripped of their DNA, from which developed eighty-one embryos. Forty-four of these were implanted into cows, resulting in eight pregnancies; five miscarried, two were induced to abort, and the remaining cow gave birth to Noah. Noah died forty-eight hours after birth, apparently from a bacterial infection.

Components in the cytoplasm govern the early stages of development by programming the genes. Therefore it is likely that if the two species are too dissimilar, then reprogramming of the genes would not be accurate. The results by White and Lanza support the hypothesis that cloning is more successful when the two species are closely related. In addition to nuclear genes, there are also genes within the mitochondria. Mitochondria are found in the oocyte cytoplasm, and therefore cloning by nuclear transfer results in the nuclear genes being derived from the donor cell and the mitochondrial genes from the recipient oocyte. Therefore, Noah was not 100 percent gaur because his mitochondrial genes were derived from the cow oocyte. In addition to this problem of "contaminating" the species genome, the mixture of mitochondrial genes and nuclear genes from different species and interaction of the nuclear genes with the oocyte cytoplasm may also disrupt the genetic programming and development of the hybrid embryo. Assuming that interspecies nuclear transfer is successful, there is also the need for a suitable surrogate mother, presumably of the same species from which the mature oocytes were procured.

Over the last fifty years, various combinations of interspecies embryo transfer have been tried. One technical requirement is the synchronization of the recipients' estrus cycle with the donor embryo, but because of variations between the species, this is not an easy task. Therefore a significant amount of knowledge is needed of the donors' and recipients' reproductive cycles and physiology. Aside from the optimal techniques and methods, the success of the transfer depends ultimately on the two species used. On the whole, embryos survive for only a short time, and therefore interspecies embryo transfer is often unsuccessful. Some mammalian embryos will develop inside a recipient from a different species from early cleavage up to the blastocyst stage but no further (see Adams, 1982). If implantation of the blastocysts into the foreign uterus does occur, it is often abnormal. Some exceptions do exist, and the embryo can develop to term, including sheep to goat and vice versa (Warwick and Berry, 1949), and horse to donkey (Allen, 1982), Przewalski's horse to domestic horse, and zebra to horse (Summers et al., 1987).

Cloning and Conservation

The live birth of Noah the gaur indicates that clones may one day be produced of endangered animals through interspecies nuclear transfer. Nevertheless, cloning is a very poor and inefficient method of reproducing and will not result in a large number of offspring. Even if offspring are created, a natural population could not be established with such a small number of genetically similar individuals because of the problems that would result from inbreeding.

If assisted reproduction is an absolute necessity, other approaches are more likely to be productive, including artificial insemination, in vitro fertilization, embryo culture and subdivision, artificial implantation, and the use of surrogate mothers (see Corley-Smith and Brandhorst, 1999).

Conclusions

This analysis suggests that present methods of cloning are unlikely to have a significant role in conservation. This reflects two separate factors. By definition, the resources of animals and oocytes are simply not available for endangered species, let alone those that are already extinct. Additionally, the differences between species make it very difficult to identify other species that are available in sufficient numbers and sufficiently close in cell and reproductive biology to be suitable as oocyte donors and recipient females. Cloning is difficult, and attempts are rarely successful. Considering the significant problems faced in cloning nonendangered animals, it is unlikely that this procedure will become a beneficial tool in the conservation of species.

Acknowledgments

The ideas discussed in this chapter were developed during research funded by the Biotechnology and Biological Sciences Research Council and the Ministry of Agriculture of Fisheries and Food.

References

Adams, C. E. 1982. Egg transfer in carnivores and rodents, between species and to ectopic sites. In *Mammalian egg transfer,* ed. C. E. Adams, 54–57. Boca Raton, Fla.: CRC Press.

Allen, W. R. 1982. Embryo transfer in the horse. In *Mammalian egg transfer,* ed. C. E. Adams, 149–151. Boca Raton, Fla.: CRC Press.

Baguisi, A., E. Behboodi, D. T. Melican, J. S. Pollock, M. M. Destrempes, C. Cammuso, et al. 1999. Production of goats by somatic cell nuclear transfer. *Nature Biotechnology* 17(5):456–461.

Campbell, K. H., J. McWhir, W. A. Ritchie, and I. Wilmut. 1996. Sheep cloned by nuclear transfer from a cultured cell line. *Nature* 380:64–66.

Corley-Smith, G. E. and B. P. Brandhorst. 1999. Preservation of endangered species and populations: A role for genome banking, somatic cell cloning and androgenesis? *Molecular Reproduction and Development* 53:363–367.

Dominko, T., M. Mitalipova, B. Haley, Z. Beyhan, E. Memili, and N. First. 1998. Bovine oocyte as a universal recipient cytoplasm in mammalian nuclear transfer. *Theriogenology* 49(1):385.

Kato, Y., T. Tani, Y. Sotomaru, K. Kurokawa, J. Kato, H. Doguchi, et al. 1998. Eight calves cloned from somatic cells of a single adult. *Science* 282:2095–2098.

Lanza, R. P., J. B. Cibelli, F. Diaz, C. T. Moraes, and P. W. Farin. 2000. Cloning of an endangered species (*Bos gaurus*) using interspecific nuclear transfer. *Cloning* 2(2):79–90.

Mitalipova, M., T. Dominko, B. Haley, Z. Beyhan, E. Memili, and N. First. 1998. Bovine oocyte cytoplasm reprograms somatic cell nuclei from various mammalian species. *Theriogenology* 49(1):389–389.

Onishi, A., M. Iwamoto, T. Akita, S. Mikawa, K. Takeda, T. Awata, et al. 2000. Pig cloning by microinjection of fetal fibroblast nuclei. *Science* 289(5482):1188–1190.

Rall, W. F. 1992. Cryopreservation of oocytes and embryos: Methods and applications. *Animal Reproduction Science* 28(1–4):237–245.

Summers, P. M., A. M. Shephard, J. K. Hodges, J. Kydd, M. S. Boyle, and W. R. Allen. 1987. Successful transfer of the embryos of Przewalski's horses (*Equus przewalskii*) and Grant's zebra (*E. burchelli*) to domestic mares (*E. caballus*). *Journal of Reproduction and Fertility* 80(1):13–20.

Wakayama, T., A. F. C. Perry, M. Zuccotti, K. Johnson, and R. Yanagimachi. 1998. Full-term development of mice from enucleated oocytes injected with cumulus cell nuclei. *Nature* 394:369–374.

Warwick, B. L. and R. O. Berry. 1949. Inter-generic and intra-specific embryo transfers. *Journal of Heredity* 40:297.

Westhusin, M. E., C. R. Long, T. Shin, J. R. Hill, C. R. Looney, J. H. Pryor, et al. 2001. Cloning to reproduce desired genotypes. *Theriogenology* 55:35–49.

White, K. L., T. D. Bunch, S. Mitalipov, and W. A. Reed. 1999. Establishment of pregnancy after transfer of nuclear transfer embryos produced from the fusion of argali (*Ovis ammon*) nuclei into domestic sheep (*Ovis aries*) enucleated oocytes. *Cloning* 1:47–54.

Wilmut, I., A. E. Schnieke, J. McWhir, A. J. Kind, and K. H. Campbell. 1997. Viable offspring derived from fetal and adult mammalian cells. *Nature* 385:810–813.

Young, L. E., K. D. Sinclair, and I. Wilmut. 1998. Large offspring syndrome in cattle and sheep. *Reviews of Reproduction* 3:155–163.

Part V

Policy, Law, and Philosophy of Conservation Biology in the Age of Genomics

The final section of this book contains two essays on philosophical issues in conservation genetics. These philosophical discussions address shifting paradigms in conservation genetics, legal issues pertinent to conservation biology in general and endangered species in particular, and nonbiological problems created by the new technological infusion into conservation. Gerald Flattman, Barbara Ruskin, and Nicholas Vogt describe many legal issues involved as a result of infusion of new technology into conservation. Subjects such as legal ownership, international law, and bio-repositories are examined in chapter 19. In chapter 20 Sandra Mitchell takes a philosophical approach to the process and relevance of genomics to conservation biology, particularly exploring how "complexities of biology and human values impinge on the means by which we can create sensible policies about genomics." Mitchell suggests that biology, unlike physics, cannot be reduced to a simple set of laws. Consequently, an approach to biology called integrative pluralism is the most reasonable way to view biology in general and conservation issues in particular.

Gerald J. Flattmann Jr.,
Barbara A. Ruskin,
and Nicholas Vogt

Can Our Laws Accommodate the New Conservation Genetics?

The importance of genomics and its central role in the future of biotechnology is widely recognized. Some have gone so far as to call genomics the world's third technology revolution, after the industrial and information revolutions.[1] As discussed in detail during the symposium "Conservation Genetics in the Age of Genomics," these new genetic technologies may enable society to recoup some of the losses suffered as a consequence of the severe depletion of the world's ecosystems and modern agricultural practices. They could even allow conservation biologists to clone endangered or extinct species from somatic cells, such as skin cells, in the hope of enhancing and protecting species diversity.

From a legal perspective, these new genetic technologies pose complex and profound questions that were barely contemplated when the existing legal framework was enacted. For example, we have no laws that directly address somatic cell cloning of any species type or the resurrection of extinct species. However, we believe that the current legal framework may prove adequate. This is so because that legal framework should govern the following:

- Access to information and knowledge: who holds the keys to today's conservation "libraries" (e.g., gene resource banks and bioinformatics)
- Property rights: who owns the products of these conservation efforts and whether any of these processes or products can be patented
- Environmental protection and conservation: how conservation efforts regarding endangered and extinct species are carried out and what species may be conserved
- Public safety: what risks are acceptable and how to manage them

Nevertheless, navigating this legal framework in a manner that promotes genetic diversity and the aims of conservation will be a challenge.

In this chapter we focus on some of the prominent new technologies being used in this field that were discussed during the symposium. First, we examine whole organism, somatic cell cloning as it relates to endangered and extinct species and explain the impact of the current legal framework on these new genetic technologies. Next, we touch on the incredible potential of the new libraries of conservation genetics—gene or tissue resource banks and bioinformatics databases—and discuss the impact of our current legal framework on these new technologies.

Cloning Technology and Conservation

Whole organism, somatic cell cloning has captured the imagination and interest of conservation biologists and the general public alike. This technique, most notably achieved by the Roslin Institute in Scotland with the cloning of Dolly, set the stage for scientists to clone animals from somatic cells such as skin and hair follicles. It offers fascinating promise and potential for biologists who seek to conserve, in laboratories, genetic resources at risk of extinction.

Although expensive, somatic cell cloning is already being used by conservation biologists. For example, symposium participant Dr. Phillip Damiani and his team of scientists have enjoyed a very public success in this field with the somatic cell cloning of Noah the gaur, the first cloned endangered species.[2] Moreover, as we and others posited at the symposium, this same technology may eventually be applied to additional endangered or threatened species. For example, there is no immediately apparent basis for assuming that the same technology could not be applied to species such as the Rwandan mountain gorilla or Przewalski's horse.

This prediction has already proven correct in some cases. For example, in April 2003 scientists successfully cloned another endangered bovine species, offering powerful evidence that cloning technology can play a role in preserving and even reconstituting threatened and endangered organisms. "The clone—a cattlelike creature known as a Javan banteng, native to Asian jungles—[was] grown from a single skin cell taken from a captive banteng before it died in 1980." Furthermore, in contrast to Noah the gaur, which died two days after its birth due to infection, the first-born banteng still appears to be healthy.[3]

There is no immediate reason to assume that the same technology could not be applied to extinct species. Depending on your perspective, this potential application is either science fiction—the stuff of Michael Crichton novels—or the inevitable next step in the cloning technology revolution. We leave it to others to decide what it means to "conserve" extinct species or whether the term *conservation* can rightly be used with reference to such lost species. Instead, we examine the possible application of the current legal framework to this technology.

Application of the Endangered Species Act to New Cloning Technology: Do Cloned Species Fall Under the Endangered Species Act?

The Endangered Species Act (ESA) governs "whether any species is an endangered species or a threatened species."[4] Before assessing the impact of the ESA on cloned, endangered, and extinct species, one must first consider whether these cloned species would even fall within the ambit of the ESA.

With respect to extinct species, soon after an organism is recreated and becomes the only living member of its species, should it be identified as "endangered" and become subject to the ESA? At this time, there is no definitive answer because the courts have yet to deal with this issue.

Furthermore, interspecies hybrids and chimeras historically were not protected under the ESA because of the infamous "hybrid policy."[5] Would a somatic cell–cloned organism born to a surrogate mother, such as Noah the gaur (which has the mitochondrial DNA of the mother species) be a hybrid or chimera under the ESA? Again, there is no definitive answer.

Accordingly, it is unclear whether the ESA would even apply to cloned extinct species or cloned endangered species born to surrogate mothers.

The purpose of the ESA is to "provide a means whereby the ecosystems upon which endangered species and threatened species depend may be conserved, to provide a program for the conservation of such endangered species and threatening species, and to take such steps as may be appropriate to achieve the purposes of the treaties and conventions."[6]

Assuming that cloned, endangered, and extinct species fall under the ESA, the next question is whether cloning would promote the aims and objectives of conservation under the ESA. To answer this question, one must determine whether cloning endangered and extinct species would promote biodiversity and conservation or endanger, and promote neglect of, existing habitats, species, and ecosystems.

A case can be made that cloning endangered species would promote the objectives of the ESA because

- Cloning could forestall extinction in the short run and preserve biodiversity (in a very limited sense).
- Cloning could promote the use of genome resource banks as virtual "cryozoos" to preserve genetic diversity.
- Used together with other strategies, cloning may save species that do not breed well in captivity, such as the cheetah or giant panda.
- Cloning could represent an accommodation of interests for certain economic activities and communities, such as farmers, ranchers, and others opposed to reintroduction of predator species such as the gray wolf.
- Cloning and the inevitable interest, debate, and discussion it generates could cause the public to focus on the ultimate underlying problems, such as habitat loss.

On the other hand, a case can be made that cloning of endangered species would hinder the aims and objectives of the ESA because

- Cloning would probably distract the government and the public from saving habitats, the preservation of which may be the most reasonable way to conserve and promote species diversity (a pressing concern in the current political climate).
- Cloning would be extremely costly and would probably displace other cost-effective conservation efforts.

- If not done responsibly and across populations, cloning could decrease biodiversity through the creation of a homogeneous, less diverse gene pool—one of the major problems already accelerating extinction rates for some species.
- Cloning may be considered unethical by some segments of the government and public.
- Cloning may be unsafe.

Similarly, with respect to extinct species, the usefulness of cloning and its impact under the ESA are unknown. Cloning could lead to reintroduction of a previously lost species back into the wild, thus increasing species diversity for at least the short term. However, would reintroducing previously extinct species back into the wild ultimately threaten species diversity? And would the time and money spent resurrecting extinct species be better spent on conserving and protecting currently endangered species and their existing habitats?

Notably, experimental activities such as cloning, which might otherwise run afoul of the ESA's prohibitions on the use of endangered species, may be explicitly permitted "for scientific purposes" or "for the establishment and maintenance of experimental populations."[7] But in this regard, the use of cloning to establish or maintain experimental populations in natural habitats may prove unrealistic. For example, cloned animals could prove to be even more fragile than captive-bred animals. Or they may threaten existing wild populations.

In sum, at this time it is difficult to assess whether the use of new cloning technologies would further or hinder the objectives of the ESA.

Ownership Rights to Cloned Species

Another important question with respect to whole organism, somatic cell cloning is who owns the products and processes of this technology. Can a cloned endangered or resurrected extinct species ever be owned? To put it more metaphorically, can we place a price tag on a dodo or a leash on a saber-toothed tiger? And perhaps more importantly, is it legal or ethical to do so?

The ownership of cloned, endangered species may very well violate the ESA and other conservation laws that prohibit the ownership of such species taken from the wild.[8] But one could also argue that cloned endangered species fall under an exception to the ESA, notably the experimental population exception.[9]

The ownership of previously extinct organisms and the products or pharmaceuticals they make may present different questions than the ownership of cloned endangered species because the organisms were "taken" from the wild long ago. Thus ownership of resurrected extinct species may not offend the ESA. Notably, "ownership" in this area typically means patentability. If an extinct species can be cloned, someone will apply for a patent on it. This reality leads to the following legal question: Are resurrected extinct organisms patentable?

Surprisingly, the answer so far has been "yes." Patents on extinct, resurrected organisms already exist. In early 1997, the U.S. Patent and Trademark Office (PTO) granted

Ambergene Corporation a patent that claims ancient microorganisms recovered from amber.[10] Although scientifically there is a big difference between resurrecting microorganisms and bringing a woolly mammoth back to life, the legal principles governing patentability of such resurrected organisms should be essentially the same.[11]

The ownership rights created by patents are based on the U.S. Constitution:

> The Congress shall have the power . . . to promote the progress of science and useful arts, by securing for limited times to authors and inventors the exclusive right to their respective writings and discoveries.[12]

By statute, one needs to prove several things to obtain a patent. One must prove that the claimed invention is useful (and not a product of nature), novel, and nonobvious.[13] In exchange for a patent, the patentee must also provide to the public an adequate written description of the invention and an adequate teaching of how to make the invention so one's peers can duplicate it after the expiration of the patent without undue experimentation.[14]

With those requirements in mind, would resurrected extinct species be patentable under U.S. patent law? In other words,

- Are they products of nature?
- Are they useful?
- Are they novel?
- Are they obvious?

First, are resurrected extinct species products of nature? Superficially, the answer is "yes," and that seems to foreclose patentability. But in 1980, the U.S. Supreme Court declared that "anything under the sun that is made by man" (i.e., the product of human manipulation) is patentable.[15] Since then, isolated clones, bacteria, and other organisms—such as the transgenic "Harvard mouse"—have been found patentable.[16] Accordingly, a strong argument exists that resurrected extinct species are not products of nature because they would be the products of human manipulation.

Second, are resurrected extinct species useful? To meet the statute, "use[s] must be more than a scientific curiosity," not based on a blanket "assertion that the organism is a valuable source of genetic material for cross-breeding or genetic engineering research experiments."[17] Rather, patent applicants must show that their new species have "'real world' value."[18] Based on this standard, resurrected extinct species would probably qualify as useful. Promoting biodiversity itself may be a use that the PTO would recognize. Moreover, these species may have commercial value in the farming industry or the pharmaceutical industry.[19] These commercial uses would probably meet the utility requirement.

Third, are resurrected extinct species novel? They appear not to be at first glance. Resurrected extinct species all existed at one time in nature, and, depending on the time of extinction, many were known to humans. But their existence may represent the recovery of a long lost organism—the practice of a lost art.[20] More than 150 years ago, the U.S. Supreme Court suggested that the practice of a lost art may constitute a novel invention:

> [The discoverer of a lost art] would not literally be the first and original inventor. But he would be the first to confer on the public the benefit of the invention. He would

> discover what is unknown, and communicate knowledge which the public had not the means of obtaining without his invention.[21]

As such, the "art" of recreating a dodo or woolly mammoth today would be undoubtedly new.[22] Whether the dodo or woolly mammoth itself would be considered a novel product of that lost art is an open question.

Fourth, are resurrected extinct species obvious? It depends. Many extinct species have genetically close living relatives. Is a woolly mammoth obvious in view of an African elephant? An Asian elephant? These essentially fact-based questions may need to be decided on a case-by-case basis.

Other questions will also be raised. Once the first extinct species is cloned, are all subsequently cloned extinct species obvious in view of that first clone? Does genetic relatedness of the species cloned matter, or is the patentable invention here the cloning method itself? Although answers to these questions have yet to be given, in the near future our courts may be faced with these issues.

Application of Other U.S. and International Laws

A host of other existing laws might also affect the introduction of previously extinct species or cloned endangered species into the wild or the marketplace. The following is a nonexhaustive list of those laws. The impact of these laws on the new cloning technologies is yet to be seen:

1. The Food, Drug and Cosmetics Act (21 U.S.C. § 301 et seq.) would apply:
    - The act would have jurisdiction over drug and food products generated from the revived organism to be introduced to the marketplace.
2. Environmental protection laws, such as the National Environmental Policy Act (42 U.S.C. § 4321 et seq.), would apply:
    - These laws would affect the introduction of any extinct species into the wild or the general environment.
    - The impact of introduction of these species would need to be assessed very carefully; perhaps experimental, controlled releases would be in order.
3. Laws concerning animal treatment, such as the Animal Welfare Act (7 U.S.C. § 2131 et seq.), may apply:
    - This law typically deals with treatment of lab animals. But it arguably governs treatment and care of any animal species generated experimentally by scientists from the moment it is considered alive.
4. Animal and Plant Health Inspection Service regulations (see 6 U.S.C. § 231):
    - These regulations present considerations that should not be ignored. We should have a more worthy objective than recreating these organisms solely for our amusement. And we undoubtedly have a responsibility to the organisms we recreate.
5. Customs and importation laws would apply:
    - The U.S. Customs Service would have jurisdiction if the organism is ever traded or imported across U.S. borders.

- Fish and Wildlife and other agencies may require licenses for import and export of such organisms or their tissue.

6. The National Forest Management Act (16 U.S.C. § 1600 et seq.) and its requirements to maintain and promote biological diversity in the nation's forests would apply.
7. A web of international laws and treaties relating to migrating birds, marine mammals, ocean fish, and trade in endangered species would also apply.
8. International biodiversity laws and policies, such as the Convention on International Trade in Extinct Species of Flora and Fauna and United Nations Program Agenda 21 would apply.

In short, the existing legal framework could affect nearly every aspect of the use of somatic cell cloning technologies to further the aims of conservation biology.

Gene and Tissue Resource Banks and Bioinformatics Databases

Powerful new technologies have recently emerged with the dawn of gene or tissue resource banks and bioinformatics databases. Whether these new technologies can be used in a manner that furthers the goals of conservation and species diversity and whether the existing legal framework can accommodate the issues posed by that use is an unanswered question.

New Resources for Conservation Biologists

In his 1997 book *Biodiversity: Exploring Values and Priorities in Conservation,* Dan Perlman refers to gene resource banks as akin to the libraries of the ancient world.[23] It was in those ancient libraries that scribes and other scholars essentially determined the fate of now classic texts; those they chose to copy by hand were preserved, and the others were irretrievably lost. This is a particularly apt metaphor to use as a backdrop for contemplating the social, legal, and ethical issues posed by these new technologies.

Conservation biologists are the new genetic librarians, who may very well control the ability of future generations to access the biological samples and data that are being carefully collected, stored, and catalogued today. Moreover, in developing countries these new technologies may offer the only practical, short-term way to promote species conservation because traditional methods are too expensive or too impractical given economic and political exigencies.[24] Skin or a blood sample, which can be easily obtained and catalogued or may already reside in a museum collection, may represent a means for conservation and future reintroduction of native species. A perfect example of the potential use of a gene resource bank is the cloning of the banteng. That clone resulted from a single skin cell taken from a captive banteng before it died in 1980.

Access, Control, and Ownership Issues

Among the significant legal issues that individuals and organizations administering gene resource banks and related databases will face are access, control, and ownership issues. For example, who will be allowed to access the genes and other biological components in gene or tissue resource banks? And for what purposes will they be allowed to access those banks? How will access to biological samples from gene and tissue resource banks be reviewed or regulated? What agreements will be put in place to protect ownership of the gene resource banks and databases? And what agreements will protect rights to information or ownership of potential inventions derived from gene banks or databases?

These and other questions of access, control, and ownership probably will be approached from different perspectives depending on who is involved. As we discussed at the symposium, nonprofit institutions or universities may very well want to use different approaches than for-profit corporations or research organizations. A number of existing legal rubrics potentially govern the access, control, and ownership issues presented by these new technologies.

Copyrights

The purpose of U.S. copyright law (17 U.S.C. § 101 et seq.) is to stimulate the creation of as many works of art, literature, music, and other "works of authorship" as possible, in order to benefit the public. Copyright law protects an author's creative expressions and creative arrangements of information. But copyright law does not protect information—such as genetic information—per se. Therefore copyrights probably would not protect gene resource banks and bioinformatics databases, except to the extent that these databases represent novel or creative arrangements of the information they hold.

Trade Secrets

Trade secrecy is one possible option for protecting these new technologies and databases. One may hold as a trade secret valuable ideas and information—such as a manufacturing process or a business method—that provide competitive advantages.[25] Thus, one may hold as a trade secret a gene or tissue resource bank or a biological database and associated analytical programs.

By holding this information as a trade secret, an institution may prohibit its employees and business associates from divulging it to others and prohibit competitors from using improper means to learn it or using the information in an unauthorized manner once disclosed.[26] Accordingly, the trade secret option may make sense for ventures that plan to commercialize collected resources or databases. However, retaining genetic and other biological information as trade secrets may be contrary to the goals of conservation biology and the goals of science in general.

Patents

Patent protection (35 U.S.C. § 101 et seq.) is also an available option for these new technologies. The U.S. Court of Appeals for the Federal Circuit, which hears all appeals from patent cases, recently confirmed on a number of occasions the legality of patenting useful methods of doing business involving computer programs and databases.[27]

The Federal Circuit explained that a patent claim may include in part an abstract idea (e.g., a mathematical algorithm, a computer program, or an information database) as long as it is applied in a practical manner to produce "a useful, concrete and tangible result."[28] But a mathematical algorithm itself—in the abstract—is not patentable subject matter.[29] In practical terms, this means that one can seek patent protection for methods or systems and even for products such as "a computer-readable storage medium" (e.g., a diskette or CD-ROM) that include formerly unpatentable information as long as they are applied to produce a useful and tangible result.

Many interpret the Federal Circuit's decisions to be broad enough to protect computer-assisted methods and systems of any kind, including bioinformatics systems, methods, and products. Notably, the PTO has recently issued numerous patents to universities and companies in the area of computer-assisted bioinformatics methods and databases.[30]

Will those who acquire patent protection for their gene resource banks and bioinformatic systems use it to promote species diversity? A patent gives its owner a limited monopoly on an invention. A patent is viewed as an agreement between the inventor and the public; in return for making and fully disclosing the invention to the public, the public grants the inventor the right to prevent others from making, using, selling, or offering to sell the invention for a limited period of time (usually seventeen years). Again, we leave you to consider whether patenting these new technologies will promote or hinder the aims and objectives of conservation genetics.

Contracts

Finally, the use of ordinary contacts between users and providers of these new genomic technologies to control access to collections and databases is another available option. This option is not mutually exclusive of the other options. Ultimately, contractual access to gene banks and bioinformatics databases may represent the best solution—from a conservation standpoint—for regulating control and ownership of these new technologies. This is true regardless of whether services and information are provided freely to the public, whether inventions will ultimately be patented, or whether gene or tissue resource banks and databases are held as trade secrets by a company, which subsequently licenses its products and services. Contractual provisions can be drafted to suit the specific facts of each case and can be tailored to secure the sometimes competing interests of rewarding innovation and promoting biodiversity and conservation.

Conclusion

The new genetic, biological, and informational technologies available to conservation biologists are already posing complex questions scarcely contemplated when the existing legal framework was enacted. Although the current legal framework may be adequate, navigating it in a manner that promotes genetic diversity and the aims of conservation will be a challenge. This is especially true in view of the speed of technological advances in the biological sciences and the boom of for-profit companies working in the areas of cloning and bioinformatics.

Notes

1. John Mattick, director of the Australian Genome Facility, Brisbane, editorial, *Science*, March 27, 1998.
2. See Lanza, R. P., B. L. Dresser, and P. Damiani. 2000. Cloning Noah's art. *Scientific American* 283:84–89.
3. See Scientists create healthy clone of endangered species. *The Atlanta Journal Constitution*, April 8, 2003, available at http://www.accessatlanta.com/ajc/news/science/0403/08cloning.html.
4. See 16 U.S.C. § 1533.
5. See 16 U.S.C. § 1539.
6. See 16 U.S.C. § 1531(b).
7. See 16 U.S.C. § 1539(a)(1)(A).
8. See 16 U.S.C. § 1538(a)(1)(d) ("It is unlawful . . . to possess" any such species).
9. Moreover, the processes or methods by which these endangered species are cloned present an entirely different question with respect to ownership. Depending on whether these processes or methods meet the requirements of U.S. patent law, scientists may be entitled to patent rights over their discoveries of novel methods of cloning. See U.S. Patent No. 6,258,998 ("Method of cloning porcine animals") and U.S. Patent No. 6,011,197 ("Method of cloning bovines using reprogrammed non-embryonic bovine cells").
10. See U.S. Patent No. 5,593,883 ("Ancient microorganisms").
11. See Rohrbaugh, M. L. 1997. The patenting of extinct organisms: Revival of lost arts. *American Intellectual Property Law Association Quarterly Journal* 25:371.
12. U.S. Constitution, Article 1, § 8, cl. 8.
13. See 35 U.S.C. §§ 101–03.
14. See 35 U.S.C. § 112.
15. *Diamond v. Chakrabarty*, 447 U.S. 303, 309 (1980) (citations omitted).
16. See U.S. Patent No. 4,736,866 ("Transgenic non-human mammals").
17. Rohrbaugh, M. L. 2005. The patenting of extinct organisms: Revival of lost arts. *American Intellectual Property Law Association Quarterly Journal* 25:388 (citing "*Brenner v. Manson*, 383 U.S. 519, 534–35 . . . (1966) (explaining that utility requires the demonstration of a specific benefit; an object of research is not inherently 'useful')").
18. Ibid. at 388 and n. 82 (citing "*Nelson v. Bowler*, 626 F.2d 853, 856 . . . (C.C.P.A. 1980) (requiring that one skilled in the art must be able to use the invention 'in a manner which provides some immediate benefit to the public')").
19. Ibid. at 388–389.
20. Ibid. at 397–407.

21. *Gaylor v. Wilder*, 51 U.S. (10 How.) 477, 497 (1850).
22. But see Jiron, D. M. 2001. Patentability of extinct organisms regenerated through cloning. *Virginia Journal of Law and Technology* 6:9, 45. Jiron argues that patenting formerly extinct species would not be novel under the "lost art" doctrine on the basis that "regenerated organisms do not meet all of the requirements for qualifying as 'lost art.'"
23. Perlman, D. L. and G. Adelson. 1997. *Biodiversity: Exploring values and priorities in conservation.* New York: Wiley-Blackwell.
24. See *"Dolly the Sheep" cloning methods could hold key to breakthrough in conserving animal genetic diversity*, Food and Agriculture Organization of the United Nations News, December 18, 1997, available at http://www.fao.org/news/en/.
25. See *N. Atl. Instruments, Inc. v. Haber*, 188 F.3d 38, 44 (2d Cir. 1999); *Roton Barrier, Inc. v. Stanley Works*, 79 F.3d 1112, 1116–17 (Fed. Cir. 1996).
26. See *Gable-Leigh, Inc. v. N. Am. Miss*, No. 01-01019 MMM (SHX), 2001 WL 521695 (C.D. Cal. Apri5l 13, 2001).
27. See *State St. Bank & Trust Co. v. Signature Fin. Group, Inc.*, 149 F.3d 1368 (Fed. Cir. 1998); *AT& T Corp. v. Excel Communications, Inc.*, 172 F.3d 1352 (Fed. Cir. 1999).
28. *State St. Bank & Trust Co.*, 149 F.3d at 1373–75 (citations omitted).
29. Ibid. at 1373.
30. See U.S. Patent No. 6,334,099 ("Methods for normalization of experimental data"), U.S. Patent No. 6,553,317 ("Relational database and system for storing information relating to biomolecular sequences and reagents"), U.S. Patent No. 6,484,183 ("Method and system for providing a polymorphism database"), U.S. Patent No. 6,229,911 ("Method and apparatus for providing a bioinformatics database"), and U.S. Patent No. 5,930,154 ("Computer-based system and methods for information storage, modeling and simulation of complex systems organized in discrete compartments in time and space").

Sandra D. Mitchell

20

The Import of Uncertainty

As new technologies develop, new questions for ethics and social policy are generated. As our understanding of the consequences of human action changes, we need to revise our view on how to implement the values we hold.

> What is most worrying about the present discourse is the near exclusion of an open scientific debate with clearly stated arguments. The arguments are highly emotional and fed by motives which are far removed from the actual issues on biotechnology. Diverse organizations use the debate for increasing their political influence without any useful result for the population. A democratic process functioning in an enlightened atmosphere seems a pipe dream.[1]

A complex web of issues surrounds the development and use of scientific knowledge in crafting policies that satisfy our sometimes conflicting desires and interests. Our assessment of the promises and perils of genomics in the twenty-first century involves the desire to preserve the biosphere so that human life can continue. Specifically we need to meet our moral obligations to the lives (human and animal) that populate the planet today by improving health and eliminating famine, to promote freedom and democracy and to increase our knowledge of the nature of nature.

In this chapter I will investigate how the complexities of biology and human values impinge on the means by which we can create sensible policies about genomics. I will consider this issue not from a scientific point of view but from a philosophical perspective. My past research has focused on the implications of biological complexity on epistemological issues.[2] That is, if complex systems such as organisms, social insect colonies, or ecosystems engage multiple, interacting (and not always linearly related) causal factors, then how should our scientific models and explanations represent that complexity? Philosophers of science have tended to look at physics and its search for the four fundamental laws as the means for understanding science in general. However, for the biological sciences the reduction of our knowledge to a few simple laws is not credible. I have defended an alternative picture of scientific practice, namely what I call integrative pluralism. Multiple theories, models, and explanations are generated for the variety of causal processes operating in complex biological systems. However, in the effort to explain a particular, local, and concrete event, these multiple approaches must be integrated to account for the singular combination of causes and historical circumstances responsible for that event.

Acknowledging complexity in nature has implications not just for our scientific theories and explanations but also for the content of and means by which we make rational policy decisions about interventions into that complexity. I will explore two different domains of complexity and describe how they affect our deliberations. Formulating rational policy requires two types of input. The first is the value that agents want to promote by a policy: What are we trying to accomplish with a policy concerning genetic modifications of plants and animals? To this question there may be a variety of answers, including the increased quality of health of our citizens, the protection of the environment for future generations, and the elimination of poverty and starvation in the world. The other input is the set of facts that describe the consequences of all the alternatives being considered. What will happen if we increase the agricultural yield of rice or the rate of growth for farmed salmon by genetic means? What are the implications if we don't do these things? What will happen if we transform goats into biofactories for the production of pharmaceutical products?

At a more general policy level we might rightly investigate the consequences of introducing a ban on human cloning or on genetically modified food. What are the consequences of no regulations for their development and production, or the consequences for any of a number of types of regulation that could be introduced? The logic is that if we know what we want (our values) and we know what we get if we adopt different actions (the facts), then we are in a position to know what we should do, how we should proceed. Do a cost–benefit analysis and maximize values. However, this simple account of a policy decision masks two sorts of complexities, corresponding to the two types of input. That is, sources of uncertainty for policy derive first from the complexity of biological systems and second from the pluralism of values held by agents for whom and by whom the policy is made.

Let's look first at the complexity of the causal structures of biological systems that preclude simplistic accounts of the consequences of genetic interventions. For example, the most widely used transgenic pest-protected plants express insecticidal proteins derived from the bacterium *Bacillus thuringiensis* (Bt). Genetically modified Bt varieties of corn, potatoes, and cotton are often thought of as a single intervention. But Bt toxins vary genetically and biochemically.[3] Furthermore, the consequences of the use of Bt plants vary in their impact on other organisms and wild types and with respect to their pesticide-reducing benefits. Replacing regular corn with Bt corn provides insect resistance but does not modify the use of externally applied chemicals because these are generally not used for the job Bt does for corn. However, the use of Bt potatoes does reduce the amount of pesticide needed for healthy crops.[4] So what sort of policy should be adopted for Bt plants? Bt potatoes are successful at resisting a variety of beetles. Their use can reduce the need for externally applied chemical insecticide. Yet the genetic modification may also have consequences for nontarget insects, induce an escalation in adaptations to evolve nonresistant parasites, or have pleiotropic effects on the modified plant itself or the landscape of genetic diversity in the wild, among other complications. The waves of consequent effects are broad, and at least some of the interactions between intricately connected parts of the ecosystem are still unknown. Because the consequences of the same genomic strategy for different agricultural plants are different, say for Bt corn and Bt potatoes, even if we understood all the implications in the case of Bt potatoes, these would not necessarily apply directly to the case of Bt corn.

Even in this brief excursion into the consequences of genetic modification, we can detect two types of uncertainty on the fact side of the policy equation that arise from biological complexity. These two sources of uncertainty have differing import for policy. The first is the uncertainty of the cascading effects of interventions on the features of the ecosystem (including on human beings by way of toxins and allergens) that we can specify and of which we have some understanding. These include consequences for the known nontarget species that interact with the transgenic organisms and the wild species that could interbreed with the transgenic organisms. Because we don't know precisely what the consequences of widespread release of transgenic organisms would be, it is difficult to assign a precise probability to the risk of these consequences. More research will improve our understanding and correspondingly our assessment of the real risks involved.

There is a second type of biological complexity, namely the consequences of what Norman Myers called the "unknown unknowns."[5] Myers recounts our recent experience of becoming aware of the degradation of the environment caused by global warming only after the damage was done. What we did had dire effects we were not aware of at the time. Given the chaotic nature of some environmental interactions, where harmful effects can be amplified, irreversible, and unpredictable in principle, there will be a class of consequences to which we could never assign a quantitative risk measurement. Indeed, once the process has begun, it may be impossible to recover from the harm.

We cannot simply plug in probabilities for some of the consequences of genetic interventions, whether they are known or unknown. How, then, do we deal with these types of factual uncertainty in making sensible policy decisions?

- *Acknowledge and manage the known risks*. Policymakers would like neat, certain answers to questions of risk so that an easily enforceable policy can be made. However, for the reasons given earlier, fixed probability assignments cannot reflect our scientific knowledge, or the nature of the uncertainty is such that it cannot be reflected by fixed probabilities in these situations. We cannot pretend that there is certainty when there is not, and we cannot hold out for certainty when it is not going to be found. In an interview in March 2000, Edwin Rhodes, aquaculture coordinator for the National Marine Fisheries Service, said he was surprised to hear that the Food and Drug Administration was overseeing the environmental review regarding the new, genetically modified salmon that has an introduced foreign gene that keeps its growth hormone continually rather than cyclically operant. Rhodes said the National Marine Fisheries Service, not the Food and Drug Administration, had the expertise to make decisions on such things as whether genetically modified fish should be grown in net pens. "We have to have absolute certainty that transgenic fish do not interact with wild stocks," Rhodes said.[6] We will never have certainty; to make our policy depend on it is a mistake.

- *Continue to investigate the unknown consequences on known factors* such as human allergies, nontarget species, and nontargeted phenotypic expression. Because this will be an ongoing process of discovery, we need to be in a position to update our policy relative to what we discover. This is in stark contrast to other environmental policy commitments, such as the "no surprises" clause in the Multiple Species Conservation Program that is being implemented in San Diego county and heralded as a model for the entire country.[7] There the rule is that once an agreement is made between local government and developers on what lands are to be set aside, nothing that may be discovered in the future can overturn that decision. This type of policy, though perhaps optimal from the developers' point of view, does not reflect the type of scientific process involved in understanding endangered and threatened species and their habitats. Rather, the current uncertainty combined with the incremental and fallible knowledge gained with continued scientific investigation requires us to develop adaptive management.[8] We need to update, overturn, and amend particular decisions as our knowledge of the consequences changes.
- *For the "unknown unknowns," if there is reason to suspect that the environmental harm will be great and irreversible, then a precautionary stance is appropriate.* Article 15 of the Biodiversity Convention states,

> In order to protect the environment, the precautionary approach shall be widely applied by states according to their capabilities. Where there are threats of serious or irreversible damage, lack of full scientific certainty shall not be used as a reason for postponing cost-effective measures to prevent environmental degradation.

This principle is lauded by environmentalists who fear the possible, though unproven, damaging effects on the environment of new technologies. It is simultaneously dismissed by those who take it as an unjustified prohibition to innovation.[9]

The intent of the precautionary principle is to shift the legal burden of proof from regulators, having to prove with "full scientific certainty" that harm would be done in order to ban a procedure, to those who want to introduce some new technology. Under the principle, products might be prohibited unless there is "full scientific certainty" that no harm could be done. Interpreting the principle with an emphasis on the proof of harm or proof of no harm invokes a type of positivism that is entirely unwarranted. Given the nature of the complexity I have described, there can be no scientific certainty either way. Rather than pass the burden of unattainable proof from one party to the other, we must acknowledge and manage the inescapable uncertainty. By *not* adopting some transgenic practices that may *not* induce irreversible harm, we are at risk of not reaping the benefits of those technologies out of unjustifiable fear, just as we are risking harmful consequence by adopting a new technology that does render irreversible harm. A precautionary approach is indicated by the nature of the possible harm. If the possible harmful consequence is both serious and irreversible, then clearly more caution is warranted than if it is minor and reversible. But caution here must reflect the nature of the uncertainty. It does not entail inaction. Rather, it suggests a more flexible policy response than total ban or complete hands-off nonregulation.

How can policy procedures take science seriously when the science invokes uncertainty with respect to future consequences of new technologies? A fixed policy—ban genetically modified organisms (GMOs) for all time—is an inappropriate response to a dynamically growing scientific base of understanding. If our policies are to reflect what we know about the world, as they should, and our knowledge is changing rapidly, then our means for policy amendment should be sensitive to those dynamics. If our policies are not so constructed, we will be stuck with the problem of having to decide once and for all on the basis of information that is ephemeral. The alternative to a flexible management response, given the impermanence of our understanding, is to have policy ignore science altogether. Because there is not a univocal accounting of the precise consequences of genetic interventions on health, productivity, biodiversity, and so on, a silencing of scientific input into policy is all too likely to ensue. This approach would be a very bad thing. Uncertain knowledge is still better than ignorance.

Uncertainties of the first type, the consequences on the known variables in a complex ecosystem, suggest a policy procedure that endorses continued investigation and is flexible enough to take into account our changing knowledge and adjust our actions appropriately. This will entail close monitoring of new technologies, with ongoing feedback from scientific analyses. Monitoring is not easy, especially given the variety of genomic interventions and variability of local environmental impact.[10] For instance, as we saw earlier, what is true for Bt potatoes is not true for Bt corn.

Other approaches to managing uncertainty in environmental policy have been proposed. However, I believe they are inadequate to address the special circumstances of genetic modifications with potentially irreversible effects. For example, Robert Costanza and Laura Cornwell suggested implementing what they call the 4P approach to scientific uncertainty.[11] The four *P*s stand for "precautionary polluter pays principle." The suggestion is to introduce an assurance bonding system that requires those who want to introduce a potentially harmful new product to commit resources up front to offset the potentially catastrophic future effects. It is a flexible system insofar as portions of the bond would be returned if and when it can be demonstrated that the suspected worst-case damage has not occurred.

This economic approach shifts both the burden of proof and the cost of uncertainty from the general public to the agent who wants to introduce the new technology, bypassing a need for specific safety standards to be established. However, it seems to me that while this might work for harmful human health consequences of introducing a new agent into the environment, it is not necessarily appropriate for the type of environmental consequences on biological diversity and ecosystem health at stake in the case of GMOs. That is, for harmful human health effects of a new product, the innovator can compensate the humans whose health is damaged. But how much money would be needed to offset the irreversible, cascading effects on the ecosystem? Rather than going full steam ahead and being prepared to pay up if there is harm, I think a slower, monitored pace with the option to stop if the harm is looking more probable is better fit to this situation. Thus, where the potential for harm is serious and irreversible and where there is no current procedure for reducing the uncertainty, then a worst-case precautionary stance is the appropriate response.

Let's turn to the other half of the policy equation: the value input. In a society that permits and endorses value pluralism, how can we generate fair policies that respect that diversity of views? Again we can distinguish between two sorts of complexity; the first is constituted by the multiple values that any one individual or social group might consistently hold at one time and that cannot all be met in a given strategy, action, or policy. We want to improve the quality of our lives, but we also want to preserve biological diversity for the future. Sometimes an action that promotes one value is in opposition to another value. This is familiar territory for the ethicist. Isaiah Berlin stated,

> In ordinary experience . . . we are faced with choices between ends equally ultimate, and claims equally absolute, the realization of some of which must inevitably involve the sacrifice of others.[12]

Moral conflict can be ameliorated if an individual can rank some values as more important than others, thus allowing a lexicographic decision procedure that follows the hierarchy. Plausibly, medical health is seen to be more important than better-tasting, cheaper, or more environmentally noninvasive food. Most people do place health fairly high up in significance, which is probably why there is not much public outcry about genomics for pharmaceutical purposes but a lot about genetically modified food. We try to satisfy as many of the most important moral values we hold as possible, and when we face a conflict we find resolution by analysis and ordering, by trading off the lesser values for the more significant ones.

The second type of complexity concerns conflicts not between values held by an individual but between different people promoting conflicting values. The agents involved in making social policy about genetically modified foods and those affected by such policies may hold very different values that are fundamentally at odds. This type of deep moral conflict characterizes the debates on abortion, for example, where all parties may agree to all the same scientific facts but have diametrically opposed interpretations of the moral significance of those facts.

Stuart Hampshire's conception of bare or procedural justice gives clues on how to manage this type of ethical complexity. He suggested that the core concept for a fair or just outcome in such a situation is to require equal and fair dealing between proponents of rival conceptions of the good; that is, it requires mutual respect:

> Mutual respect seems to be necessary to keep open the possibility of resolving, on a *moral* basis, any significant dispute about public policy that involves fundamental moral conflict. If citizens do not practice mutual respect as they try to come to agreement on a morally disputed policy, or as they try to live with the disagreement that remains after the disputed policy is adopted, they are forced to turn to nonmoral ways of dealing with moral conflict. They are driven to count on procedural agreements, political deals, and threats of violence—all of which obviously stand in the way of

> moral deliberation. The underlying assumption is that we should value reaching conclusions through reason rather than force, and more specifically through moral reasoning rather than through self-interested bargaining. Nevertheless people holding such divergent moral values are still equal participants in a democratically structured decision process.[13]

Such deep differences are evident in the policy debates on transgenic foods. For example, some adopt the view that "Nature and all that is natural is valuable and good in itself; all forms of biotechnology are unnatural in that they go against and interfere with Nature, particularly the crossing of natural species boundaries; all forms of modern biotech are therefore intrinsically wrong."[14] Or, to quote Prince Charles in reference to genetically modified food, "I happen to believe that this kind of genetic modification takes mankind into realms that belong to God, and to God alone."[15] The other side of the debate includes those who argue that only by using genetically modified foods will we be able to feed undeveloped nations or treat vitamin A deficiency effectively. "If we value the ethic of 'to each according to need' . . . then the introduction of GM crops on a large scale would be a moral imperative. This is because GM crops might produce more food, or more employment income with which to obtain food, for those who need it most urgently. More food for the hungry, unlike tomatoes with a longer shelf-life, is a strong ethical counterweight to set against the concerns of the opponents of GM crops."[16] How can we make common cause between Prince Charles and Monsanto?

Although tolerance and respect acknowledge that there is room for many moral views, they do not entail that *any* view is to be considered. In particular, some moral stances are based on confusion, and one of the main philosophical tools for resolving moral disagreement is to clarify the views and the assumptions on which they rest in the hopes of moving positions closer together.

For example, the "crossing the species boundary is unnatural and thus immoral" argument is really at odds with our understanding of evolutionary processes. If you were to accept that position, you would also have to believe that the evolution of all new species was in some sense unnatural and immoral. Because anyone would be hard pressed to reject all the results of evolution, this grounding for what is "natural" as a basis for what is morally permissible won't hold up under scrutiny.

Even after screening out the inconsistent or confused positions, there will remain genuine moral differences that must be acknowledged and perhaps accommodated. There are well-known problems with attempting to compare, measure, and resolve different valuations of goods or outcomes. Kenneth Arrow demonstrated formally that the collective preferences of groups cannot always be determined from the individual preferences of their members.[17] Thus there is no algorithm for reaching consensus, but the reasoning is that if all internally consistent views are given a fair hearing (i.e., if Hampshire's bare procedural principle of justice is adopted), and we all agree that such a process is a fair one, then the policy decisions that result should be acceptable. Importantly, they should be acceptable even to those who "lose" in the end.

It may well be that there are moral differences between individuals that cannot be negotiated into some collective decision. Even moral and political views that counsel tolerance find it unsavory to tolerate intolerance. These bottom-line differences will have to find satisfaction through dividing the domains in which people can comfortably live or through a political process. However, it does not seem to be the case that the different moral stances participating in the debates on genetically modified food are of this character. One indication is that there is very little objection to the use of genetic engineering in the production of medicines, but there is major public dismay at the use of genetic engineering in food products. For most, the moral objection cannot be to all genetic technologies and therefore does not seem to fit the bottom-line profile. In the case of GMOs, mediation and resolution may be possible.

There are steps we can take to encourage mediation between the diverse views concerning genetically modified foods. The first step to accommodating moral pluralism is to let into the discussion the different moral stakeholders. As we saw earlier, this is not anyone with a view but anyone with a coherent, consistent moral position. Thus there is a weeding out of the merely politically expedient, of those paying merely lip service to a view who do not act on the view, and of those voicing one-issue slogans without accepting the consequences of that view. Second, there are ways of framing disagreements that acknowledge moral conflict and ambiguity while preserving our ability to continue to live together and make joint policy. David Wong articulated principles for achieving this.[18]

- Act on one's moral position in a way that minimizes potential damage to one's broader relationship to others who have opposing positions.
- Other things being equal, select issues that minimize the opportunity for serious disagreement. In this case, we want to decide policies on a fairly specific rather than general level. If we pose the options as banning all GMOs or having no regulation at all, there are greater chances of polarization and disagreement. A case-by-case framework is dictated by the nature of the diversity and complexity of the consequences of different proposed interventions.
- Adopt a willingness to bridge differences. This permits domains of agreement to be achieved by removing the "us versus them" attitude.

How can we create an environment in which we can engage in scientifically informed, respectful dialogue? How do we reconfigure the debate to take account of both biological complexity and moral pluralism for policy decisions on GMOs? The current climate for decisions about genomic policy is teeming with hype, innuendo, and unbridled emotion. This is obviously not a good situation for clear thinking. For example,

> The reason that governments treat GMOs differently from other kinds of introduced organisms (like biocontrol agents or new garden plants) is because of fear and uncertainty. Most of the fears are completely unjustified but not all. . . . The worst way to deal with fear and uncertainty is by denial (e.g., by saying that GMOs are just like any other organisms so there is nothing to be concerned about). This is an example of what we might call the "trust me I'm a doctor" syndrome.[19]

Evidence of distrust and disrespect for science abounds, with the introduction of labels such as "Frankenfish."[20] By acknowledging and acting with respect for all parties involved one can hope to raise the level of the debate. But that means there is responsibility on all sides. First of all, citizens have to recognize the complexity of the scientific issues. The same holds for policymakers, who also would prefer not only simple consequences but also certainty, and if not certainty then at least accessibility to quantitative risk analysis. We cannot demand proof and nothing but the proof in a context of biological complexity and uncertainty. Rather, we have to develop adaptive policy mechanisms for management to take into account the scientific understanding of complex, irreversible processes. In addition, scientists have an obligation to present to the public and to policymakers the full story of complexity and to avoid soundbite science. In a context of political polarization and public distrust, more education has to be done. Citizens and policymakers have to hear the real story. Overenthusiastic, oversimplified science will just increase the distrust of new technology.

The overarching concern is that, given the thorniness of making policy in the midst of these complications, how can we avoid a complete stalemate? How can we let both the scientific knowledge we do have about the risks and benefits of genetic interventions and the moral values we endorse guide our decisions? It is irrationalism (rejection of science) and tyranny (rejection of democracy). Each of the sources of complexity generates uncertainty that contributes ingredients to the recipe for a reasonable, democratic process by which to reach some perhaps provisional consensus on genomics. Indeed, there are concrete things to be done to move from the current context, which is rife with emotional intensity and hype, to one of reflection and respect.

Postscript

I presented this paper at the conference "Genetic Resources for the New Century," sponsored by the San Diego Zoological Society in 2000. In May 2003, the Pew Stakeholder Forum on agricultural biotechnology concluded its two-year study and issued a report.[21] The goal was to produce "consensus on a package of regulatory reforms described in sufficient detail to enable an agreement on implementation. That package was to address animal and plant applications of biotechnology, public health and environmental concerns, and the regulatory system at the U.S. Department of Agriculture (USDA), the Food and Drug Administration (FDA), and the Environmental Protection Agency (EPA). Forum members agreed to outcomes, principles, and components for a regulatory system for agricultural biotechnology that protects public health and the environment."[22] In many ways the design of the Stakeholder Forum instantiated the types of respect-based procedures for decision making under conditions of moral pluralism discussed earlier. The body was constructed to be representative of all parties that had an interest in agricultural biotechnology and its consequences. All parties agreed to participate in good faith and were given equal opportunities to be heard. They reached consensus that policy in this area needs to be adaptive, responding to increasing scientific knowledge and better techniques for monitoring. Indeed, they agreed to the following:

> In order to protect public health and the environment without unduly burdening the development of innovative, productive, and sustainable agricultural practices, Forum

members agreed that a set of fundamental building blocks, or components, needs to be in place in the regulatory system. These components include adequate legal authority, adequate resources, a safety-driven approach to risk assessment, and appropriate risk management. Furthermore, in order to ensure continuous improvement, and to build and maintain public confidence in the regulatory system, members agreed that the system must be adaptive, efficient, equitable, transparent, and participatory.[23]

What is disappointing is that some participants deemed the process to have failed because they could not reach a consensus of sufficient detail to be of immediate aid to policymakers.[24] But there was much to applaud in this effort. The group agreed to meet a year or so after their final report of May 2003 in order to see whether more areas of agreement could be reached. (This might be more successful if the analysis in the *Washington Post* is correct that the current political climate in which the Bush administration filed a complaint against the European Union biotech moratorium was not conducive to broad-ranging agreement.) Another possible reason for the failure was attributed to Monsanto's refusal to agree to give the FDA authority to conduct mandatory safety assessments of biotech crops. Monsanto denied responsibility.[25] There is insufficient data in the public record to give a detailed analysis of where the Stakeholder's Forum might have gone astray. I cautiously conclude that their willingness to continue to meet in the future and the broad areas of consensus that were reached by this body suggest that this type of structure does hold the possibility for negotiated agreement to counter the complexities of the biological and social worlds that policy on new biotechnologies exhibits.

Notes

1. European Federation of Biotechnology. 1999. *Focus on future issues in biotechnology.* Task Group on Public Perceptions of Biotechnology and the European Molecular Biology Organisation, April 7–9.
2. Weingart, P., S. D. Mitchell, P. J. Richerson, and S. Maasen, eds. 1997. *Human by nature: Between biology and the social sciences.* Mahwah, N.J.: Erlbaum; Mitchell, S. 2000. Dimensions of scientific law. *Philosophy of Science* 67:242–265; Mitchell, S. 2002. Integrative pluralism. *Biology and Philosophy* 17:55–70; Mitchell, S. D. 2003. *Biological complexity and integrative pluralism.* Cambridge: Cambridge University Press.
3. Committed on Genetically Modified Pest-Protected Plants. 2000. *Genetically modified pest-protected plants: Science and regulation.* Washington, D.C.: National Academies Press, p. 109.
4. Ibid., p. 111.
5. Myers, N. 1995. Environmental unknowns. *Science* 269:358–360.
6. Kaesuk Yoon, C. 2000. Altered salmon lead the way to the dinner plate, but rules lag. *New York Times,* May 1.
7. Jasny, M., J. Reynolds, and A. Notthoff. 1997. *Leap of faith: Southern California's experiment in natural community conservation planning.* New York: Natural Resources Defense Council.
8. Mangel, M., L. M. Talbot, G. K. Meffe, M. T. Agardy, D. L. Alverson, J. Barlow, et al. 1966. Principles for conservation of wild living resources. *Ecological Applications* 6(2):338–362.

9. Chairman Nick Smith of the Subcommittee on Basic Research for the Committee on Science for the One Hundred Sixth Congress, second session. 2000. *Seeds of opportunity: An assessment of the benefits, safety, and oversight of plant genomics and agricultural biotechnology.* Washington, D.C.: U.S. Government Printing Office.
10. Diamond, E. 1999. Genetically modified organisms and monitoring. *Journal of Environmental Monitoring* 1:108N–110N.
11. Costanza, R. and L. Cornwell. 1992. The 4P approach to dealing with scientific uncertainty. *Environment* 34:12–20.
12. Berlin, I. 1969. Two concepts of liberty. In *Four essays on liberty,* p. 168. Oxford: Oxford University Press.
13. Gutmann, A. and D. Thompson. 1990. Moral conflict and political consensus. *Ethics* 101(1):64–88, p. 77.
14. EFB Task Group on Public Perception of Biotechnology. 1999. *Ethical aspects of agricultural biotechnology,* p. 10. Brussels: European Federation of Biotechnology.
15. Seeds of disaster. Opinion page, *Telegraph,* June 13, 1998.
16. Nuffield Council on Bioethics. 1999. *Genetically modified crops: The ethical and social issues,* p. 68. http://www.nuffieldbioethics.org/go/ourwork/gmcrops/introduction.
17. Arrow, K. 1951. Social choice and individual values. New York: Wiley.
18. Wong, D. B. 1992. Coping with moral conflict and ambiguity. *Ethics* 102(4):763–784.
19. Crawley, M. 1994. Long term ecological impacts of the relations of GMOs. In *Pan European Conference on the Long Term Ecological Impacts of GMOs,* ed. P. J. van der Meer and P. M. Schenklears, 31–50. Working document 14. Strasbourg: Council of Europe Press.
20. *Time Magazine* 155(9), March 6, 2000.
21. Neuhäuser, S. 2003. *Genetically modified organisms: The transatlantic conflict is picking up speed.* Washington, D.C.: Office of Science and Technology. http://www.ostina.org/V/print/biotech_print.pdf.
22. Ibid., p. 3.
23. Ibid., p. 31.
24. Schnepf, R. 2003, June 9. Genetically engineered soybeans: Acceptance and intellectual property rights. *Food Chemical News* 45(17):CS2.
25. Ibid., p. CS4.

Further Reading

Part I: Perspectives on the Union of Conservation and Genetics

Review Articles

Avise, J. C. 2007. The history, purview, and future of conservation genetics. In *Conservation biology: Evolution in action,* ed. S. P. Carroll and C. W. Fox, 10–26. Oxford: Oxford University Press.

Cronin, M. A. 2007. Limitations of molecular genetics in conservation. *Nature* 447:638–639.

Forest, F., R. Grenyer, M. Rouget, T. J. Davies, R. M. Cowling, D. P. Faith, et al. 2007. Preserving the evolutionary potential of floras in biodiversity hotspots. *Nature* 445:757–760.

Kohn, M. H., W. J. Murphy, E. A. Ostrander, and R. K. Wayne. 2006. Genomics and conservation genetics. *Trends in Ecology and Evolution* 21:629–637.

Leinonen, T., R. B. O'Hara, J. M. Cano, and J. Merilä. 2008. Commonness, population depletion and conservation biology. *Trends in Ecology and Evolution* 23:14–19.

Mills, L. S. and M. E. Soulé. 2006. A brief history of the role of genetics in conservation. In Conservation genetics, ed. F. Allendorf and G. Luikart, 115–131. New York: Blackwell.

Ridenhour, B. J. 2005. Stress and adaptation in conservation genetics. *Journal of Evolutionary Biology* 18:750–755.

Smith, T. B. and G. Grether. 2008. The importance of conserving evolutionary process. In *Conservation biology: Evolution in action,* ed. S. P. Carroll and C. Fox, 115–131. Oxford: Oxford University Press.

Vernesi, C., M. W. Bruford, G. Bertorelle, E. Pecchioli, A. Rizzoli, and H. C. Hauffe. 2008. Where's the conservation in conservation genetics? *Conservation Biology* 22:802–804.

Wang, J. 2005. Conservation biology: Ecosystem recovery enhanced by genotypic diversity. *Heredity* 95:183–184.

Wayne, R. and P. Morin. 2004. Conservation genetics in the new molecular age. *Frontiers in Ecology and the Environment* 2.2:89–97.

Books

Carroll, S. P. and C. W. Fox. 2006. *Conservation biology: Evolution in action.* Oxford: Oxford University Press.

Caughley, G. and A. Gunn. 1996. *Conservation biology in theory and practice.* Cambridge, Mass.: Blackwell Science.

Frankel, O. H. and M. E. Soulé. 1981. *Conservation and evolution.* Cambridge: Cambridge University Press.

Gaston, K. J. and J. I. Spicer. 2004. *Biodiversity: An introduction.* Oxford: Blackwell Science.

Groom, M., G. K. Meffe, and C. R. Carroll. 2007. *Principles of conservation biology.* 3rd ed. Sunderland, Mass.: Sinauer.

Mills, L. S. 2006. *Conservation of wildlife populations: Demography, genetics, and management.* Oxford: Blackwell.

Primack, R. B. 2004a. *Essentials of conservation biology.* Sunderland, Mass.: Sinauer.

Primack, R. B. 2004b. *A primer of conservation biology.* Sunderland, Mass.: Sinauer.

Pullin, A. S. 2002. *Conservation biology*. Cambridge: Cambridge University Press.

Soulé, M. E. 1986. *Conservation biology: The science of scarcity and diversity.* Sunderland, Mass.: Sinauer.

Van Dyke, F. 2008. *Conservation biology: Foundations, concepts, applications.* Berlin: Springer.

Part II: Conservation Genetics in Action: Assessing the Level and Quality of Genetic Resources in Endangered Species

Review Articles

Beebee, T. J. 2005. Conservation genetics of amphibians. *Heredity* 95:423–427.

Frankham, R. 1995. Conservation genetics. *Annual Review of Genetics* 29:305–327.

Frankham, R. 2003. Genetics and conservation biology. *Comptes Rendus Biologies* 326:S22–S29.

Frankham, R. 2005a. Conservation biology: Ecosystem recovery enhanced by genotypic diversity. *Heredity* 95:183–184.

Frankham, R. 2005b. Ecosystem recovery enhanced by genotypic diversity. *Heredity* 95:183.

Frankham, R. 2005c. Stress and adaptation in conservation genetics. *Journal of Evolutionary Biology* 18:750–755.

Leblois, R., A. Estoup, and R. Streiff. 2006. Genetics of recent habitat contraction and reduction in population size: Does isolation by distance matter? *Molecular Ecology* 15:3601–3615.

Michaux, J. R., O. J. Hardy, F. Justy, P. Fournier, A. Kranz, M. Cabria, et al. 2005. Stress and adaptation in conservation genetics. *Journal of Evolutionary Biology* 18:750–755.

Phillimore, A. B. and I. P. Owens. 2006. Are subspecies useful in evolutionary and conservation biology? *Proceedings. Biological Sciences* 273:1049–1053.

Strasburg, J. L. 2006. Conservation biology: Roads and genetic connectivity. *Nature* 440:875–876.

van Oppen, M. J. and R. D. Gates. 2006. Conservation genetics and the resilience of reef-building corals. *Molecular Ecology* 15:3863–3883.

Vernesi, C., M. W. Bruford, G. Bertorelle, E. Pecchioli, A. Rizzoli, and H. C. Hauffe. 2008. Where's the conservation in conservation genetics? *Conservation Biology* 22(3):802–804.

Woodworth, L., M. Montgomery, D. Briscoe, and R. Frankham. 2002. Rapid genetic deterioration in captive populations: Causes and conservation implications. *Conservation Genetics* 3:277–288.

Books

Allendorf, F. and G. Luikart. 2006. *Conservation and the genetics of populations.* Oxford: Blackwell.

Avise, J. C. 1994. *Molecular markers, natural history, and evolution.* New York: Chapman & Hall.

Avise, J. C. and J. L. Hamrick, eds. 1996. *Conservation genetics: Case histories from nature*. New York: Chapman & Hall.

Frankham, R., J. D. Ballou, and D. A. Briscoe. 2002. *An introduction to conservation genetics.* Cambridge: Cambridge University Press.

Part III: Saving Genetic Resources

Reviews

Zbinden, J. A., C. R. Largiadèr, F. Leippert, D. Margaritoulis, and R. Arlettaz. 2007. Back to the future: Museum specimens in population genetics. *Trends in Ecology and Evolution* 22:634–642.

Part IV: Genomic Technology Meets Conservation Biology

Kohn, M. H., W. J. Murphy, E. A. Ostrander, and R. K. Wayne. 2006. Genomics and conservation genetics. *Trends in Ecology and Evolution* 21:629–637.

Contributors

GEORGE AMATO, American Museum of Natural History
JUDITH A. BLAKE, The Jackson Laboratory
SYDNEY BRENNER, The Molecular Sciences Institute
REBECAH BRYNING, Zoological Society of San Diego
RICHARD CAHOON, Cornell Center for Technology, Enterprise, and Commercialization
LEONA G. CHEMNICK, Center for Reproduction of Endangered Species
WILLIAM CONWAY, Wildlife Conservation Society
ANGÉLIQUE CORTHALS, Ambrose Monell Cryo Collection
ROB DESALLE, Sackler Institute of Comparative Genomics
NORMAN C. ELLSTRAND, Department of Botany and Plant Science
GERALD J. FLATTMANN JR., Fish & Neave
JAMES P. GIBBS, Department of Environmental and Forest Biology
PAUL Z. GOLDSTEIN, Florida Museum of Natural History
ROBERT HANNER, Coriell Institute for Medical Research
PHILIP W. HEDRICK, Department of Biology, Arizona State University
MARLYS L. HOUCK, Center for Reproduction of Endangered Species, Zoological Society of San Diego
ROBERT C. LACY, Department of Conservation Biology, Daniel F. & Ada L. Rice Center
CATHI LEHN, Cleveland Metroparks Zoo
PATRICK E. MCGUIRE, University of California, Davis
ANNE MCLAREN, Wellcome/CRC Institute
SANDRA D. MITCHELL, Department of History and Philosophy of Science, University of Pittsburgh
STEPHEN R. PALUMBI, Department of Biological Sciences, Stanford University, Hopkins Marine Station
LESLEY PATERSON, Roslin Institute
CALVIN O. QUALSET, Agriculture and Natural Resources, University of California, Davis
DEBORAH L. ROGERS, University of California, Davis
BARBARA A. RUSKIN, Fish & Neave
OLIVER A. RYDER, Center for Reproduction of Endangered Species, Zoological Society of San Diego
NICHOLAS VOGT, Fish & Neave
VITALY VOLOBOUEV, MNHN Zoologie
IAN WILMUT, Roslin Institute

Index

adaptive variation
  adaptive genetic variation, 43, 48–52
agriculture
  animals
    artificial insemination (AI), 199–200
  crop transgenes in natural populations, 187–195
  hybridization between engineered crops and wild plants
    consequences of, 187–195
  National Agricultural Library, 189
  plants
    food crops (table), 189
    herbicide, tolerance for, 188, 192–193
    natural hybridization between food crops and wild relatives (table), 190
    risk assessment on new products, 194
  transgenic crops versus traditionally improved crops, 192
animal conservation
  assisted reproduction, 199–203
  comparative study of development, 183
  disease, 6
  whole genome sequences of animals, 171
animal diversity
  animals cared for in captivity, 160
  animals conserved in their natural habitats, 160
  collections of biological samples, 160
  global, 160–165
  institutional collections, 160
  learning through careful scientific study, 160
animal records keeping system
  ZIMS software package, 162–163
applied wildlife management, 6
aquariums; *See:* zoos and aquariums
areas of endemism, 18

biodiversity
  archiving and storage, 115–120
  assessing, 174–176
  collecting sources of DNA, 17
  information access, 118–120
  species loss, 58
  versus ecological and evolutionary processes, 58
biodiversity conservation
  global program of diversity science, 130
  role of cryopreserved cell and tissue collections, 131–138
  systematics
    characters, highly specific genetic, 131
biodiversity crisis, 113, 141–156
  assessments, 154
  continuing losses of biological diversity, 141–156
  losses of ex situ reserves, 147–148
  opportunities to mitigate, 153–154
  predominant problems
    abandonment, 150
    accidents and natural disasters, 149–150
    crime, 150
    inadequate distribution policies, 151
    insufficient, reduced, or fluctuating funding, 148–149
    insufficient taxonomic expertise, 151
    international impediments, 151–152
    undervaluation of collections, 149
    war, 150
biological materials
  living, 132
  managing in the university environment, 164–165
  transferring biological samples to researchers, 164
biological species concept (BSC), 100–103
biological systems, complexity of, 225–227
Biomaterials Action Plan, 161
Biomaterials Banking Advisory Group (BBAG), 161
biotechnology
  genetic change in captive populations, 60–62
  products of crop biotechnology, 187–195
breeding, 4–6, 11–13, 68–76. *See also:* reproduction
  captive species, 4–6, 11–12
    avoiding, 41
    inbreeding depression, 6, 18, 44, 47
  effective population size, 6, 18
  genetics, 11
  inbred populations, 59
  inbreeding, 5, 41, 44–45, 59
    avoiding, 41
    inbreeding depression, 6, 18, 44, 47
    Wright's inbreeding coefficients, 7, 19

breeding *(continued)*
maintaining genetic variation, 41
programs, 11–13, 49, 60–76
related individuals, 45
strategies, 68–73, 76
bushmeat kill, 26–27

captive populations, 11–12, 58–77
causes of mortality, 60
evolution in, 60
genetic change through biotechnology, 60–62
genetic drift, 60
genetic management, 58–77
inherited disorders, 47
used to educate people, 59
used to restore wild populations, 59
captive species breeding, 4, 11–12
causes of mortality
captive populations, 60
wild populations, 60
CCTCs (cryopreserved cell and tissue collections), 131–138
capabilities of, 132–134
chromosomal analysis, 132–134
collections of nucleic acids and proteins, 132
conservation strategies, 134
genome banks, 133
genome database on global scale, 138
organizing to global standard, 137–138
planned replenishment of, 135–136
potential, 137–138
role of, 131–138
source of living biological materials, 132–138
study of unexplored species diversity, 136–137
supposed weaknesses, 134
true genetic variability of the species, 135
working strategy, 136–137
cellular systems, 183
character fixation
phylogenetic species, 100–101
chromosomal analysis, 133–134
cladistic diversity, 7, 9, 18
climate change, 29–39
cloning, 15, 147
bringing back from extinction, 61
cloned species, 214
cloning procedures, 204–205
cloning technology, 214
closely related species, 207
conservation, 214
conservation and cloning, 204–208
embryo splitting, 201–202
endangered species, 204–298, 214–216
creating copies of, 204–208
for and against objectives of ESA, 215–216
extinct species, 214
creating copies of, 204–208
if recreated, whether ESA applies, 214
inducing quiescence, 205
low success rate, 204–205
nuclear transfer, 201–208
between species, 206–207
ownership rights, 216–218
potential for assisting conservation, 207–208
preserving endangered species, 61
resources needed, 205–206
science behind, 205
somatic cell, 204, 214
collections. *See:* collections, laws governing; museums and herbaria; zoos and aquariums
collections, laws governing
access to information, 213
bioformatics databases, 219
gene and tissue resource banks, 219
comparative biology
structure and evolution of biological entities, 183–184
complexity in nature, impact on policy, 225–227
conservation biology, 7, 35–38
adaptations of genome technology to, 174–176
biological collections
support for genetic diversity in natural populations, 146
challenges, 5–6
conserving by saving genetic resources, 113–166
context dependency, 16–17
continuing losses of biological diversity, 141–156
crisis discipline, 3
economic, social, and cultural factors, 18
genetic analysis of species and populations, 9
genetics, 3–4
genomics and methods of data acquisition, 4
genomics technologies most pertinent to, 169–173
genomics technologies with biggest impact, 174–176
genomic technology, 167–208
incorporating genomic technology, 176
incorporating molecular information, 1–18
intraspecific genetic diversity within and between populations, 107
markers, 5
matchmaking, 11–12
mathematical advances in conservation theory, 3
molecular evolutionary genetic techniques, 5
new resources, 219
noninvasive, 14–15
northeastern beach tiger beetle, 96–103
preserving ecological or geographic areas that support endangered organisms, 10
relative merit and priority of genetics, 5
relevance of genetics to demographic issues, 5
role of collections, 113–166

role of genomics in, 169–176
satellite-based imaging of areas, 3
species boundaries, 4, 7
subspecies, 6
unit designation, 6
use of global positioning systems (GPSs), 3
wildlife conservation, 30–31, 83–93, 97
conservation forensics, 4, 15–16
conservation genetics, 1–18, 34–111, 213–222
breeding genetics, 11
centralized DNA barcodes, 15–16
combining genetic methods with ecological and landscape approaches, 4
components of, 101–103
conservation geneticists, 3–4
context dependency, 16–17
economic challenges, 4
enhanced by genomic techniques, 167–208
erosion of genetic resources, 106–111
evolving of field, 34–38
expansion of, 3–18
extinction crisis, 25–31
genetic analysis, 9–12, 15, 35
genetic change through biotechnology, 60–62
genetic data, 96–103
genetic distances between species versus those between populations, 107
genetic resources in endangered species, 39–111
genetic signals of population ecology, 88
genetic threats to endangered species, 11–16
genomics, potential of, 1
genotyping technology, 4
high-throughput DNA sequencing, 4
incorporating both pattern and process, 7–9
intraspecific genetic diversity within and between populations, 107
legal issues, 213–222
low-tech, 62–76
management decisions, 4
merging pedigree analysis with molecular genetics, 74–76
molecular genetic analysis, 35
neutral, detrimental, and adaptive variation in, 41–54
new insights into, 184
new studies in endangered species, 41–42
pattern in, 4
philosophical issues, 211–233
phylogenetic species, 101–103
political boundaries, 4
role of genetics in conservation, 53–54
roles of (in table), 5
scope of, 4–11
using unconventional sources of tissue, 14
conservation priority areas, 9
conservation programs, 161
conservation strategies
breeding, 68–73
conserving by saving genetic resources, 113–166
cryopreserved cell and tissue collections (CCTCs), 134
Frozen Zoo at San Diego Zoo, 124–130
in the future, 17, 30–31
protecting global genetic resources, 111
rescuing species while ignoring population loss, 107
conservation theory, 3
conservation units, 6, 7, 100
delineation of, 14
diagnosing in nature
character-based approach, 8–9
tree-based approach, 8–9
contracts, 220–221
controversies over transgenic products, 187–195
Convention on International Trade in Endangered Species of Wild Fauna and Flora (CITES), 14, 17–18
copyrights, 220–221
cryopreservation, 113–138
correct temperature and backup freezer, 117
cryopreserved cell and tissue collections. *See* CCTCs
gametes and embryos, 61
ideal treatment of tissue samples, 117
living biological materials, 132
curation
of animals, 160
of biological material obtained from animals, 160
data acquisition, 4, 169–174
high throughput approaches, 170–171
data analysis, 169–170
databases
animal records, 162–163
gene discovery using, 160–170
genome database on global scale, 138
global, 181
International Nucleotide Sequence Database Collaboration, 181
legal issues, 219
model organism databases, 182
on genetic distances between and within species, 110
on species and aggregate loss of genetic diversity, 109
on species surviving only in cultivation or captivity, 108
on species that have become extinct or imperiled, 108
sequence data, 181
data storage (archiving), 169–170
demographics, 5–10, 14
demographic and fine-grained genetic approaches, 6

demographics *(continued)*
  demographic factors, 5
  intersecting demographics and genetics, 9
detrimental genetic variation, 44–47
diversity
  life strategies of marine creatures, 93
DNA
  barcoding, 171, 174
    centralized DNA barcodes, 15–17
  detecting DNA polymorphisms, 173–176
  diagnostics, 12–14
  extracting viable DNA from specimens, 116
  genetic resource collections, 142–143
  molecular basis of inheritance, 181
  registries, 16
  sequence, 62
    sequence information from extinct species, 147
  sequencing technologies, 171–173
  sequencing, high-throughput, 4
  using unconventional sources, 15
documentation
  publishing results of work, 182
  world-standard recordkeeping system, 161

ecological communities
  results of evolutionary processes, 58
ecological genetics
  intraspecific genetic diversity within and between populations, 107
ecological processes
  genetic signals of population ecology, 88
  versus biodiversity, 58
ecology, 4, 6, 10–11, 16, 26, 36–38
ecosystems
  biodiversity, 115
  broken, 1
  marine, 93
  protecting, 28
  species loss, 58
endangered organisms
  preserving ecological or geographic areas, 9–10
endangered populations, 4
endangered species, 4–6, 9–16, 27–28
  adaptive variation, 48
  conserving by saving genetic resources, 113–166
  creating copies of endangered or extinct species, 204–208
  examination of genetics, 42
  genetic resources, 124–130
  genetic studies, 41
  humans putting recreation first, 97
  inherited disorders, 47
  managing, 4, 11
  migratory shorebirds, 97
  new studies, 41–42
  northeastern beach tiger beetle, 96–103
    effect of barrier beaches on, 97
  population size, 42
Endangered Species Act (ESA), 17–18, 214–215
  hybrid policy, 215
endangering wildlife, 12–14, 27–28
environmental challenges
  adaptive variation, 48
  economic approach, 228–229
  precautionary approach to, 227–228
environmental destruction, 26–27
  habitat destruction, 49
ethics, 163–165, 225–233
evolution
  captive populations, 60
  cryopreservation of gametes and embryos, 61
  estimating evolutionary parameters, 41
  evolutionary novelty, 18
  MHC evolution, 49
  of genomes, 179
  stopping evolution, 58–77
  structure and evolution of biological entities, 183–184
  study of endangered populations and species, 4
evolutionarily significant unit (ESU), 6, 18
evolutionary biology, 7
  intraspecific genetic diversity within and between populations, 107
  population genetics, 4–7
  systematics, 5
evolutionary processes
  versus biodiversity, 58
extinction, 3, 5, 12, 41
  cloning extinct species, 214
  crisis, 25–31, 34–38, 106–107, 110
    background of, 28
  extinct organisms, 15
  introducing individuals from related populations, 46
  over coming decades, 202
  population extinction, 106–108
  species extinction, 106–108

forensics and dead DNA, 14–16
future challenges and trends
  better use of observed associations, 42
  climate change, 29–30
  conservation, 17–18
  genetics role in conservation, 53
  merging pedigree analysis with molecular genetics, 74–76
  pattern and process, 7–8

genes, 15, 42
  crop transgenes in natural populations, 187–195
  gene and tissue resource banks, 219
  gene diversity, 58
  gene flow, 107

gene flow, crop to wild, 187–195
biologically significant or not, 190
Gene Ontology (GO) Consortium, 182
gene pool preservation, 115
gene sequences, 39, 171, 179–183
genetic differentiation among populations, 107
genetic resource collections, 142–143
involved in adaptation, 44
MHC, 49
neutral polymorphic, 42
of mutant phenotypes, 181
genetic data, 96–103
genetic differences
between-species variation, 108
within-species variation, 108
genetic diversity
biological collections
support for diversity in natural populations, 146
losses of, due to population extinction, 108
losses of, due to species extinction, 108
results of evolutionary processes, 58
genetic drift, 42, 45–46, 50, 107
captive populations, 60
changes in allele frequencies, 60
loss of polymorphism, 60
reduction in molecular variation, 42
genetic engineering, 61
enhancing prospects for survival, 61
genetic factors, 5
genetic load, 44–45
genetic management
captive populations, 58–77
genetic markers
molecular, 42
neutral, 43
identify species, evolutionary significant units, management units, 43
variety of classes of, 108
genetic resource collections
acknowledgment in publications, 152–153
ancient fungi trapped in glacial ice, 147
appropriate use of, 152
banking of genetic resources, 124–130, 147–130
better communication
between users and holders, 155
characterization of, 154–156
communication, 152
contributions to restoration, 145
endowments, 155
ex situ genetic resource collections, 143–147
cessation of natural evolution, 145
grantmakers focus on obtaining and maintaining genetic material, 154
increased efficiency of maintenance and use, 154–156
losses of, worldwide, 141–156
networking and collaboration, 155–156
new technology, 146–147
organelles, 142–143
organizations (table of), 144
policy, 153
research, 146
varied uses, 145
genetic resources, 39, 113, 142–143
banking of, 124–130
disaster mitigation
agricultural resources, 147
rice gene bank, 147
endangered species, 124–130
erosion of, 106–111
fossils from lake sediments, peat bogs, or buried trunks, 147
fossils from ocean floors or glacial ice, 147
germplasm resources, 124
global genetic resources
loss of genetic diversity, 107
range collapse, 107
importance to agriculture and food production, 115
living or nonliving species outside natural habitats, 142–143
living species in natural habitats, 142
loss of populations, effect on, 107–111
loss of species, effect on, 107–111
nature and types of, 141–156
saving genetic resources, 113–166
genetic threats to endangered species, 11–16
genetic variation, 5
between/among populations, 11
declining, 35
examining and characterizing, 34
within a population, 11
genetically modified foods, 231
genomes, 169–170, 179–181. *See also:* sequencing
analysis of multiple genome sequences, 180
comparative biology techniques for, 179
evolution of, 179
genome banks, 133
genome comparative analysis, 181
genome database on global scale, 138
genome mapping technologies, 183
human genome, 25
sequencing of, 169–170
whole genome sequences of animals, 171
whole genome sequences of plants, 171
genomics, 25
archiving and storage, 116
comparative method, 183–184
conservation biology, 170–176
conservation genetics, 179–184
data acquisition, analysis, and storage, 169–170

genomics (*continued*)
  genomic studies, 115
  genotyping, 4
  high-throughput methods of data acquisition, 4
  policies about, 225–233
  refining language of, 182
  role in conservation biology, 169–176
  scope of (figure), 169–170
  sequencing complete genomes, 180–183
  sociogenomics, 169
  technology
    data acquisition, storage, and analysis, 3
  working definition of, 169
genomic technology
  conservation biology, and, 167–208
genotypes
  genetic resource collections, 142–143
global cooperation, 161

habitat
  collections of living species in natural habitats, 142
  habitat threat, 6
  saving wildlife habitats, 31
  shore-line habitats, 97
herbaria. *See:* museums and herbaria
herbicide, tolerance for, 188, 192–193
holism versus reductionism, 34–38
Human Genome Project, 179
hybridization, 12
  between engineered crops and wild plants, 187–192
    aggressive weeds, 191
    extinction of endangered species, 191
    hybridization rate, 191
hybrid policy of Endangered Species Act, 215

inbreeding, 5, 41, 44–45, 59
  depression, 6, 18, 44, 47
informatics, 171
inheritance
  genetic variability, 179
  molecular basis of, 181
integrative pluralism
  approach to biology and conservation issues, 211
International Nucleotide Sequence Database Collaboration, 181
International Species Information System (ISIS), 161
invertebrate conservation, 96–103

land management
  intraspecific genetic diversity within and between populations, 107
laws that apply to conservation genetics, 213–222
  access to information and knowledge, 213
  access, control, and ownership issues
    contracts, 220–221
    copyrights, 220–221
    patents, 220–221
    trade secrets, 220–221
  bioformatics databases, 219
  environmental protection and conservation, 213
  gene and tissue resource banks, 219
  international, 218–219
  property rights, 213
  public safety, 213
  U.S., 213–219
legal issues in conservation genetics, 211–233
life history, 1, 5, 11

management
  applied wildlife management, 6
  biological samples
    Biomaterials Banking Advisory Group (BBAG), 160–161
    conservation programs, 160–161
    Zoological Information Management System (ZIMS), 160–161
  captive populations, 11, 58–77
  critical areas, 4
  endangered species, 4, 11
  genetic, 58–77
  ocean neighborhoods, 93
  pedigree, 63–68
  single species, 11
management unit (MU), 5, 14, 19, 100
marine conservation
  effect of industry and technology on oceans, 82
  efforts on a smaller scale, 83
  marine reserves, 90–91
  ocean neighborhoods, 82–93
marine ecosystems
  diversity, 93
marine populations
  chaotic patchiness, 89–90
  effect of hurricanes on, 98
  genetic studies, 89
marine reserves
  scale of coastal populations, 90–91
markers, 4–5, 13, 16, 41–44, 53
  alleles, 42, 49
  DNA, 174
  genetic, 43, 53
  molecular, 43–44
  molecular genetic, 42
  neutral genetic, 43
  neutral, 53
  neutral molecular, 43–44
  neutral nuclear, 43–44
material for taxon being studied
  storage techniques, 132
Material Transfer Agreements (MTAs), 163–165
  managing biological materials in the university environment, 164–165
  transferring biological samples to researchers, 164

microbes
  having whole genome sequence, 180
  microbial communities, environmental sampling of, 175
microsatellites, 4, 53
minimum viable population size, 19
minisatellites, 4
molecular biology
  archiving and storage, 116
  refining language of, 182
  system approaches in, 183
molecular genetic markers
  estimating fundamental parameters in conservation, 42
molecular studies, 115
  formalin-fixed specimens, 116
molecular variation
  increase in due to mutation, 42
  reduction in due to genetic drift, 42
moral conflict, mutual respect, 229
museums and herbaria, 15–17, 37, 59, 101–103, 113–121. *See also:* zoos and aquariums
  archiving and storage, 115–120
  biological resources, 113
  collections
    Ambrose Monell Cryo Collection (AM-CC), 118–120
    biological, 117–118
    classic, 113
    documentation, 113
    frozen tissue, 113
    genetic resource, 141–156, 142–143
    living tissue, 125
    organizing to a global standard, 137–138
    reference collections to backup molecular data generated, 117
    role of, 113–166
    specimens from mammalian taxa, 125
    use of DNA from museum specimens of extant species, 147
  databasing and inventory, 118–119
  degradation of materials, 117
  genetic resources, 113
  specimens
    cultures, 117
    digital representation of, 120
    extracting DNA from specimens, 116
    formalin-fixed specimens, 116
    ideal treatment of tissue samples, 117
    voucher specimens, 117
mutation, 107
  genes of mutant phenotypes, 181
  increase in molecular variation, 42
  rate of compared to crop-to-wild gene flow rate, 191

natural imperative, 19
nested clade analysis, 7
neutral variation
  genetic, 42–44
  molecular, 42
  indicator of detrimental and adaptive variation, 52–53

ocean neighborhoods
  demographic differences, 88
  implications of size, 83, 90–93
  management strategies, 93
oceans
  coastal versus open sea, 83–87
  effect of industry and technology, 82
  fossils from ocean floors or glacial ice, 147
  ocean habitats, 83–87
  oceanography, 83
  planktonic phases, 85
  protection and restoration, 82–93
  red tides and dead zones, 83
  scale, 82–83
ocean wildlife
  coastal fish and invertebrates, 83
  independent populations now intermingling, 88
  larvae, 83–93
  limpets, 91
ontologies
  precise language for scientists, 182
  terms and classifications, 182
organ systems, 183
organelles
  genetic resource collections, 142–143
organisms
  cellular systems, 183
  collections of, 17
  composition of metapopulations, 175
  endangered, 17
  expanding knowledge, of through genomics, 179
  germplasm resources, 124
  having a whole genome sequence, 179–183
  organ systems, 183
  tissue systems, 183
ownership of animals and biological samples of same, 160–165
  legal and ethical issues, 163–165
  ownership rights, 216–218
  responsibilities of ownership, 163–165

paraphyletic, 19
patents, 220–221
pattern and process, 5–9
PCR (bridge polymerase chain reaction), 172
pedigree, 11
  analysis
    information, 6–7, 62–76
    merging with molecular genetics, 74–76
  management, 63–68
phenotype, 179
  genes of mutant phenotypes, 181
  of transgene, 192–193

philosophical issues in conservation genetics, 211–233
phylogenetic species, 100–103
  character fixation in, 101
  phylogenetic species concept (PSC), 100–103
plants
  comparative study of development, 183
  disease, 6
  hybridization between engineered crops and wild plants
    consequences of, 187–195
  plant seed banks, 15
  whole genome sequences of plants, 171
  wild sunflowers versus crop sunflowers, 188
policy issues in conservation genetics, 211–233
polymerase chain reaction. *See* PCR
population
  allelic composition of populations, 10
  effective population size, 73
  fitness declining over time, 44
  minimum viable population size, 6–7
    population viability analysis (PVA), 7
  population aggregation analysis (PAA), 16, 19
  population dynamics, 7
  population genetics, 4–7, 9
    analysis of molecular variance (AMOVA), 7
    coalescence theory-based analyses, 7
    problems when uninformed, 12–14
  population loss, 107
  viability analysis, 19
population biology, 3
population genetics, 4–7, 9, 41
populations
  captive, 1, 11–12
  crop transgenes in natural populations, 187–195
  declining in numbers, 35
  detrimental traits, 46–47
  diversity of genes, 58
  effective population size, 73
  endangered, 4–6
  extinction, 106–107
  genetic resource collections, 142–143
  geographically separated, 107
  impact of loss of populations, 110
  inbred, 46
  migration of alleles from one to another by pollen or seed, 187–192
    spread of beneficial alleles, 187–192
  natural, 1, 6
    breeding effective population size, 6
    gene flow, 6
    genetic variation, 6
    inbreeding depression, 6, 18
    intrapopulation and genealogical problems, 6
    minimum viable population size, 6
  new environmental challenges, 48
  peripheral and isolated, 107
  polymorphic, 99–101
  population loss, 107–108
  results of evolutionary processes, 58
  small, 28, 42, 46–47
process and pattern, 5–9

reductionism versus holism, 34–38
reintroducing animals into the wild, 28–29
  negative consequences, 47
  northeastern beach tiger beetle, 97–98
reproduction, 199–203
  alleviation of human infertility, 199
  artificial insemination (AI), 199–200
  cryopreservation of eggs, 200
  cryopreservation of embryos, 200–201, 206
  cryopreservation of fertilized eggs, 200–201
  cryopreservation of sperm, 200
  embryo splitting, 201
  embryo transfer to surrogate mother, 201
  nuclear transfer cloning, 201–208
  regulation of human fertility, 199
  superovulation, 200, 205
reproductive science, 25–31
restoration
  genetic resource collections
    contributions to, 145
  northeastern beach tiger beetle
    ecological considerations, 98–99
    genetic considerations, 98–101
    logistical considerations, 98–99
restoration ecology
  intraspecific genetic diversity within and between populations, 107
risk assessment on new crops, 194
RNA, 113, 172

selection, 107
sequencing
  analysis of multiple genome sequences, 180
  automated high-throughput, 170–171
  bridge polymerase chain reaction (PCR), 172
  complete genomes, 171, 179–183
  DNA sequencing technologies, 171–173
  emulsion PCR, 172
  Human Genome Project, 179
  loss of hybridization signal approach, 173–176
  new technologies, new algorithms, 180
  parallelization of DNA sequencing approach, 171–172
  Quake approach, 172
  Sanger dideoxy approach, 171
  Solexa approach, 172
  whole genome sequences of animals, 171
  whole genome sequences of plants, 171
sociogenomics, 169
species
  cetacean, 10
  characters, highly specific genetic, 131
  coastal waters, 85–87

- concepts delineated by topological criteria and patterns of character fixation, 100
- discovering, describing, and classifying, 131
- DNA from museum specimens of extant, 147
- documenting as yet unknown, 115
- endangered, 9, 35
  - genetic threats, 11
  - precision and quantity of data on, 3–6
- endangered populations, 9
- extinction, 3, 106–107
- genetic characterization of, 132
- genetic resource collections, 142–143
- identifying, 43
- incipient, 18
- loss, 58, 107–108
- marginalized, 1
- phylogenetic, 100–103
- phylogenetic relationships between, 131
- range collapse, 107
- recognition, 39
- results of evolutionary processes, 58
- study of unexplored species diversity, 136–137
- survival, 25
  - species survival program (SSP), 19
- technologies having impact on conservation, 174–176
- units, 6
- whale species, 10

species boundaries
- between recently diverged species
  - general systematics approaches: character based, 7
  - general systematics approaches: tree based, 7
- problems, 4

species concepts
- biological species concept (BSC), 100–103
- phylogenetic species concept (PSC), 100–103

species survival programs (SSPs), 161

subspecies, 6, 19

systematic biology, 7

systematic research
- scarcity of material to carry out genetic characterization of species, 132

systematics, 3, 5–7, 19
- boundary between recently diverged species
  - general systematics approaches: character based, 7
  - general systematics approaches: tree based, 7
- molecular, 131
- molecular systematics programs, 116
  - arching and storage, 116
- nucleic acids, proteins, and chromosomes, 131
- phylogenetic inference methods, 102
- problems, potential, 12–14

systems
- cellular, 183
- organ, 183
- system approaches in molecular biology, 183
- tissue, 183

taxonomy
- characterizing a taxon, 136
- classifications, 182
- higher taxomic levels, 9
- mammalian taxa specimens, 125
- nucleotide substitutions separating taxa, 108
- taxonomy advisory groups, 161

techniques, methods
- allozymes, 4
- amplified fragment length polymorphism (AFLP), 4
- applied to systematics, 131
- cladistic, 9, 100–103
- cryogenic, 15
- detecting DNA polymorphisms, 173
- DNA sequencing, 4
- genomic, 167
- inter-simple sequence repeat (ISSR), 4
- making examination of genetics feasible, 42
- microarrays, 4
- microsatellite, 4, 12
- minisatellite, 4
- molecular evolutionary genetic, 5
- phylogenetic, 100–103
- population genetics
  - analysis of molecular variance (AMOVA), 7
  - coalescence theory-based analyses, 7
- quantitative trait mapping, 4
- random amplification of polymorphic DNA (RAPD), 4
- restriction fragment length polymorphism (RFLP), 4
- screening techniques for genetic variability, 4
- single nucleotide polymorphism (SNP) analysis, 4
- systematics
  - high-throughput sequencing, 4
  - SNP analysis, 4
- systems or population control, 5

technology
- cloning, 214
- genome mapping, 183
- genomics, 1, 3
  - data acquisition, storage, and analysis, 3
  - species survival, 25
- genotyping, 4
- legal issues, 213–218
- population biology, 3
- systematics, 3
- ZIMS software package: animal records, 162–163

tissues
- genetic resource collections, 142–143
- systems, 183

trade secrets, 220–221

variation
- adaptive, 41–43, 48–53
- detrimental, 41–47, 52–53
- DNA, 48

variation *(continued)*
genetic, 41–47
MHC, 48–49
molecular, 42, 46
neutral, 42–44, 52–53
viruses
genetic resource collections, 142–143
having whole genome sequence, 180

whales and whaling, 11
wildlife conservation, 30–31
coastal, 97
ocean life, 83–93
saving wildlife habitat, 31
wild populations
addax, 59
amphibians, 110
antelopes, 27
apes, 26–27, 161, 214
bandicoots, 59
beetles, 59
big cats, 14, 26–27, 161
birds, 26–27, 110
bison, 29
bonobos, 63
causes of mortality, 60
condors, 59, 63
cormorants, 27
coyotes, 12
cranes and storks, 26, 63
crocodiles, 26
crows, 59
deer, 59
eagles, 27
elephants, 26–27
ferrets, 59
fish, 27, 110
goats, 63–64
golden lion tamarins, 59
honeycreepers, 27
horses, 59, 214
iguanas, 27
kestrels, 29
kingfishers, 63
lakeside daisy, 14
larvae
DNA tags, 86
patterns of dispersal, 85–87
macaques, 63
mammals, 27, 110
monkeys, 26
northeastern beach tiger beetle, 96–103
oryx, 29, 59, 63–64
parrots, 12–13, 27, 37, 63
peccaries, 26
penguins, 27
peregrin falcons, 59
plants, 27
pythons, 27
rails, 63
red wolves, 12, 49–52
reptiles, 110
restoring, 59
rhinos, 27
seals, 27
snails, 59
tapirs, 26
terns, 27
tortoises, 27
toucans, 26
tuatara, 12
turtles, 26
wild sunflowers, 188
Wright's inbreeding coefficients, 7, 19

Zoological Information Management System (ZIMS), 162–163
zoos and aquariums, 11
Association of Zoos and Aquariums, 160
breeding programs, 63–76
collections
genetic resource, 124–30, 141–156,
of biological samples, 160–161
role of, 113–166
educating people, 59
Frozen Zoo at San Diego Zoo, 124–130
acquisition strategy, 126
current holdings, 124
DNA preparations, 127–128
fibroblast cell lines, 125
living tissue collection, 125, 127
mammals represented in (table of), 126
nucleic acid preparations, 127–128
recent efforts to expand, 128–129
specimens from mammalian taxa, 125
global zoo directory, 145
inbreeding, 59
institutional collections
animal diversity, 160
storage and distribution of biomaterials, 161
transferring biological samples to researchers, 164
use of artificial insemination, 199–200